CARDIAC ELECTROPHYSIOLOGY, CIRCULATION, AND TRANSPORT

DEVELOPMENTS IN CARDIOVASCULAR MEDICINE

Reiber, J.H.C., Serruys, P.W., Slager, C.J.: Quantitative coronary and left ventricular cineangiography. ISBN 0-89838-760-4.
Fagard, R.H., Bekaert, I.E., eds.: Sports cardiology. ISBN 0-89838-782-5. DICM 52.
Reiber, J.H.C., Serruys, P.W., eds.: State of the art in quantitative coronary arteriography. ISBN 0-89838-804-X. DICM 53.
Roelandt, J., ed.: Color doppler flow imaging. ISBN 0-89838-806-6. DICM 54.
van de Wall, E.E., ed.: Noninvasive imaging of cardiac metabolism. ISBN 0-89838-812-0. DICM 55.
Liebman, J., Plonsey, R., Rudy, Y., eds.: Pediatric and fundamental electrocardiography. ISBN 0-89838-815-5. DICM 56.
Higler, H., Hombach, V., eds.: Invasive cardiovascular therapy. ISBN 0-89838-818-X. DICM 57.
Serruys, P.W., Meester, G.T., eds.: Coronary angioplasty: a controlled model for ischemia. ISBN 0-89838-819-8. DICM 58.
Tooke, J.E., Smaje, L.H., eds.: Clinical investigation of the microcirculation. ISBN 0-89838-833-3. DICM 59.
van Dam, Th., van Oosterom, A., eds.: Electrocardiographic body surface mapping. ISBN 0-89838-834-1. DICM 60.
Spencer, M.P., ed.: Ultrasonic diagnosis of cerebrovascular disease. ISBN 0-89838-836-8. DICM 61.
Legato, M.J., ed.: The stressed heart. ISBN 0-89838-849-X. DICM 62.
Safar, M.E., ed.: Arterial and venous systems in essential hypertension. ISBN 0-89838-857-0. DICM 63.
Roelandt, J., ed.: Digital techniques in echocardiography. ISBN 0-89838-861-9. DICM 64.
Dhalla, N.S., Singal, P.K., Beamish, R.E., eds.: Pathophysiology of heart disease. ISBN 0-89838-864-3. DICM 65.
Dhalla, N.S., Pierce, G.N., Beamish, R.E., eds.: Heart function and metabolism. ISBN 0-89838-865-1. DICM 66.
Dhalla, N.S., Innes, I.R., Beamish, R.E., eds.: Myocardial ischemia. ISBN 0-89838-866-X. DICM 67.
Beamish, R.E., Panagia, V., Dhalla, N.S., eds.: Pharmacological aspects of heart disease. ISBN 0-89838-867-8. DICM 68.
Ter Keurs, H.E.D.J., Tyberg, J.V., eds.: Mechanics of the circulation. ISBN 0-89838-870-8. DICM 69.
Sideman, S., Beyar, R., eds.: Activation metabolism and perfusion of the heart. ISBN 0-89838-871-6. DICM 70.
Aliot, E., Lazzara, R., eds.: Ventricular tachycardias. ISBN 0-89838-881-3. DICM 71.
Schneeweiss, A., Schettler, G.: Cardiovascular drug therapy in the elderly. ISBN 0-89838-883-X. DICM 72.
Chapman, J.V., Sgalambro, A., eds.: Basic concepts in doppler echocardiography. ISBN 0-89838-888-0. DICM 73.
Chien, S., Dormandy, J., Ernst, E., Matrai, A., eds.: Clinical hemorheology. ISBN 0-89838-807-4. DICM 74.
Morganroth, J., Moore, E. Neil, eds.: Congestive heart failure. ISBN 0-89838-955-0. DICM 75.
Heintzen, P.H., Bursch, J.H., eds.: Progress in digital angiocardiography. ISBN 0-89838-965-8.
Scheinman, M., ed.: Catheter ablation of cardiac arrhythmias. ISBN 0-89838-967-4. DICM 78.
Spaan, J.A.E., Bruschke, A.V.G., Gittenberger, A.C., eds.: Coronary circulation. ISBN 0-89838-978-X. DICM 79.
Bayes de Luna, A., ed.: Therapeutics in cardiology. ISBN 0-89838-981-X. DICM 81.
Visser, C., Kan, G., Meltzer, R., eds.: Echocardiography in coronary artery disease. ISBN 0-89838-979-8. DICM 80.
Singal, P.K., ed.: Oxygen radicals in the pathophysiology of heart disease. ISBN 0-89838-375-7. DICM 86.
Iwata, H., Lombardini, J.B., Segawa, T., eds.: Taurine and the heart. ISBN 0-89838-396-X. DICM 93.
Mirvis, D.M., ed.: Body surface electrocardiographic mapping. ISBN 0-89838-983-6. DICM 82.
Morganroth, J., Moore, E.N., eds.: Silent myocardial ischemia. ISBN 0-89838-380-3. DICM 88.

CARDIAC ELECTROPHYSIOLOGY, CIRCULATION, AND TRANSPORT

Editors

Samuel Sideman, D.Sc.
R.J. Matas Winnipeg Professor of Biomedical Engineering
The Julius Silver Institute
Department of Biomedical Engineering
Head
Cardiac System Research Center
Technion–Israel Institute of Technology
Haifa, Israel
Distinguished Visiting Professor
Department of Biomedical Engineering
Rutgers University
Piscataway, New Jersey

Rafael Beyar, M.D., D.Sc.
Associate Professor
The Julius Silver Institute
Department of Biomedical Engineering
Associate Head
Cardiac System Research Center
Technion–Israel Institute of Technology
Haifa, Israel

André G. Kléber, M.D.
Professor
Department of Physiology
University of Bern
Bern, Switzerland

Springer-Science+Business Media, B.V.

Library of Congress Cataloging-in-Publication Data

Cardiac electrophysiology, circulation, and transport / editors.
Samuel Sideman, Rafael Beyar, André G. Kléber.
p. cm. — (Developments in cardiovascular medicine ; v. 121)
Proceedings of the 7th Henry Goldberg Workshop, May 13–17, 1990, organized jointly by Technion-Israel Institute of Technology, Haifa, Israel, and the University of Berne, Berne, Switzerland, in cooperation with Rutgers University, N.J. USA.
Includes index.
ISBN 978-1-4613-6737-6 ISBN 978-1-4615-3894-3 (eBook)
DOI 10.1007/978-1-4615-3894-3
1. Heart—Physiology—Congresses. 2. Electrophysiology--Congresses. 3. Heart—Pathophysiology—Congresses. 4. Heart--Metabolism—Congresses. 5. Coronary circulation—Congresses.
I. Sideman, S. II. Beyar, Rafael. III. Kléber, André G.
IV. Tekniyon, Makhon tekhnologi le-Yisra'el. V. Universität Bern.
VI. Rutgers University. VII. Henry Goldberg Workshop (7th : 1990 : Bern Switzerland) VIII. Series.
[DNLM: 1. Electrophysiology—congresses. 2. Heart—physiology--congresses. 3. Myocardium—metabolism—congresses. W1 DE997VME v. 121 / WG 202 C26453 1990]
QP112.5.E46C37 1991
612.1'7—dc20
DNLM/DLC
for Library of Congress 91-7013
CIP

Printed on acid-free paper.

Library of Congress Cataloging-in-Publication Data

Cardiac electrophysiology, circulation, and transport / editors.
Samuel Sideman, Rafael Beyar, André G. Kléber.
p. cm. — (Developments in cardiovascular medicine ; v. 121)
Proceedings of the 7th Henry Goldberg Workshop, May 13–17, 1990, organized jointly by Technion-Israel Institute of Technology, Haifa, Israel, and the University of Berne, Berne, Switzerland, in cooperation with Rutgers University, N.J. USA.
Includes index.
ISBN 978-1-4613-6737-6 ISBN 978-1-4615-3894-3 (eBook)
DOI 10.1007/978-1-4615-3894-3
1. Heart—Physiology—Congresses. 2. Electrophysiology--Congresses. 3. Heart—Pathophysiology—Congresses. 4. Heart--Metabolism—Congresses. 5. Coronary circulation—Congresses. I. Sideman, S. II. Beyar, Rafael. III. Kléber, André G.
IV. Tekniyon, Makhon tekhnologi le-Yisra'el. V. Universität Bern. VI. Rutgers University. VII. Henry Goldberg Workshop (7th : 1990 : Bern Switzerland) VIII. Series.
[DNLM: 1. Electrophysiology—congresses. 2. Heart—physiology--congresses. 3. Myocardium—metabolism—congresses. W1 DE997VME v. 121 / WG 202 C26453 1990]
QP112.5.E46C37 1991
612.1'7—dc20
DNLM/DLC
for Library of Congress 91–7013
CIP

Printed on acid-free paper.

DEDICATED

To those we love,
who support and encourage this endless search for the secrets of life.

To our students and colleagues
who share the excitement of this voyage to the unknown.

In Memory of
Janet Goldberg Ruberman

TABLE OF CONTENTS

List of Contributors xi
Preface xix

Modeling of Electrical Activity and Propagation

1 A Cellular Activation Model Based on Macroscopic Fields 3
Robert Plonsey and Adriaan van Oosterom

2 Modeling of Cellular and Intercellular Propagation 13
Dan Adam and Samuel Sideman

3 Anisotropic Conduction and Inhomogeneity in Refractoriness 29
Frans J.L. van Capelle and Michiel J. Janse

4 Different Mechanisms for Reentry in the Epicardial Border Zone of Healing Canine Infarcts: The Roles of Functional and Anatomical Conduction Block 37
Adam E. Saltman, James Coromilas, Bernd Waldecker, Stephen M. Dillon and Andrew L. Wit

5 Effects of Intercellular Resistive Coupling on Action Potential Initiation 47
Ronald W. Joyner and Rosemarie C. Tan

6 Reentry of Cardiac Excitation: A Simulation Study 63
Yoram Rudy and Weilun Quan

7 Conduction Velocity Variability in the Ischemic Myocardium as the Cause of Reentrant Arrhythmias 73
Shlomo A. Ben–Haim and Yoram Palti

Propagation of the Cardiac Impulse

8 Determinants of Action Potential Transfer From Cell to Cell 83
Robert Weingart

9 The Effect of the Finite Extracellular Resistance on Impulse Conduction in the Ventricular Myocardium 91
André G. Kléber and Gan–Xin Yan

10 The Effects of Potassium on Anisotropic Conduction 101
Lucas V.A. Boersma, Josep Brugada, Charles J.H.J. Kirchhof and Maurits A. Allessie

11 Effect of Ryanodine and Other Modulators of Intracellular Calcium on Impulse Propagation During Simulated Ischemia 109
Leonard S. Gettes and Toru Maruyama

12 Low Dimensional Chaos and the Transition From Rhythmic to Arrhythmic Behavior in Cardiac Tissue 115
José Jalife and Dante R. Chialvo

13 Fractal Mechanisms in Cardiac Electrophysiology 125
Ary L. Goldberger

Cell Physiology and Pathophysiology

14 Cardiac Sodium Channel Kinetics 137
Harry A. Fozzard

15 Regulation of Cationic Flux Through Cardiac Sarcoplasmic Reticulum 145
Qi–Yi Liu and Harold C. Strauss

16 Voltage Clamp Tests of a Model of Calcium Handling by Myocardial Cells 155
Per Arlock, Bjorn Wohlfart and Mark I.M. Noble

17 Modulation of Ventricular Impulse Initiation and Repolarization by Sympathetic Innervation 169
Michael R. Rosen, Susan F. Steinberg, Gabriella Malfatto and Tove S. Rosen

18 Relating Membrane Receptors to Drugs 179
Arnold M. Katz

19 Excitation–Contraction Coupling in Stunned Myocardium 185
Hideo Kusuoka, William C.Rose and Eduardo Marban

20 Biochemical Mechanisms Contributing to Arrhythmogenesis During Ischemia 195
Peter B. Corr

21 The Control of Cardiac Adenine Nucleotides: The Discovery and Potential Function of High–Energy Oligomeric Derivatives of ATP 205
John Mowbray

22 On the Contractile Mechanism in Cardiac Muscle 215
Gerald H. Pollack

Myocardial Mechanics

23 Modeling the Effect of Contractility on Distribution of Coronary Flow 229
Pieter Sipkema and Nicolaas Westerhof

24 Intramyocardial Pressure Revisited 237
Nicolaas Westerhof

25 On the Mechanism of Transmural Myocardial Compression and Perfusion 245
Rafael Beyar, Reuven Kamminker, Dan Manor, Riva ben Ari and Samuel Sideman

Circulation and Coronary Flow

26 A Model for Disturbed Coronary Artery Flow with Phonocardiographic Verification 261
Walter Welkowitz, Metin Akay, Jin–zhao Wang, John Semmlow, and John Kostis

27 Left Ventricular Pressure Versus Contractility as Causes for Pulsatile Coronary Flow 273
Erik Kouwenhoven and Jos A.E. Spaan

28 Are Flow Transport Capacities and Metabolic Requirements Matched Regionally in the Myocardium? 281
James B. Bassingthwaighte, James H. Caldwell, Gary Raymond, Keith Kroll and Gary V. Martin

29 Diminished Tolerance to Ischemia of Hypertrophied Hearts 295
Luc H.E.H. Snoeckx, Will A. Coumans, Ger J. van der Vusse and Robert S. Reneman

30 Capillary Growth in Normal and Hypertrophic Heart 307
Olga Hudlicka and Margaret D. Brown

Myocardial Metabolism and Energetics

31 Myocardial Blood Flow and Oxidative Metabolism Findings in Normal and Diseased Myocardium 319
Heinrich R. Schelbert

32 Oxidative Substrate Metabolism in Reperfused Myocardium 333
Rene Lerch, Gunter Goerge and Richard Benzi

33 The Role of Driving Forces in Myocardial Energy Metabolism and Transport Processes 341
Helmut Kammermeier

34 The Effect of Exogenous Lactate and Pyruvate on Triacylglycerol Turnover in the Normoxic Heart 351
Ger J. van der Vusse and Monique J.M. de Groot

General Discussion

35 General Discussion: Propagation and Arrhythmias 365
Muscle Function and Coronary Flow 369
CA^{2++} and the Membrane 370
Metabolism and Transport 372
Closing Remarks 374
Samuel Sideman and Rafael Beyar: Moderators

Index 375

LIST OF CONTRIBUTORS

Dan Adam, D.Sc., Julius Silver Institute, Heart System Research Center, Department of Biomedical Engineering, Technion–IIT, Haifa, 32000, Israel.

Metin Akay, Ph.D., Department of Biomedical Engineering, Rutgers University, Piscataway, NJ 08854–0909 USA.

Maurits A. Allessie, M.D., Ph.D., Department of Physiology, Biomedical Center, University of Limburg, P.O.B. 616, 6200 MD Maastricht, The Netherlands.

Per Arlock, Ph.D., Department of Zoophysiology, University of Lund, Lund, Sweden.

James B. Bassingthwaighte, M.D., Ph.D., Center for Bioengineering, University of Washington WD–12, Seattle WA 98195 USA.

Riva ben Ari, B.Sc., Julius Silver Institute, Heart System Research Center, Department of Biomedical Engineering, Technion–IIT, Haifa, 32000, Israel.

Shlomo A. Ben–Haim, M.D., D.Sc., Cardiovascular Research Group, Rappaport Family Institute, Technion–IIT, Haifa, 31096, Israel.

Richard Benzi, M.D., Cardiology Center, University Hospital, CH–1211 Geneva 4, Switzerland.

Rafael Beyar, M.D., D.Sc., Julius Silver Institute, Heart System Research Center, Department of Biomedical Engineering, Technion–IIT, Haifa, 32000, Israel.

Lucas V.A. Boersma, M.D., Department of Physiology, Biomedical Center, University of Limburg, P.O.B. 616, 6200 MD Maastricht, The Netherlands.

Margaret D. Brown, Ph.D., Department of Physiology, University of Birmingham, Birmingham B15 2TT, UK.

Josep Brugada, M.D., Department of Physiology, Biomedical Center, University of Limburg, P.O.B. 616, 6200 MD Maastricht, The Netherlands.

James H. Caldwell, M.D., Cardiology, University of Washington, VA Medical Center, 1660 Columbian Way, ZB–20, Seattle, WA 98108 USA.

Dante R. Chialvo, M.D., Department of Pharmacology, SUNY Health Science Center, 766 Irving Avenue, Syracuse, NY 13210 USA.

James Coromilas, M.D., Departments of Pharmacology and Medicine, College of Physicians and Surgeons of Columbia University New York, NY 10032 USA.

Peter B. Corr, Ph.D., Cardiovascular Division, Washington University Sch. of Medicine, 660 South Euclid Avenue, Box 8086 St. Louis, Missouri 63110 USA.

Will A. Coumans, Chem. Tech., Department of Physiology, University of Limburg, P.O. Box 616, 6200 MD Maastricht, The Netherlands.

Stephen M. Dillon, Ph.D., Departments of Pharmacology and Medicine, College of Physicians and Surgeons of Columbia University, New York, NY 10032 USA.

Uri Dinnar, D.Sc., Julius Silver Institute, Department of Biomedical Engineering, Technion–IIT, Haifa, 32000, Israel.

Harry A. Fozzard, M.D., Cardiac Electrophysiology Laboratories, University of Chicago, Chicago, IL 60637 USA.

Leonard S. Gettes, M.D., Division of Cardiology, CB #7075 Burnett–Womack Bldg., University of North Carolina, Chapel Hill, NC 27599–7075 USA.

Gunter Goerge, M.D., Cardiology Center, University Hospital, CH–1211 Geneva 4, Switzerland.

Ary L. Goldberger, M.D., Department of Medicine, Beth Israel Hospital, Harvard Medical Sch., 330 Brookline Ave., Boston, MA 02215 USA.

Monique J.M. de Groot, M.Sc., Department of Physiology, University of Limburg, P.O.B. 616, 6200 MD Maastricht, The Netherlands.

Olga Hudlicka, M.D., Ph.D., D.Sc., Department of Physiology, University of Birmingham, Birmingham B15 2TT, UK.

José Jalife, M.D., Department of Pharmacology, SUNY Health Science Center, 766 Irving Avenue, Syracuse, NY 13210 USA.

Michiel J. Janse, M.D., Department of Experimental Cardiology, Academic Medical Center, Nijbergdreef 9, Amsterdam, The Netherlands.

Ronald W. Joyner, M.D., Ph.D., Todd Franklin Cardiac Research Laboratory, The Children's Heart Center, Department of Pediatrics, Emory University, Atlanta, GA 30323 USA.

Helmut Kammermeier, M.D., Department of Physiology, Klinikum, RWTH Aachen, 5100 Aachen, Federal Republic of Germany.

Reuven Kamminker, B.Sc., Julius Silver Institute, Heart System Research Center, Department of Biomedical Engineering, Technion–IIT, Haifa, 32000, Israel.

Arnold M. Katz, M.D., Cardiology Division, Deptartment of Medicine, University of Connecticut, Farmington, CT 06032 USA.

Charles J.H.J. Kirchhof, M.D., Department of Physiology, Biomedical Center, University of Limburg, P.O.B. 616, 6200 MD Maastricht, The Netherlands.

André G. Kléber, M.D., Department of Physiology, University of Bern, Bühlplatz 5, CH–3012, Bern, Switzerland.

John Kostis, M.D., Department of Cardiology, UMDNJ – Robert Wood Johnson, Medical Sch., New Brunswick, NJ USA.

Erik Kouwenhoven, M.Sc., University of Amsterdam, Department of Medical Physics, Academic Medical Centre, Meibergdreef 15, 1105 AZ Amsterdam, The Netherlands.

Keith Kroll, Ph.D., Center for Bioengineering, University of Washington WD–12, Seattle WA 98195 USA.

Hideo Kusuoka, M.D., Ph.D., Division of Cardiology, Department of Medicine, Johns Hopkins University, Sch. of Medicine, 725 N. Wolfe St., Baltimore, MD 21205 USA.

Rene Lerch, M.D., Cardiology Center, University Hospital, CH–1211 Geneva 4, Switzerland.

Qi–Yi Liu, Ph.D., Department of Pharmacology, Duke University Medical Center, Durham, NC 27706 USA.

Gabriella Malfatto, M.D., Istituto di Clinica Medica Generale, e Terapia Medica, Universita di Milano, Centro di Fisiologia Clinica, e Ipertensione, Via Francesco Sforza, 35, 20122 Milano, Italy.

Dan Manor, M.Sc., Julius Silver Institute, Heart System Research Center, Department of Biomedical Engineering, Technion–IIT, Haifa, 32000, Israel.

Eduardo Marban, M.D., Ph.D., Division of Cardiology, Department of Medicine, Johns Hopkins University, Sch. of Medicine, 725 N. Wolfe St., Baltimore, MD 21205 USA.

Gary V. Martin, M.S., Cardiology, University of Washington, VA Medical Center, 1660 Columbain Way, ZB–20, Seattle WA 9810895 USA.

Toru Maruyama, M.D., Division of Cardiology, CB #7075 Burnett–Womack Bldg., University of North Carolina, Chapel Hill, NC 27599–7075 USA.

John Mowbray, Ph.D., Biochemistry Department, University College London, Gower St., London WC1E 6BT, UK.

Mark I.M. Noble, D.Sc., M.D., Ph.D., FRCP, FESC, Academic Unit of Cardiovascular Medicine, Charing Cross & Westminster Medical Sch., 17 Horseferry Rd., London SW1P 2AR, UK.

Yoram Palti, M.D., Ph.D., Rappaport Family Institute, Technion–IIT, Haifa, 31096, Israel.

Robert Plonsey, Ph.D., Department of Biomedical Engineering, Duke University, Durham, NC 27706 USA.

Gerald H. Pollack, Ph.D., Bioengineering WD–12, University of Washington, Seattle, WA 98195 USA.

Weilun Quan, Ph.D., Department of Biomedical Engineering, Case Western Reserve University, Cleveland, OH 44106 USA.

Gary M. Raymond, M.S., Center for Bioengineering, WD–12, University of Washington, Seattle WA 98195 USA.

Robert S. Reneman, M.D., Ph.D., Department of Physiology, Cardiovascular Research Institute Maastricht, University of Limburg, P.O. Box 616, 6200 MD Maastricht, The Netherlands.

William C. Rose, Ph.D., Division of Cardiology, Department of Medicine, 116, Johns Hopkins University, Sch. of Medicine, 725 N. Wolfe St., Baltimore, MD 21205 USA.

Michael R. Rosen, M.D., Departments of Pharmacology & Pediatrics, College of Physicians & Surgeons of Columbia University, 630 West 168 St., New York, NY 10032 USA.

Tove S. Rosen, M.D., Department of Pediatrics, College of Physicians & Surgeons, of Columbia University, 630 West 168 St., New York, NY 10032 USA.

Yoram Rudy, Ph.D., Department of Biomedical Engineering, Case Western Reserve University, Cleveland, OH 44106 USA.

Adam E. Saltman, M.D., Ph.D., Departments of Pharmacology and Medicine, College of Physicians & Surgeons, of Columbia University, 630 West 168 St., New York, NY 10032 USA.

Heinrich R. Schelbert, M.D., Division of Nuclear Medicine and Biophysics, Department of Radiological Sciences, UCLA School of Medicine University of California, Los Angeles, CA 90024 USA.

John Semmlow, Ph.D., Department of Biomedical Engineering, Rutgers University, Piscataway, NJ 08854–0909 USA.

Samuel Sideman, D.Sc., Heart System Research Center, The Julius Silver Institute, Department of Biomedical Engineering, Technion–IIT, Haifa, 32000, Israel.

Pieter Sipkema, Ph.D., Laboratory for Physiology, Free University of Amsterdam, Amsterdam, The Netherlands.

Luc H.E.H. Snoeckx, Ph.D., Department of Physiology, University of Limburg, P.O. Box 616, 6200 MD Maastricht, The Netherlands.

Jos A.E. Spaan, Ph.D., University of Amsterdam, Dept. of Medical Physics, Academic Medical Centre, Meibergdreef 15, 1105 AZ Amsterdam, The Netherlands.

Susan F. Steinberg, M.D., Departments of Medicine & Pharmacology, College of Physicians & Surgeons, of Columbia University, 630 West 168 St., New York, NY 10032 USA.

Harold C. Strauss, M.D., Department of Medicine, Duke University Medical Center, Bell Building #345, Durham, NC 27706 USA.

Rosemarie C. Tan, B.S.B.E., Todd Franklin Cardiac Research Laboratory, The Children's Heart Center, Department of Pediatrics, Emory University, Atlanta, GA 30323 USA.

Frans J.L. van Capelle, Ph.D., Laboratory of Experimental Cardiology, Academic Medical Center, Interuniversity CArdiology Institute, Amsterdam 1105 AZ, The Netherlands.

Ger J. van der Vusse, Ph.D.,
Department of Physiology, Cardiovascular Research Institute Maastricht, University of Limburg, P.O. Box 616, 6200 MD Maastricht, The Netherlands.

Adriaan van Oosterom, Ph.D., Laboratory of Medical Physics and Biophysics, University of Nijmegen, The Netherlands.

Bernd Waldecker, M.D., Departments of Pharmacology and Medicine, College of Physicians & Surgeons of Columbia University, New York, NY 10032 USA.

Jin–zhao Wang, Ph.D., Department of Biomedical Engineering, Rutgers University, Piscataway, NJ 08854–0909 USA.

Robert Weingart, Ph.D., Department of Physiology, University of Berne, Buhlplatz 5, 3012 Berne, Switzerland.

Walter Welkowitz, Ph.D., Department of Biomedical Engineering, Rutgers University, Piscataway, NJ 08854–0909 USA,

Nicolaas Westerhof, Ph.D., Laboratory for Physiology, Free University of Amsterdam, Amsterdam, The Netherlands.

Bjorn Wohlfart, M.D., Department of Clinical Physiology, University Hospital, Lund, Sweden.

Andrew L. Wit, Ph.D., Departments of Pharmacology and Medicine, College of Physicians & Surgeons of Columbia University, New York, NY 10032 USA.

Gan–Xin Yan, M.Sc., Department of Physiology, University of Bern, Bühlplatz 5, CH–3012, Bern, Switzerland.

Proceedings of the 7th Henry Goldberg Workshop
on Activation, Circulation and Transport in the Cardiac Muscle,
May 13–17, 1990, organized jointly by
Technion–Israel Institute of Technology, Haifa, Israel,
and the University of Berne, Berne, Switzerland,
in Cooperation with Rutgers University, NJ, USA.

Scientific Advisory Committee

Prof. J.I.E. Hoffman
University of California, San Francisco, CA, USA

Prof. Eliezer Kaplinski
Meir Hospital, Kfar Saba, Israel

Prof. Robert Reneman
University of Limburg, Maastricht, The Netherlands

Prof. M.L. Weisfeldt (Chair)
Johns Hopkins University Hospital, Baltimore, MD, USA

Prof. Andrew Wit
College of Physicians & Surgeons, New York, NY, USA

Organizing Committee

Prof. James Bassingthwaighte
University of Washington, Seattle, WA, USA

Assoc. Prof. Rafael Beyar (Co–Chair)
Technion–Israel Institute of Technology, Haifa, Israel

Prof. André G. Kléber (Chair, Local Committee)
University of Berne, Berne, Switzerland

Prof. Robert Plonsey
Duke University, Durham, NC, USA

Prof. Samuel Sideman (Chair)
Technion–Israel Institute of Technology, Haifa, Israel

Prof. Walter Welkowitz
Rutgers University, Piscataway, NJ, USA

The 7th Henry Goldberg Workshop was held
under the auspices of

The American Heart Association
and
The Israel Heart Association

Sponsors

Mr. Henry Goldberg
Julius Silver Institute of Biomedical Engineering, Technion–IIT
Max und Elsa Beer–Brawand Foundation, University of Bern
Bayer AG
Swiss National Science Foundation
Zyma AG Nyon

Co–Sponsors

Knoll (Switzerland) AG
F. Hoffmann – La Roche AG
Medtronic, Switzerland AG
Sandoz AG
Loeb AG
Sanofi AG
Sigma Tau AG
Ciba–Geigy AG
Glaxo AG

THE EDITORS

Samuel Sideman, D.Sc., R.J. Matas/Winnipeg Professor of Biomedical Engineering, is Chairman of the Department of Biomedical Engineering, Director of the Julius Silver Institute of Biomedical Engineering, and Head of the Cardiac System Research Center of the Technion–Israel Institute of Technology.

Born in Israel (1929), he received his B.Sc. and D.Sc. from the Technion and his M.Ch.E. from the Polytechnical Institute of Brooklyn. On the faculty of the Technion since 1957, he served as Dean of Faculty, Dean of Students, and Chairman of Chemical Engineering, was a Visiting Professor at the University of Houston and CCNY, a Distinguished Visiting Professor at Rutgers University, NJ, and is a Visiting Professor of Surgery (Bioengineering) at UMDNJ, USA.

His interests include transport phenomena, with particular emphasis on the analysis and simulation of the cardiac system. He has authored and co–authored over 250 scientific publications and co–edited 9 books, A recipient of a number of professional awards and citations, on the editorial board of some major scientific journals, he is a member of a number societies, and Fellow of the American Institute of Chemical Engineering and the New York Academy of Science.

Rafael Beyar, M.D., D.Sc., is an Associate Professor in the Department of Biomedical Engineering and is Associate Head of the Cardiac Research Center at the Technion.

Born in Israel (1952), he received his M.D. from Tel Aviv University and obtained his D.Sc. in Biomedical Engineering from the Technion–Israel Institute of Technology. In the Julius Silver Institute, Department of Biomedical Engineering, Technion–IIT since 1984, he was (1985 to 1987) at the Division of Cardiology, Johns Hopkins University Hospital, Baltimore.

His interests include simulation of the cardiovascular system, resuscitation, and circulatory assist. He is a member of medical and engineering societies, and a recipient of a number of institutional and national excellence awards. He has authored and co–authored over 60 scientific publications and is co–editor (with Prof. Sideman) of 6 books on the analysis and simulation of the cardiac system.

André G. Kléber, M.D. is a Professor Physiology in the Department of Physiology, University of Bern, Switzerland.

Born in Switzerland (1944), he received his M.D. form the University of Bern (1970). From 1970–1980 he passed his clinical training in Internal Medicine and Cardiology at various Swiss Universities, and as post–doctoral fellow he worked at the Department of Physiology, University of Bern, Switzerland and at the Department of Clinical Physiology, University of Amsterdam, The Netherlands.

His interests include the electrical and ionic changes occurring during myocardial ischemia and the cellular mechanisms of cardiac impulse conduction. He has authored and co–authored a number of important publications related to these fields, received two national awards and serves as a member of the editorial board of 2 international journals.

PREFACE

The cardiac system represents one of the most exciting challenges to human ingenuity. Critical to our survival, it consists of a tantalizing array of interacting phenomena, from ionic and molecular transport, membrane channels and receptors through cellular metabolism, energy production to fiber mechanics, microcirculation, electrical activation to the global, clinically observed function, which is measured by pressure, volume, coronary flow, heart rate, shape changes and responses to imposed loads and pharmaceutical challenges. It is a complex interdisciplinary system requiring the joint efforts of the life sciences, the exact sciences, engineering and technology to understand and control the pathologies involved. The Henry Goldberg Workshops were set up to address these challenges. Briefly, our goals are:

1. To foster interdisciplinary interaction between scientists from different areas of cardiology, identify missing links, and catalyze new questions.
2. To relate microscale cellular phenomena to the global, clinically manifested cardiac function.
3. To relate conceptual modeling and quantitative analysis to experimental and clinical data.
4. To encourage international cooperation so as to disseminate medical and technological know how and lead to better understanding of the cardiac system.

Today we celebrate the tenth birthday of a dream come true, a dream to bring together the diversified expertise in the various fields of science, engineering and medicine, to relate to the numerous interactive parameters and disciplines involved in the performance of the heart. This gathering of outstanding scientists from all over the world is a testimonial to international cooperation and highlights the pleasure of joining hands in the pursuit of the secrets of life.

The outstanding success of the past workshops has encouraged their continuation. The first Henry Goldberg Workshop, held in Haifa in 1984, introduced the concept of interaction between mechanics, electrical activation, perfusion and metabolism, emphasizing imaging in the clinical environment. The 2nd Workshop, in [illegible], discussed the same parameters with a slant towards the [illegible]. The 3rd Henry Goldberg Workshop, held in the USA at Rutgers University in [illegible], highlighted the transformation of the microscale subcellular dysfunctions to macroscale activity and performance.

PREFACE

The cardiac system represents one of the most exciting challenges to human ingenuity. Critical to our survival, it consists of a tantalizing array of interacting phenomena, from ionic microscopic transport, membrane channels and receptors through cellular metabolism, energy production to fiber mechanics, microcirculation, electrical activation to the global, clinically observed, function, which is measured by pressure, volume, coronary flow, heart rate, shape changes and responds to imposed loads and pharmaceutical challenges. It is a complex interdisciplinary system requiring the joint efforts of the life sciences, the exact sciences, engineering and technology to understand and control the pathologies involved. The Henry Goldberg Workshops were set up to address these challenges. Briefly, our goals are:

1. To foster interdisciplinary interaction between scientists from different areas of cardiology, identify missing links, and catalyze new questions.
2. To relate microscale cellular phenomena to the global, clinically manifested cardiac function.
3. To relate conceptual modeling and quantitative analysis to experimental and clinical data.
4. To encourage international cooperation so as to disperse medical and technological know how and lead to better understanding of the cardiac system.

Today we celebrate the 7th birthday of a dream come true; a dream to bring together the diversified expertise in the various fields of science, engineering and medicine, to relate to the numerous interactive parameters and disciplines involved in the performance of the heart. This gathering of outstanding scientists from all over the world is a testimonial to international cooperation and highlights the pleasure of joining hands in the pursuit of the secrets of life.

The outstanding success of the past workshops has encouraged their continuation. The first Henry Goldberg Workshop, held in Haifa in 1984, introduced the concept of interaction between mechanics, electrical activation, perfusion and metabolism, emphasizing imaging in the clinical environment. The 2nd Workshop, in 1985, discussed the same parameters with a slant towards the control aspects. The 3rd Henry Goldberg Workshop, held in the USA at Rutgers University in 1986, highlighted the transformation of the microscale activation phenomena to macroscale activity and performance, relating electrophysiology, energy metabolism and cardiac mechanics. The 4th Henry Goldberg Workshop in 1987 continued the effort to elucidate the interactions between the various parameters affecting cardiac performance, with emphasis on the ischemic heart. The 5th Workshop, held in Cambridge, UK, in 1988, dwelt on the effects of inhomogeneity of the cardiac muscle on its performance in health and disease. The 6th Workshop, aimed to highlight the role of new modern imaging techniques, which allow to gain more insight into local and global cardiac performance in cardiac research and clinical practice. The present 7th Workshop aims at in–depth exploration of the basic microlevel phenomena that affect the cardiac system in health and disease.

It is with great pleasure that we acknowledge here those who have helped make this meeting a reality. Special thanks are due to Mr. Julius Silver and Ms. Dinny Winslow (Silver) of New York, for their personal support and continued friendship which inspired and shaped our goals and made it all possible. Thanks are also due to the Women's Division of the American Society for the Technion, who encouraged us with their unshakable trust and provided the means to start the Heart System Research Center. Personal thanks go to a friend and collaborator, Prof. Walter Welkowitz of Rutgers University, who was the first to learn of our wild dreams and has, ever since, helped us to materialize them. Particular thanks go to our host, Prof. Andre Kléber, of the University of Berne, who set the environment to make this meeting a most enjoyable encounter. Last but not least, our warm hearted thanks go to our sponsors and supporters, and particularly to Mr. Henry Goldberg and his wife Viola, for their generosity and kindness which made these Henry Goldberg Workshops significant milestones in cardiac research.

Samuel Sideman

My greetings to all the participants of this important meeting which deals with the forefront of medical sciences. We are very proud that the Technion participates in an effective way in advanced interdisciplinary areas such as the cardiac system.

I would like to thank all the sponsors and supporters for their help and congratulate the organizers on their initiative. I hope you will all enjoy useful deliberations as well as social contacts that are so important for international scientific cooperation.

Max Reis
President, Technion–IIT

Modeling of Electrical Activity and Propagation

CHAPTER 1:

A CELLULAR ACTIVATION MODEL BASED ON MACROSCOPIC FIELDS

Robert Plonsey and Adriaan van Oosterom

ABSTRACT

A model of cellular activation of normal ventricular tissue is proposed, and the sources and fields of this model evaluated. The results agree with macroscopic field measurements. Since the model assumes a maximal superposition of elemental sources, there is little room for alternative hypotheses.

INTRODUCTION

There is little information describing normal activation of cardiac tissue on a cellular level because of the difficulty in making measurements at such a small size scale. The work of Spach et al. [1] has documented the possibly complex pathways which depends on the detailed structure, particularly under abnormal conditions. In this paper a model of cellular activation is proposed which is consistent with a source description established from macroscopic measurements. While such a model could be thought to overlook details which have only a small effect on cardiac sources, the main and most significant characteristics are likely to be emphasized. This paper brings up-to-date an earlier effort to infer microscopic behavior from well-accepted macroscopic considerations [2].

MACROSCOPIC ACTIVATION PATTERN

Scher and Young [3] were the first to investigate the macroscopic pattern of normal ventricular activation. They determined the location of sequential surfaces of the activation wavefront (isochrones) showing a generally uniform progression from endocardium to epicardium. Since it has also been established that cardiac fibers are oriented parallel to the endo– and epi–

cardium [4], clearly the macroscopic activation proceeds in a cross–fiber direction. This particular orientation of the isochronous surfaces is not surprising. In the first place, activation is initiated by the conduction system along a broad endocardial surface. Secondly, since the propagation velocity along the fiber axis is perhaps three times that in the cross–fiber direction [5], the stable orientation of an activation wavefront is, in fact, perpendicular to the fiber axis. Underlying the macroscopic observation of uniform propagation is the existence of numerous interconnections between each cardiac cell and both its lateral and axial neighbors [6]. These intercellular junctions have been shown to be of low resistance [7] and to a first approximation permit the treatment of normal cardiac tissue as a continuum (syncytium).

For simplicity we assume the cardiac fibers to be parallel and axially extensive, so that variations in the axial direction can be neglected. A cross-sectional view of a segment of tissue is shown in Fig. 1; consistent with cardiac histology, fiber shape is depicted as irregular. For simplicity, the extensive vascular system has not been included.

Normal, transverse, plane wave propagation in the x direction is assumed. Successive isochrones are shown in Fig. 1 and numbered from 1 to 5. Spacing between isochrones corresponds to a temporal change of approximately 0.02 ms, based on a cell diameter of 10 μm and a propagation velocity of 0.5 m.s^{-1}.

SOURCE MODEL

Our proposed source model assumes synchronous activation of all cell membrane elements lying in the same isochrone. This idea is illustrated in Fig. 1 where we have emphasized two isochrones separated by a differential in time Δt, hence these surfaces are also separated spatially by the designated

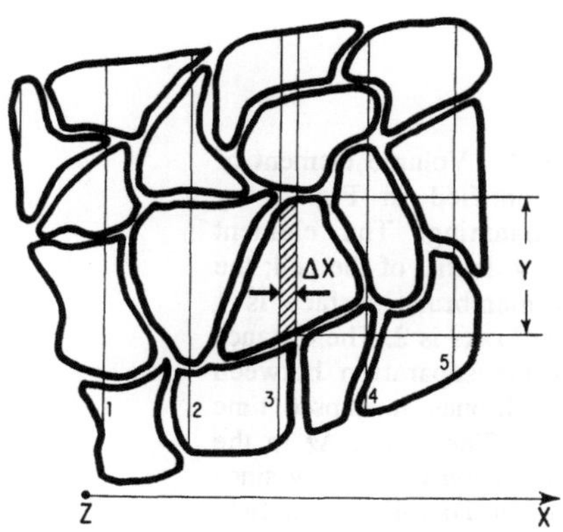

Figure 1: Cross–section of cardiac muscle fiber bundle whose axis is in the z direction. A uniform plane wave is propagating in the transverse x direction and sequential isochrones are labeled 1, 2, 3, 4, 5.

differential $\Delta x = \theta \Delta t$, where θ is the velocity. All membrane elements lying between these surfaces, regardless of its shape or with which cell its is associated, are assumed to behave synchronously. The implications regarding the electrical sources which are generated will be examined by a consideration of the prototypical cell identified in Fig. 1 by cross–hatching. An element of this cell is also shown in perspective in Fig. 2 for an axial distance of Δz.

The source arising from cellular activation has been shown to be a double–layer lying in the cell membrane and oriented normal thereto, whose strength is given by [8].

$$\boldsymbol{\tau}_s = (\Phi_i \sigma_i - \Phi_e \sigma_e)\, \mathbf{n} \tag{1}$$

where (Φ_i, Φ_e) are the potentials just inside and outside the membrane; (σ_i, σ_e) are the intracellular, extracellular conductivities, and $\mathbf{n}$ (a unit dimensionless vector) is the outward surface normal. In Eq. (1) $\boldsymbol{\tau}_s$ has the units of $A \cdot mm^{-2}$ and constitutes a dipole *surface* density (i.e. a double–layer). A consequence of our source model is that all membrane elements lying between the surfaces Δx must be the site of the same dipole density source given by Eq. (1). The membrane surfaces labeled 1 and 2 in Fig. 2 have, therefore, the same uniform double–layer source. The extracellular field generated by these membrane sources is given by the expression

$$\Phi(P) = (1/4\pi\sigma_e) \cdot \int \boldsymbol{\tau} \cdot \nabla(1/r)\,dS \tag{2}$$

where the surface integral is taken over the membrane surfaces.

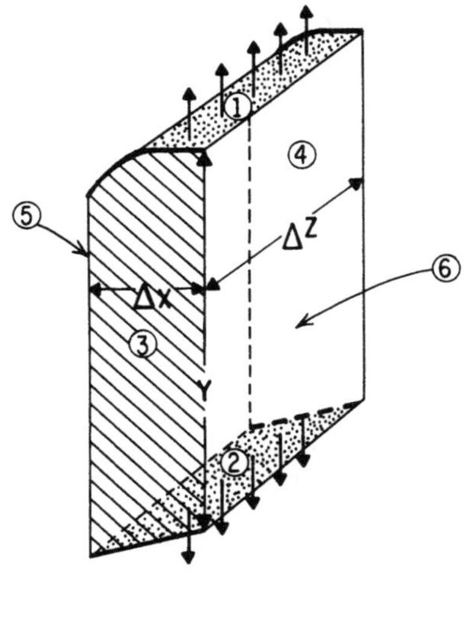

Figure 2: Volume element of cell identified in Fig. 1 by cross–hatching. The element height, y, is that of the cell; the upper membrane surface is 1 and the lower is 2. The distance Δx is the separation between two isochrones at elapsed time Δt apart. The extent Δz in the axial direction is arbitrary since we assume no variation of field quantities in this direction.

The volume element described in Fig. 2 is shown bounded by six surfaces. If each surface held a uniform double–layer source of the same strength as that on the membranes (given in Eq. (1)) then, using Eq. (2) to evaluate the field would show that it is zero at any extracellular field point, hence demonstrating the well known fact that the field generated by a closed uniform double–layer is zero. Consequentially, the field from the active membrane surfaces 1 and 2 can also be evaluated from the negative of the same uniform double layer on surfaces 3, 4, 5 and 6. The sources arising from an entire tissue require the superposition of cellular elements such as shown in Fig. 2. When this is done, source elements on 3 and 4 will be exactly canceled by adjoining cell elements, since these double–layer strengths are determined by identical membrane strengths. However, cancellation will not arise from surfaces 5 and 6 since adjoining elements lie in the direction of propagation, the x direction, and their membrane source strengths will be slightly different. In fact the resulting source per unit length, i.e. the axial source density, is simply the rate of change with respect to x of the right hand side of Eq. (1), and this is described, in the limit as $\Delta x \rightarrow 0$, by

$$\vec{\tau}_v = [\, \partial(\Phi_i \sigma_i - \Phi_e \sigma_e) / \partial x \,]\, \vec{a}_x \qquad (3)$$

which constitutes a *volume* dipole source oriented in the direction of propagation (i.e. the x direction). (In Eq. (3) $\vec{a}_x$ is a unit dimensionless vector in the x direction). This volume dipole source fills each cell but since the cells are densely packed one could assume the entire tissue filled with the source density given by Eq. (3) provided the strength is multiplied by the volume fraction, f, occupied by the cells. A graphical description of these sources is given in Fig. 3. Consistent with the model, the dipole density along any isochrone is uniform, but a variation along successive isochrones may arise according to Eq. (3).

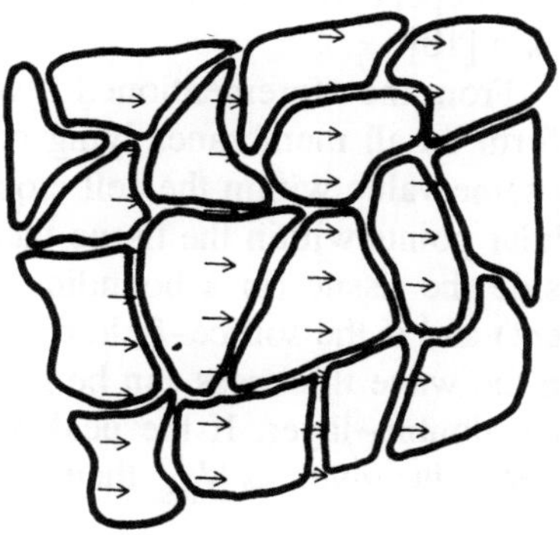

Figure 3: An equivalent source distribution, oriented in the direction of propagation, for the tissue described in Fig. 1.

7

MACROSCOPIC FIELD EXPRESSIONS

The source described by Eq. (3) can be evaluated from a description of the transmembrane action potential $V_m(t)$ based on the following assumptions:

(a) Negligible variation of $V_m(t)$ throughout the tissue (during the rising phase).

(b) Uniform propagation velocity.

(c) Applicability of the linear–core–conductor model.

The validity of these assumptions should be reasonably satisfied in normal tissue. Probably the main challenge to their validity arises from the presence of discrete intercellular junctions. When the resistance magnitude is not sufficiently low, the treatment of cardiac tissue as a continuum, as done here, is invalidated.

From the linear core conductor model [9] we have

$$\Phi_i = r_i / (r_i + r_e) \cdot V_m \qquad (4)$$

$$\Phi_e = - r_e / (r_i + r_e) \cdot V_m \qquad (5)$$

where r_i, r_e are resistances for a unit axial length (x in Fig. 1) for a unit cross–section (i.e. yz in Fig. 1). Substituting Eq. (4) and Eq. (5) into Eq. (3) and evaluating the extracellular field from these sources results in

$$\Phi_e(P) = 1/4\pi \cdot r_e / (r_e + r_i) \int_V \partial / \partial x(V_m) \, \vec{a}_x \cdot \nabla(1/r) dV \qquad (6)$$

where V is the total volume and the factor f has been absorbed in the coefficient on the right hand side of Eq. (6). While the linear core conductor model is normally applied to a one–dimensional preparation, it should be noted that in syncytial tissue, as assumed here, Eq. (4) and Eq. (5) applies to plane wave propagation in any direction. However, if the medium is anisotropic then r_i and r_e depend on the direction of propagation as described by Muler and Markin [18].

From the aforementioned it will be seen that V_m is not only defined as uniform on all membranes lying on an isochrone but also is defined to have this same value within the cell cross–section. If Eq. (6) is evaluated at extracellular points within the tissue then Eq. (5) is returned. If the field point lies outside the tissue (in a bounding uniform and isotropic medium of infinite extent) and if the source–field distance is large compared to the wave–width, then the wave thickness can be neglected and the dipole source density is an actual double–layer. If the peak value of V_m is designated V_{peak} while its resting value given as V_{rest} then

$$\Phi_e = V_{wave}\, \Delta\Omega/4\pi \tag{7a}$$

where

$$V_{wave} = (V_{peak} - V_{rest})\; r_e / r_e + r_i \tag{7b}$$

and $\Delta\Omega$ is the solid angle subtended by the activating wave at the fieldpoint. If the wavefront is smooth with a large radius of curvature it may be divided into planar (triangular) elements to each of which the aforementioned expressions apply approximately, with the final result obtained by superposition.

MACROSCOPIC FIELD EVALUATION

Within the Cardiac Tissue

For field points on opposite sides of the wavefront Eq. (7) can be applied with $\Delta\Omega$ essentially $\pm 2\pi$. Accordingly, the maximum voltage obtained in a bipolar measurement is V_{wave}, and one obtains the experimental result of 40 mV [10–12]. A comparison with the theoretical value arising from this model depends on values of r_e and r_i. In Table 1 we show a list of values reported in the literature.

Table 1: A list of resistance values reported in the literature.

	Clerc [13]	Roberts & Scher [11]	Epstein & Foster [14]	Kléber & Riegger [15]
r_e^l	160 Ωcm	855 Ωcm	383 Ωcm**	252 Ωcm
r_i^l	575 Ωcm	291 Ωcm	319 Ωcm**	221 Ωcm
r_i	5181 Ωcm	1678 Ωcm	—	—
r_e	424 Ωcm	1247 Ωcm	1250 Ωcm	—
V_{wave}*	8 mV	43 mV	—	—

* Assumes V_m = 100 mV; ** assumes the axial resistance ratio of Kléber and Riegger [15]; r_e, r_i – transverse intracellular and interstitial resistivities; r_e^l, r_i^l – axial intracellular and interstitial resistivities.

In addition to the desired quantities of r_e, r_i, which express transverse intracellular and interstitial resistivities, we have also included axial intracellular and interstitial resistivities (superscript l) as a means of evaluating the quality of data. Neither Roberts and Scher [11] or Clerc [13] are completely consistent, but the former is somewhat better. Using the data of Roberts and Scher [11] yields a value of V_{wave} of 43 mV, in good agreement with measurements.

Body Surface Potential

Equation (7) has been used to calculate potentials at the torso surface taking into account actual body, heart, and lung geometry as well as the conductivities of these organs. The inhomogeneities may be taken into account through evaluation of secondary sources (which depends on the discontinuity in conductivity). Using the boundary element method for treating the inhomogeneous volume conductor, V_{wave} = 40 mV, $\sigma_{heart}/\sigma_{lung} = 5$, $\sigma_{blood}/\sigma_{heart} = 3$, van Oosterom and Huiskamp [16] found excellent agreement between computed and measured ECG's both in waveform and magnitude. Since no arbitrary scale factor was introduced, and the geometry used was that found from MRI images, this provides additional support for the choice of V_{wave} = 40 mV and, indirectly supports further the microscopic model proposed here.

DISCUSSION

A proposed cellular model of activation is evaluated here by examining its macroscopic fields and comparing the results with experiment. While good agreement is achieved we note that this depends on tissue resistances which are not unequivocally determined. Beyond this uncertainty is the possibility of modifying the model in ways which do not substantially change the source magnitude or orientation. Zig–zag propagation which makes small angles with the x axis, for example, cannot be ruled out. On the other hand zig–zag propagation that is substantially in the axial direction (z direction in Fig. 1), as suggested by Clerc [13] and Geselowitz and Miller [17], would substantially affect the generated fields and can therefore be excluded. As a general observation one notes that a maximal superposition of sources in the outward direction is achieved here. Since this is just sufficient to produce the observed macroscopic behavior alternative hypotheses which result in some cancellation of source elements or in source directions that are not outward would probably be unsatisfactory. The model advanced here is based on the view of cardiac tissue as continuous and syncytial, where uniform plane–wave propagation in any direction is described by the linear–core–conductor model for that direction [18].

Acknowledgement

The research was supported by the National Institutes of Health through grant HL40609, from the Heart Lung and Blood Institute. The author (RP) expresses his appreciation to Prof. H.K. Boom for making available resources of the Division of Biomedical Engineering at the University of Twente at which this work was done while on sabbatical leave from Duke University, and for his helpful discussions.

DISCUSSION

Dr. Janse: You assume in your model that throughout the ventricular wall the fibers are parallel to the epicardium. I thought that the fibers make a curve of 180° from epicardium to endocardium.

Dr. Plonsey: Part of the discrepancy is that I have interpreted the data a little differently. All that I know about fiber direction I get from the work of Streeter et al. *[4]*. If the fibers lie parallel to the endocardium, then as we go toward the endocardium from the epicardium, my understanding is that the fibers continue to be parallel even though their direction is rotating through 180°, so they are always parallel. My model is a very simple model because I have ignored this rotation.

Dr. Rudy: First a comment related to Dr. Janse's question. The fibers in the thickness of the myocardial wall remain parallel to the endocardial and epicardial surfaces and their rotation is limited to this plane. A recent paper by Frazier et al. *[Circ Res 1988;65:135–146]* deals with certain aspects of the "rotational anisotropy" and its effects on transmural activation.

My question is to clarify whether the analysis is based on the properties of a plane wave and is therefore limited to normal propagation of a planar wavefront across the ventricular wall, or is it applicable to point, ectopic, stimulation?

Dr. Plonsey: I am dealing here solely with normal propagation in normal cardiac tissue, so I am excluding such things as ectopic stimuli. In fact, I also exclude the behavior in ischemic tissue because the idea of treating cardiac muscle as a continuum begins to break down when the junctional resistances begin to get high due to decoupling of cells. Your comments are right.

Dr. Bassingthwaighte: It seems to me that there is incorporated into r_e and r_i an analog to intercellular communication in the lateral direction; I am not sure whether that was included in r_e or r_i. It could be either one I suppose. But, as you go from endocardium to epicardium, there are some changes in fiber direction, and in the degree of effective intercellular communication. That would mean that the isochrones would be unequally spaced across the endocardium to epicardium. How does that work out?

Dr. Plonsey: If there were some changes in the value of r_e and r_i, between the endocardium and epicardium then it would certainly mean that the strength of the source would be nonuniform depending on how this particular ratio was affected. But the model is a very simple minded model and it essentially assumes that there is no substantial change in those particular values. A kind of supporting macroscopic evidence of that is the fact that one can model the electrocardiogram so the comparison is with measurements at the body surface. One can model the elec–

trocardiogram using a uniform double layer, assuming that the double layer strength is uniform at the very start to the very end. One can get quite a good comparison of the magnitude and wave shape on that basis. This suggests that whatever changes (in r_e and r_i) might be happening during activation, it is not having a big effect, at least as reflected at the body surface.

REFERENCES

1. Spach MS, Miller WT, Dolber P, Kootsey JM, Sommer JR, Moser CE Jr: The functional role of structural complexities in the propagation of depolarization in the atrium of the dog. *Circ Res* 1982;50:175–191.
2. Plonsey R: An evaluation of several cardiac activation models. *J Electrocard* 1974;7:237–244.
3. Scher A, Young AC: Ventricular depolarization and the genesis of the QRS. *Ann NY Acad Sci* 1957;65:768–778.
4. Streeter DD Jr, Spotnitz HM, Patel DP, Ross J Jr, Sonnenblick, EH: Fiber orientation in the canine left ventricle during diastole and systole. *Circ Res* 1969; 24:339–347.
5. Spach MS, Miller WT, Miller–Jones E, Warren RB, Barr RC: Extracellular potentials related to intracellular action potentials during impulse conduction in anisotropic canine cardiac muscle. *Circ Res* 1979;45:188–204.
6. Sommer JR, Scherer B: Geometry of cell and bundle appositions in cardiac muscle: light microscopy. Am. *J Physiol* 1985;248:H792–H803.
7. Chapman RA, Fry CH: An analysis of the cable properties of frog ventricular myocardium. *J Physiol* 1978;283:263–282.
8. Plonsey R: The formulation of bioelectric source–field relationships in terms of surface discontinuities. *J Franklin Inst* 1974;297:317–324. (Also reprinted in Pilkington T, Plonsey R: *Engineering Contributions to Biophysical Electrocardiography*, IEEE Press, New York, 1982).
9. Plonsey R, Barr RC: *Bioelectricity: A Quantitative Approach*. Plenum Press, New York, 1988.
10. Van Oosterom A: Cell models – macroscopic source descriptions. In: *Comprehensive Electrocardiology*, Macfarlane PW, Lawrie TDV (eds), Pergamon Press, New York, pp 155–179, 1989a.
11. Roberts DE, Scher AM: Effect of tissue anisotropy on extracellular potential fields in canine myocardium in situ. *Circ Res* 1982;50:342–351.
12. Corbin LV, Scher AM: The canine heart as an electrocardiographic generator. *Circ Res* 1977;41:58–67.
13. Clerc L: Directional differences of impulse spread in trabecular muscle from mammalian heart. *J Physiol* 1976;255:335–346.
14. Epstein BR, Foster KR: Anisotropy in the dielectric properties of skeletal muscle. *Med Biol Eng Comp* 1983;21:51–55.
15. Kléber AG, Riegger CB: Electrical constants of arterially perfused rabbit papillary muscle. *J Physiol* 1987;385:307–324.
16. Van Oosterom A, Huiskamp GJ: The effect of torso inhomogeneities on body surface potentials quantified using "tailored" geometry. *J Electrocard* 1989;22:53–72.
17. Geselowitz DB, Miller WT: Active electric properties of cardiac muscle. *Bioelectromagnetics* 1982;3:127–132.
18. Muler AL, Markin VS: Electrical properties of anisotropic nerve–muscle syncytia–II Spread of flat front of excitation. *Biophysics* 1977;22:536–541.

CHAPTER 2:

MODELING OF CELLULAR AND INTERCELLULAR PROPAGATION

Dan Adam and Samuel Sideman

ABSTRACT

The activation of the myocardium is a very complicated phenomenon which depends on cell–membrane processes, as well as on the loading conditions within the cell itself and in its neighboring cells. The complex interactions between the membrane voltage sources and the cell loading in the nonuniform anisotropic three–dimensional (3D) arrays of cells that make up the left ventricular (LV) myocardium are difficult to study. Simulations allow some insight into this 3D process, but consume extremely large computational resources, when required to account for detailed cellular structure and cell–to–cell relationships. The present report describes our attempts to generate normal and pathological ventricular myocardial activation patterns by combining two models: a detailed two–dimensional (2D) small tissue model and a global 3D ventricular model. While the 2D model responds to the physiological parameters of the membrane and allows descriptions of inter– and intracellular processes, the 3D ventricular model produces patterns of activation of the whole myocardium. Combining the two models allows simulation of global activation patterns under different physiological conditions and cellular level intervactions.

INTRODUCTION

The electrical activation process of the myocardium is a very complex 3D process [1–3], which is practically impossible to measure and difficult to analyze. Finite element procedures [4] which can better approximate the real system as well as faster, analytically–based, approaches [5–7] have been suggested. These studies however, involve simplifications and usually compromise by neglecting the cellular level membrane activation properties, relating to assumed average local transport properties, and introducing geometrical consideration and simplification of the 3D heart so as to reduce the computation load.

Clearly, a procedure which will combine basic membrane data with the global scale activation process is highly desired. In this approach a 2D tissue preparation is activated and the microscale activation impulses are allowed to propagate into the 3D ventricles. This approach has been used in physiological experiments [3] and in modeling studies [8], but with limited success, since the extension from 2D to 3D changes the process drastically.

A different approach is suggested here, involving the incorporation of two different models, a 2D myocardial cell based tissue model and a 3D global model of the myocardium of both ventricles. The 2D model is detailed, responding to normal and pathological propagation of the activation signal at that location; it is being activated by, or is activating, the 3D model, thus yielding the global patterns of the ventricular activation.

Physiological studies of the electrical propagation in a 2D myocardial tissue reveal the major parameters which determine the velocity of propagation. The 2D tissue preparation has been used extensively to study the effects of different interventions. It has been reported that the effective longitudinal velocity, along the myofibers, is approximately 3 times higher than the transverse velocity, [9, 10]. There are also measured differences in the action potential (AP) characteristics, depending on its direction of propagation [11, 12]. Ischemia is known to slow down the activation process [10], and induce reentry [13, 14], which is clinically described by ventricular tachycardia or late potentials.

When attempting to model the activation of a 2D myocardial tissue, several options are available. Models that assume continuous properties [15, 16] ignore the problem of discretization, but loose the ability to describe the inhomogeneities of the tissue. Modeling by lumping properties is another, and more common, approach. Here, the discretization, or level of lumping parameters, is of major concern. As in our previous study [17], we model here a 2D piece of myocardial tissue by lumping cellular properties into a single node of the 2D network [18], with the node generating the local AP, as defined by Beeler and Reuter (BR) [19]. The cell–to–cell interconnection is described by a single resistance, and the values of this resistance in the different directions produce the different velocities of propagation, i.e. longitudinal versus the transverse velocities. A single node per cell has been chosen to save computation time, as even a relatively small 2D sheet of cells includes some 4300 cells.

The 2D model of myocardial tissue can be used to test the effects of ischemia, which modifies membrane properties but also increases inter–cellular resistances and thus slows down conduction. As previously reported [13], the geometric shape of the ischemic region, is an important factor in initiating and maintaining reentry, since the shape of the border of the ischemic region may cause a unidirectional block [20].

Experimental data concerning the 3D spread of the activation impulse in the myocardium is rather limited. The measurements are technically difficult, and do not usually have sufficient resolution to describe the sequence of myocardial activation under various conditions. Furthermore, the spatial distribution of properties is altered and the initiation of activity is shifted in pathological circumstances.

Earlier related studies involved a 3D finite–elements model [1, 4] of the LV and the right ventricle (RV), which has been used to study global activation patterns. The studies demonstrated the effects of the spatial arrangement of myocardial fibers, as well as the distribution of conduction velocities, on the myocardial activation patterns [1]. In those studies, the myocardial volume was sectioned into layers and areas, so that different properties may be pre–assigned at each location. The propagation velocity along the myocardial fiber is assumed to be higher then in the transverse direction. The activation patterns have been found to be sensitive to changes in the location of the foci of activation, and to "pathological" slow velocities in the "ischemic" areas.

The transition from normal myoelectrical activity to arrhythmias, tachycardias and fibrillation may be studied by accounting for the appropriate mechanisms. An attempt is made here to combine two models, the 2D tissue model which can describe pathological propagation between cells, and a 3D model of the two ventricles which accounts for the inhomogeneity and anisotropy of the tissue. Such a combination of models can focus on the generation, based on physiological parameters, of myocardial potentials at the particular location where the 2D model is 'inserted', and can produce 3D activation maps [8] of the myocardium. Figure 1 is a graphic presentation of the combined models.

METHODS AND PROCEDURES

The 2D Tissue Model

The following assumptions are used in this model:

1. The 2D lumped model used here includes uniform inter–cellular anisotropic resistivity and constant electrical membrane properties.
2. The model describes a thin, anisotropic piece of myocardial tissue, e.g. an epicardial layer with a necrotic underlying subepicardium [14, 20].

Figure 1: Graphic representation of a cardiac cross–section, exposing the mid–level myocardial wall cross–section, from which the simulation results are given. A thin, 2D anisotropic section of myocardial tissue, described by lumped inter-cellular anisotropic resistivities and BR active cells, is located at the mid–LV free wall. The ischemic area is at its center.

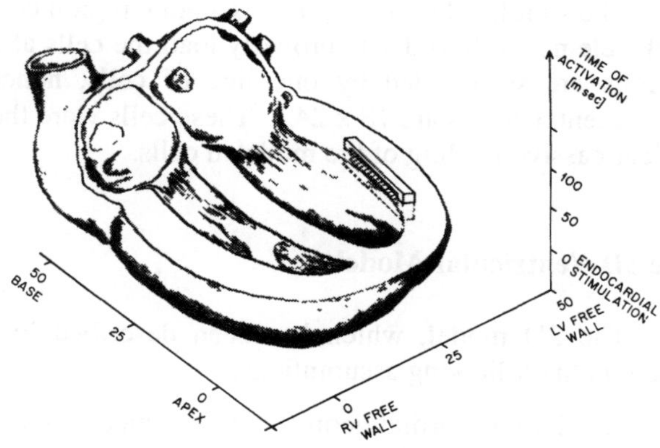

3. Longitudinal inter–cellular and different transverse inter–cellular resistances are used as lumped elements to describe the cytoplasmic and intracellular resistances. The values are modified to produce propagation velocities of 90 cm/s in the longitudinal direction and 30 cm/s in the transverse direction.
4. The interstitial medium is assumed isotropic and unbound, and its resistivity lower than the intracellular resistivity.
5. The intracellular potential is assumed equal to the transmembrane potential.
6. Each cell is characterized by a uniform membrane potential, as described by the Beeler and Reuter relationships [19], which very accurately describe the upstroke and decay of the AP.
7. Simulating the effects of regional ischemia in the 2D piece of tissue is done by increasing junctional resistances 20 times the normal values [18]. The normal ratio of longitudinal to transverse junctional (inter–cellular) resistivities (1:7) is also maintained in the ischemic area.
8. The propagation velocities are taken to have a 3:1 ratio in the longitudinal to transverse directions, and a 6:1 ratio between the normal and ischemic regions.
9. The shape of the border line between the ischemic and normal regions [20] is modified to affect the propagation between these two regions. Figure 1 represents the ischemic area within the 2D model. As shown before [17], an ischemic concave border, or even a straight one, may activate its adjacent normal tissue or be activated by it. An ischemic convex border may activate adjacent normal cells, but may also behave as an unidirectional block and may not be activated by the normal tissue.

The simulated tissue with an ischemic region consisted of 16 x 240 cells, or BR elements. In order to properly load the cells at the edges, the 2D arrays of cells are surrounded by one line of cells, hence the real sizes of the experimental sheets are 18 x 242. These cells were then distally loaded by the typical passive loading of the modeled cells.

The 3D Ventricular Model

The 3D model, which has been described in detail elsewhere [1], is based on the following assumptions:

1. The myocardial geometry is assumed to be non–contracting. The small error added, due to the myocardial twisting and change of size, may have some effect during repolarization, but not during the activation.
2. The myocardial volume is described by ellipsoids, sectioned to 98 x 78 x 98 'cells', and grouped into 4 confocal layers in the LV and 2 layers in the RV. Each layer is divided into 14 subgroups. All the 'cells' in each subgroup have similar properties. The properties are allowed to vary in a Gaussian distribution.
3. Each 'cell' is defined by an absolute refractory period (RP); an activation time (AT) during which the 'cell' may activate its neighbors; a local propagation speed, with one direction in the 3D space in which propagation velocity is 3 times the local propagation velocities in all other 14 directions.
4. The high speed direction is set in the myocardial spiral fibers orientation at that location [1]. Propagation is allowed in all directions from an active 'cell' to all its non–active neighbors.
5. The inner layer (Purkinje fibers) of the LV and RV, is characterized by higher conduction velocities, (2–4 times the velocity in the myocardial fibers) and directed mostly from apex to base. Two–way interaction is assumed between this layer and the adjoining myocardium (the more central parts of the conduction system, above the bundle branches, are not included).
6. Conduction velocity is slow when (a) an area is assumed to be ischemic or (b) when prematurity is simulated and 'cells' which are in the stage of repolarization (during the relative refractory period), are re–activated and propagate the re–activation process at a relatively slow velocity [14].

Membrane properties do not appear explicitly as parameters of the model, but are accounted for indirectly through their effect on the local conduction velocities and the refractory periods. Propagation from one 'cell'

Figure 2: Time of activation of myocardial 'cells' at the mid–level cross–section of the normal 3D model. The Z axis presents, in arbitrary units, the time of activation during one beat, after stimulation at 4 locations in the Purkinje layer. Highest peaks represent latest activation.

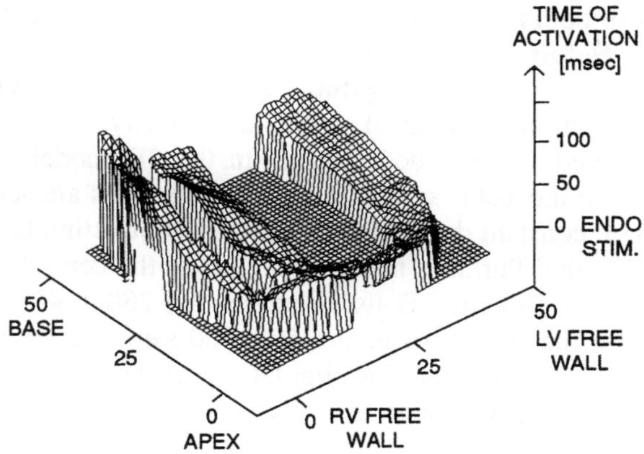

to its adjacent neighbors thus depends only on geometrical relationships. This simplification allows for a significantly reduced computational complexity.

The parameter values for all 'cells' are selected by setting the mean and S.D. of the mean, before each simulation. The normal stimulation points are located on the endocardium, within the Purkinje network. At each simulation time step, the 'cells' which are active effect their neighbors according to the rules stated above. In this manner the activity spreads to all 'cells' that are not refractory. The results of simulation of normal conditions are given, in Fig. 2, as activation times of each 'cell' in the central cross–section. Secondary activation wavefronts may appear when the 2D tissue model generates a reentry process or premature beats due to slowed conduction.

Interaction Between the 2D and 3D Models

As the emphasis in this study has been only on the activation process, it is possible to calculate the activation in the 2D tissue and the 3D ventricular model separately, and then relate the results from one to the other. Specifically, a normal beat is first simulated in each of the models. Next, pathological conditions are imposed on the 2D model, and the activation process is simulated. The activation times and the end of refractory times are transported to the 3D model, where the bordering 'cells' are activated, if they are not refractory.

The two models obviously relate to elements of different sizes. The size ratio of 3D to 2D elements is 7:1 in the longitudinal axis. The 2D tissue model is 'inserted' at a specific coordinate (x,y,z), with an orientation of choice. This is simply done by specifying 3 corner points of the 2D model in the 3D coordinate system. In the simulation results presented here, the longitudinal axis of both models are matched, thus simplifying the transfer of data.

Before each experiment, all the parameters are set in both models. The 2D model requires the gating parameters, conductivities and other geometric and electrophysiological data. Additionally, the inter–cellular resistors are calculated as described above. In the 3D model, the conduction velocities between the 'cells' and their refractory periods are selected. These parameters remain constant during the simulation. Four stimulation points, located on the endocardial Purkinje network, are set at the central cross–section plane. The simulations of the 2D tissue model run 280–350 simulation time steps (e.g. ms) for the normal case, and 600–800 steps for the case of reentry with slow ischemic conduction. The simulations of the 3D model run 60–120 simulation time steps, long enough to activate all the myocardium.

RESULTS

In the normal 2D myocardial tissue model, the complete AP is generated at each cell (node), and an excited node affects its neighbors by charging them through the resistors until they reach their own threshold and become active. The AP calculations proceed according to the BR set of equations, calculated at time resolution of 1 ms. The results are shown as a series of rectangles depicting the same 2D array of cells at consecutive (1 ms apart) intervals, after the stimulation. The color code (gray level) allows to demonstrate the spatial variations of the transmembrane potentials and, by setting a threshold, the pattern of propagation of the electrical activity can be depicted. The whole AP time course may be followed by concentrating on a single cell (node) in the 2D space over the series of rectangles.

The combined two models are tested for their capability to generate normal patterns of activation; the 2D model is 'inserted' at a given location, is activated by the normal activation process of the 3D model. After propagation within the 2D model, the 'cells' activate the 3D model 'cells' at its side. Figure 3 depicts the simulation results of this experiment. Comparison to Fig. 2, which depicts the simulation results of the 3D model alone, shows that the overall pattern of activation of the 3D model is not affected at all by the inclusion of the 2D model. The similarity between Figs. 2 and 3 is evident, taking into account the variability of different normal simulations, with somewhat different parameters.

Next, we introduce an example of using the 2D model under pathological conditions corresponding to conditions of premature ventricular contractions (PVC). Since the main purpose of this study is the combination and proper synchronization of the two models, the initiation of the premature electrical activity has been dealt here in a simplistic way. A set of 4 'cells' in

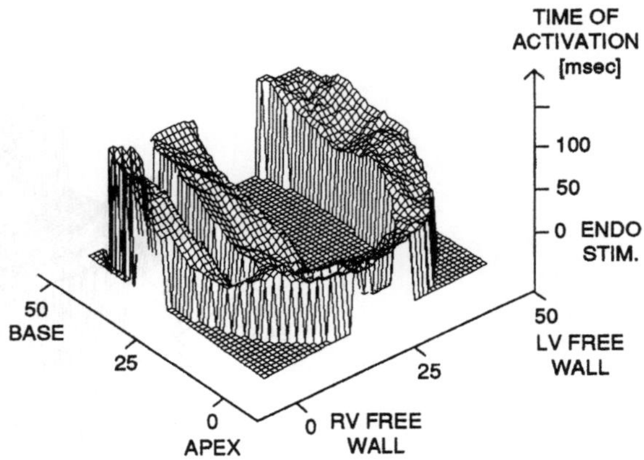

Figure 3: Time of activation (Z axis) of myocardial 'cells' at the mid-level cross-section under normal conditions. Similar to Fig. 2, but with the 2D tissue model incorporated. The 2D model activates the 3D model at the proper timing and location.

the 2D model have been stimulated by an external current source, i.e. no automaticity has been simulated. The use of an external current source generates a rapid upstroke of the AP, and spread the activity in all directions, masking the problem of modeling pacemakers, where few cells need to activate a large array of cells (see [18]). The stimulated cells in the 2D tissue model produce an elliptical pattern of activation, since the anisotropic conduction parameters impose a 3:1 conduction velocity ratio between the longitudinal and transverse directions. (This is created by a 7:1 ratio of resistances between the nodes.) Once the activity front reaches the borders of the 2D tissue model, the 3D ventricular model is activated at the adjacent border, by transferring all the required information from one model to the other. The pattern of activation in the 3D model is much more complex than in the 2D model, as the orientation of the 'fibers', and therefore the different conduction velocities, vary from layer to layer and from one location to the other. Nevertheless, a nearly elliptical spread of the second wavefront is seen in Fig. 4 (a), superimposed on the first normal activation at the central cross-section. This premature activation, initiated at the LV free wall, activates this wall (Fig. 4 (b)), and then propagates through the Purkinje network and activates the septum and then the RV free wall (Fig. 4 (c)).

The second pathology studied here is that due to the generation of a reentry loop which reactivates the whole myocardium. Thus, a single beat of ventricular tachycardia is simulated. First, a normal beat is generated in the 3D ventricular model by the normal stimulation of the Purkinje network. The activation front propagates normally until it reaches the 2D tissue. The 2D tissue model is now programmed to describe a different condition: local ischemia. The ischemic cells are allocated inter-cellular resistive values (20 times the normal values), producing conduction velocities equivalent to 1/6 of the normal velocities in both directions. An important feature here is the shape

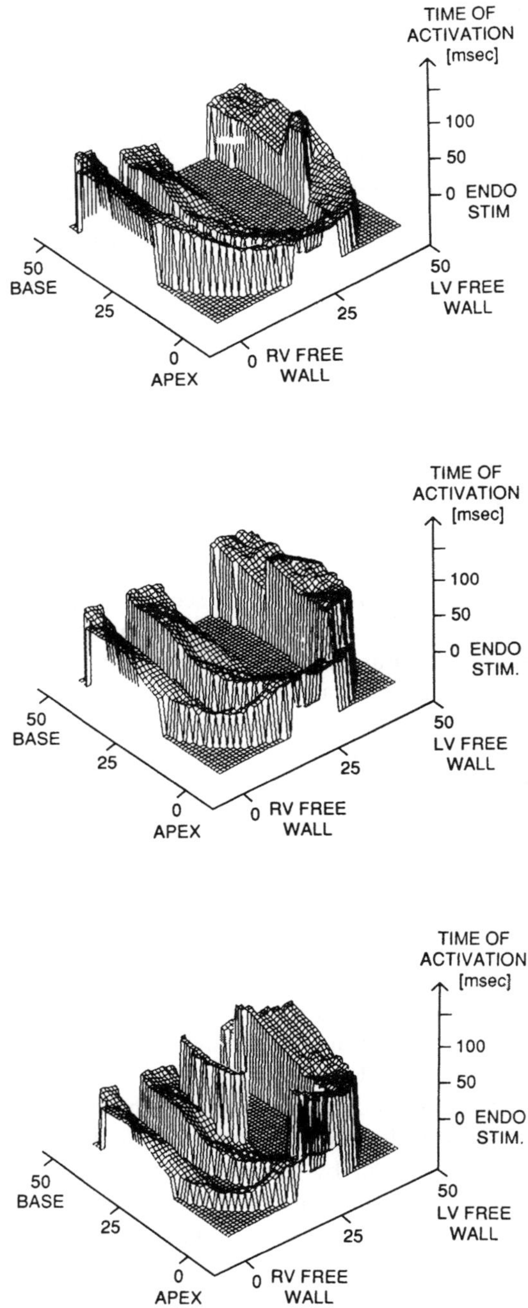

Figure 4: Time of activation (Z axis) of myocardial 'cells' at the mid–level cross–section. After the first normal activation of the whole myocardium, (3D model), the generation of patho–logical activity at the 2D tissue model spreads out through and activates the 3D model. (a) 10 ms after the second ectopic activity is initiated, (b) and (c) the same simulation but 20 and 30 ms later, respectively. Higher peaks (on the Z axis) represent longer times since the initial stimulation.

of the ischemic area, as it was found [14, 17] that an entry block, and therefore reentry activity, may be generated due to the particular shape of the ischemic area. Here, the ischemic area is made up of a rectangle with straight borders, except the left–side border (see below). Very high transverse resistances at the upper and lower borders of the ischemic area allow only for an extremely slow, hardly noticed, propagation in these directions. The straight borderline on the right allows excitation of the ischemic area by the normal tissue.

The concave shape of the left–side borderline is the cause of the entry block. As shown [17], the block is generated when each of the normal cells at this borderline is required to activate more than one ischemic cell, and is unable to do so. This result depends on the model used, here BR model, when a cell is unable to drive the membrane potential of its neighboring ischemic 'cells' above their threshold. In the example given in Fig. 5, in a tissue size of 16 x 240 cells, the left side border of the ischemic area has been set as jagged and convex. Consider that the activation front which propagates from the endocardium of the 3D model reaches the 2D healthy tissue at the lower edge of the 2D tissue piece. The activity then propagates (upwards) and around the ischemic area, until it reaches the border mentioned above (from left to right), and the entry block is complete. The activation wavefront reaches also the right side of the ischemic area, where it penetrates and propagates from right

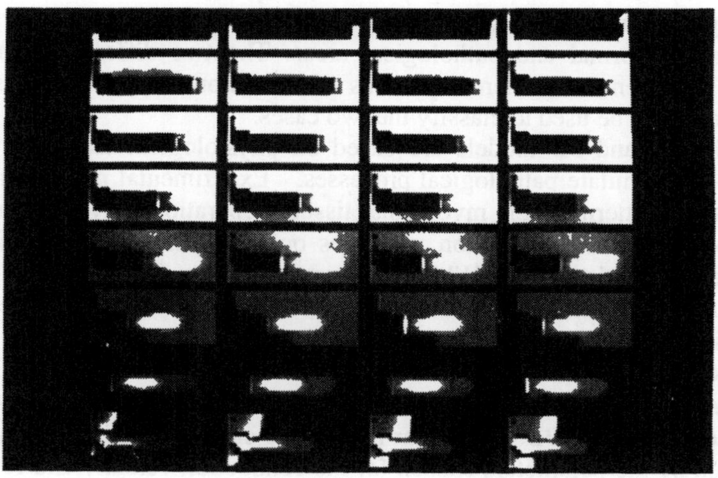

Figure 5: Effects of an area of ischemia in the 2D tissue model, with convex shape of the borderline at one end producing a unidirectional block. Model size is 16x240 cells. The normal activation is propagating upwards from the lower part of the piece of tissue. The 2D piece of tissue model is depicted several times at different time instants. The color code (gray level) demonstrates the spatial variations of transmembrane potentials.

to left. Contrary to the block encountered earlier, the ischemic cells are able to activate a healthy cell, specifically when the cells ratio is 2:1.

The reentry loop produced in this manner, creates a circulating wavefront as is clearly demonstrated in Fig. 5. Once the circulating wavefront reaches the borders of the 2D tissue model, it activates the 3D model at these locations, in all the 3D directions. However, due to the anisotropy of the 3D model, the penetration of the activation is not even at all the 2D–3D border points around the 2D tissue. This effect, shown in Fig. 6 (a), is smoothed out soon afterwards, as seen in Fig. 6 (b), and several ms later in (Fig. 6 (c).

DISCUSSION

The purpose of this study is to investigate the possibility of incorporating a 2D cell–based model of the myocardial tissue in a 3D global model of the LV and the RV. Both models account for the anisotropic properties of the tissue, specifically those that control the propagation velocity [4, 9].

The combined models are first tested by simulating normal patterns of propagation, and then used to study the effects of simulated ectopic activity and the effects of a regional area of slow conduction. An area of slowed conduction, which occurs during ischemia, may form a unidirectional block, resulting in a reentry, depending on the geometrical shape of the border between the normal and pathological areas. The two different pathologies studied here generate different patterns of activation in the ventricles; these may eventually be used to classify the two cases.

The 2D and 3D models are based on physiological data, and can be extended to simulate pathological processes. Experimental measurements of propagation patterns in 2D myocardial tissue preparations have demonstrated the dependence of conduction velocities on membrane characteristics [11] and fiber orientation [4, 9]. While the former property has been used only in the 2D model, the latter has been used in both 2D and 3D models: Membrane properties appear explicitly as parameters of the 2D tissue model, corresponding to the BR formulation, and therefore this model can be used to represent the area that generates the pathologies. In the other parts of the myocardium, i.e. the 3D ventricular model, the electrical activity is indirectly described by the parameters chosen for the conduction velocity and refractory period over a few 'cells'. Thus, only a small compromise is made on the spatial resolution of the calculations and the display of the results, while the computational complexity is reduced very significantly.

The combination of the anisotropy of the tissue, the existence of an area of slow conduction and of some kind of a conduction block [5], has been considered to be a major contributor to the generation and maintenance of reentry loops [13, 14]. This phenomenon has been demonstrated here in the 2D model, and extended to show its effect on the whole myocardium. The 2D model has been successfully used to simulate regional ischemia (i.e. regional area of increased inter-cellular resistance), and demonstrate the initiation and maintenance of a sustained reentrant activation. The initiation of the reentry loop is caused here by the unidirectional block at the border between the healthy and ischemic tissue. Reentry formation depends, therefore, on the

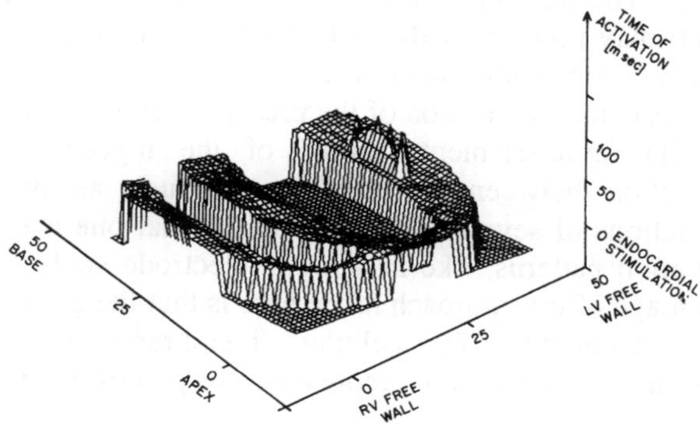

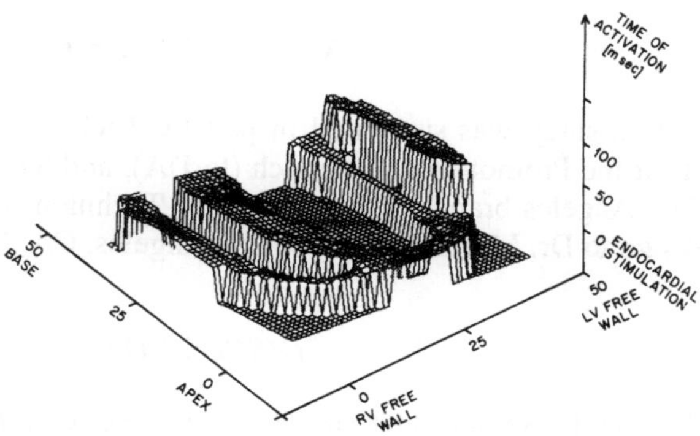

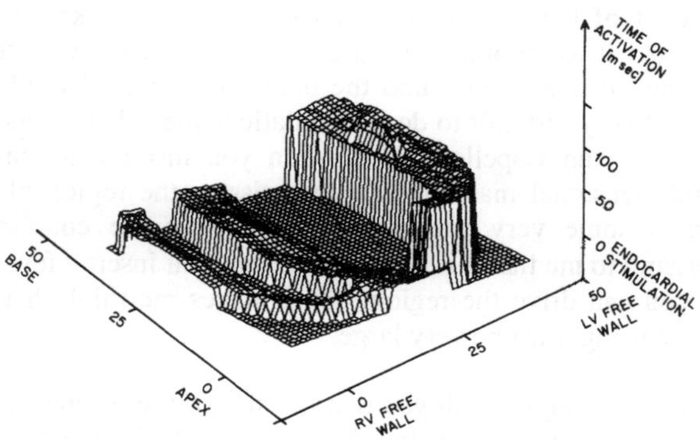

Figure 6: Time of activation (Z axis) of myocardial 'cells' at the mid-level cross-section. Similar to Fig. 4, but here the pathological conditions at the 2D model (local ischemia) generate reentry, which re-activates the whole myocardium of the 3D model. (a) 3 ms after re-activation of the 3D myocardial 'cells', (b) 10 ms later, and (c) 20 ms later.

direction of the propagating wavefront when it reaches this area. The normal activation wavefront propagates from endocardium to epicardium [3], moving perpendicular to the fiber orientation. When reaching the ischemic area (as in Fig. 5), the activation front may propagate normally through the unidirectional block, if the wavefront reaches the right side earlier. The wavefront may also deviate somewhat and reach the ischemic area in a different route, say at the left side, and be blocked. This block may initiate a sustained reentry loop. The length of the reentry loop depends on the length of the slow conducting fibers. The formation of the reentry loop describes the underlying processes which may produce body surface late potentials, which then deteriorate into sustained ventricular tachycardia.

A natural extension of the present attempt to combine the 2D model with the 3D finite–elements model of the myocardium is to automate the interaction between the models as much as possible. This will allow simulations of several cardiac beats, so that one may study sequential global activation patterns, like in a multi–electrode study of the heart. The biggest advantage of the approach taken here is that the global patterns may be studied while accounting for cellular (membrane) properties and intracellular resistances in a small but representative part of the myocardium.

Acknowledgements

This study was supported in part by Technion VPR – Henry Gutwirth Fund for the Promotion of Research (to DA), and the Technion 2000 group of the Los Angeles branch of the American Technion Society, USA. Particular thanks go to Dr. Martin Kellner of Los Angeles, CA, USA.

DISCUSSION

Dr. Joyner: I suspect that the size of the finite elements that you used for the model has a lot of bearing on the results. You have taken a Beeler–Reuter model and converted it to automatic pacing and then inserted two cells, or rather two discrete elements of this model, into the multi–dimensional field, and drive the sheet. We find that very difficult to do in simulations that I have done in collaboration with Drs. Janse and van Capelle *[21]*. When you insert a region of automaticity within a multidimensional mass of quiescent tissue, the region of automaticity stops, unless there is some very special arrangement of the coupling resistance. It is very surprising to me that this can happen; that you insert a few cells with automaticity and that you can drive the region. That makes me think that the discrete elements that you are using must be very large.

Dr. Adam: I agree with you that the size of the discrete elements of the model has a strong bearing on the results. Since, in this study, our purpose was to investigate and generate the whole myocardial pattern of activation, and not the loading effects of quiescent cells on their active neighbors, we have chosen the parameters' values to describe elements of the size of several physiological myocytes. Under these conditions, a few elements in the 2D tissue model were modified to describe automaticity and they generated enough current to activate their neighbors. It is important to note that for the 2D tissue model, the inter–elements (or intercellular) resistances are high in the transverse direction to the myofiber axis, so most of the current flows along the myofiber axis direction. When the 2D tissue is fully activated, it is large enough to activate the elements of the 3D myocardial model.

Dr. Joyner: What is the size of the discreet elements in the 2D tissue model? How many microns are the edges of those finite elements?

Dr. Adam: Each element in the 2D tissue is assumed to be 160 μ in length, i.e. the length of about two myocytes. The whole length of the 2D sheet (which included 240 elements) is about 38 mm. When this 2D model was fully activated, there were enough cells here to activate the other model.

Dr. Rudy: Was the 3D simulation of propagation in the whole heart also a discretized model based on elements of excitable membrane, i.e. Beeler–Reuter?

Dr. Adam: No, the 3D model of the two ventricles describes a large volume of many physiological cells, and therefore must include many elements. Each element is therefore very simple (with only three parameters: geometry, time of activation, refractory period) and only coarse generalized patterns of ventricular activation are generated.

Dr. Rudy: What was the size of the discretization element in your 3D tissue model? It is important, especially in view of the ability to drive the tissue.

Dr. Adam: In the 3D whole ventricles model, the size of the elements is about 1 mm^3. The ratio to the 2D tissue model is about 7:1 (when considering the longitudinal direction of the cells).

Dr. Pollack: In your 3D ventricular model, the only parameter which determines propagation velocity is a discrete delay from one element to the next. Do you think that is valid? Is your model sensitive to the difference between that sort of approach to regulate intercellular conduction or a different approach of controlling the velocity of propagation?

Dr. Adam: That is a very tough question. Let me say that while in the 3D ventricular model only a fixed delay (and geometry) determines the propagation velocity, in the 2D tissue model inserted within the 3D model, the resistance between cells, i.e. between Beeler–Reuter models, modified the conduction velocity. It is basically the time constant of the RC, the serial resistance and the capacitance of the next Beeler–Reuter model, which defines the propagation velocity of the activation. Under these conditions, the action potential amplitude and shape (e.g. under pathological conditions), have an effect on the propagation velocity. There is no intracellular propaga–

tion in this present model. We are working hard on this topic just now, trying to study propagation within the cell as compared to the effects of the gap junction; how it modifies the overall tissue conduction velocity. When propagation within the cell is studied, membrane properties are taken into account. Though the process within the cell is rapid, channel delays and other properties that determine the propagation velocity, are included in our model as they may play a role under pathological conditions.

REFERENCES

1. Adam D, Barta B: The effects of anisotropy on myocardial activation. In: *Simulation and Control of the Cardiac System, Volume III*, Sideman S and Beyar R, (eds), CRC Press, Boca Raton, 1987, pp 13–27.
2. Solomon JC, Selvester RH: Simulation of measured activation sequence in the human heart. *Am Heart J* 1973;85:518–523.
3. Spach MS, Barr RC: Analysis of ventricular activation and repolarization from intramural and epicardial potential distributions for ectopic beats in the intact dog. *Circ Res* 1975;37:830–843.
4. Adam D, Sideman S: Aberrant conduction patterns in and around an ischemic zone. In: *Analysis and Simulation of the Cardiac System – Ischemia*, Sideman S and Beyar R (editors), CRC Press/FL, 1989, pp 189–199.
5. Barta E, Adam D, Salant E, Sideman S: 3–D ventricular myocardial electrical excitation: a minimal pathways model. *Annals Biomed Eng* 1987;15:443–456.
6. Barta E, Sideman S: 3–D simulation of model of the electrical activation patterns in the ischemic heart. In: *Analysis and Simulation of the Cardiac System – Ischemia*, Sideman S and Beyar R (editors), CRC Press/FL, 1989, pp 175–188.
7. Barta E, Sideman S: 3D simulation of electrical activation patterns in the ischemic left ventricle. *Annals Biomed Eng*, 1991;in press.
8. Okajima M, Fujino T, Kobayashi T, Yamada K: Computer simulation of the propagation process in excitation of the ventricles. *Circ Res* 1968;23:203–211.
9. Clerc L: Directional differences of impulse spread in trabecular muscle for mammalian heart. *J Physiol* (London) 1976;255:335–346.
10. Kléber AG, Janse MJ, Wilms–Schopmann FJG, Wilde AAM, and Cornel R: Changes in conduction velocity during acute ischemia in ventricular myocardium of the isolated porcine heart. *Circulation* 1986;73:189–198.
11. Spach MS, Miller WT, Geselowitz DB, Barr RC, Kootsey JM, Johnson EA: The discontinuous nature of propagation in normal canine cardiac muscle. Evidence for recurrent discontinuities of intracellular resistance that affect the membrane currents. *Circ Res* 1981;48:39–54.
12. Spach MS, Miller WT, Dolber PC, Kootsey JM, Sommer JR, Mosher CE: The functional role of structural complexities in the propagation of depolarization in the atrium of the dog. *Circ Res* 1982; 50: 175–191.
13. Allessie MA, Bonke FIM, Schopman FJG: Circus movement in rabbit atrial muscle as a mechanism of tachycardia. III. The "leading circle" concept. *Circ Res* 1977;41:9–18.
14. Lammers WJ, Wit AL, Allessie MA: Effects of anisotropic conduction on properties of functional reentrant circuits. In: *Simulation and Modeling of the Cardiac System: From Cellular Activation to Muscular Activity*, Sideman S, Beyar R (eds), Martinus Nijhoff, The Hague, 1988, pp 133–150.

15. Plonsey R, Barr RC: Current flow patterns in two-dimensional anisotropic bisyncytia with normal and extreme conductivities. *Biophys J* 1984;45:557-571.
16. Geselowitz DB, Miller WT: A bidomain model for anisotropic cardiac muscle. *Ann Biomed Eng* 1983;11:191-206.
17. Adam D, Hoffman N, Sideman S: On the generation of re-entrant activity by a lumped 2-Dimensional model. In: *Imaging, Analysis and Simulation*, Sideman S, Beyar R (eds), Freund Publ., London, 1990, in press.
18. Roberge FA, Vinet A, Victorri B: Reconstruction of propagated electrical activity with a two-dimensional model of anisotropic heart muscle. *Circ Res* 1986;58:461-475.
19. Beeler GW, Reuter H: Reconstruction of the action potential of ventricular myocardial fibres. *J Physiol* 1977;268:177-210.
20. Ursell PC, Jardiniere PI, Albala A, Fenoglio JJ, Wit AL: Structural and electrophysiological changes in the epicardial border zone of canine myocardial infarcts during infarct healing. *Circ Res* 1985;56:436-451.
21. Joyner RW, van Capelle FJL: Propagation through electrically coupled cells: how a small SA node drives a large atrium. *Biophys J* 1986;45:1157-1164.

CHAPTER 3:

ANISOTROPIC CONDUCTION AND INHOMOGENEITY IN REFRACTORINESS

Frans J.L. van Capelle and Michiel J. Janse

ABSTRACT

Regional differences in refractoriness can set the stage for the occurrence of reentrant tachycardias. They may be due to regional differences of the intrinsic membrane properties, but conduction characteristics may also be of importance, especially when the wavefront abruptly changes its direction, switching from longitudinal to transverse propagation or vice versa.

INTRODUCTION

One of the prerequisites for the genesis of reentrant arrhythmias is the existence of a sufficiently large area of unidirectional block. This occurs as a consequence of differences in refractory periods between adjacent areas. This may be caused by intrinsic differences in membrane properties in various regions of the heart, or else it may accompany abrupt changes in the direction of propagation in an anisotropic medium. Here, we briefly discuss these different mechanisms of inhomogeneous conduction.

ANISOTROPIC CONDUCTION

The characteristics of anisotropic conduction were studied in a computer model that has been described in detail elsewhere [1, 2]. Briefly, it consists of up to 650 excitable elements which can be connected in different ways in a two–dimensional matrix. The behavior of the different elements is determined by a set of parameters that can be adjusted to trim their electrophysiological properties (e.g. the duration of the refractory period). The geometrical arrangement of the elements, together with the values of the coupling resistances is specified interactively using a graphic editor. We simulated anisotropic conduction in a rectangular sheet by connecting the elements with different resistances in the horizontal and vertical directions. Figure 1 shows a circus movement of the activation wave in a sheet of 41 x 10 identical excitable elements. The elements (indicated as small '+' signs in the sheet) were coupled to their immediate neighbors in a square lattice. The vertical coupling resistances were four times as large as the horizontal ones, resulting

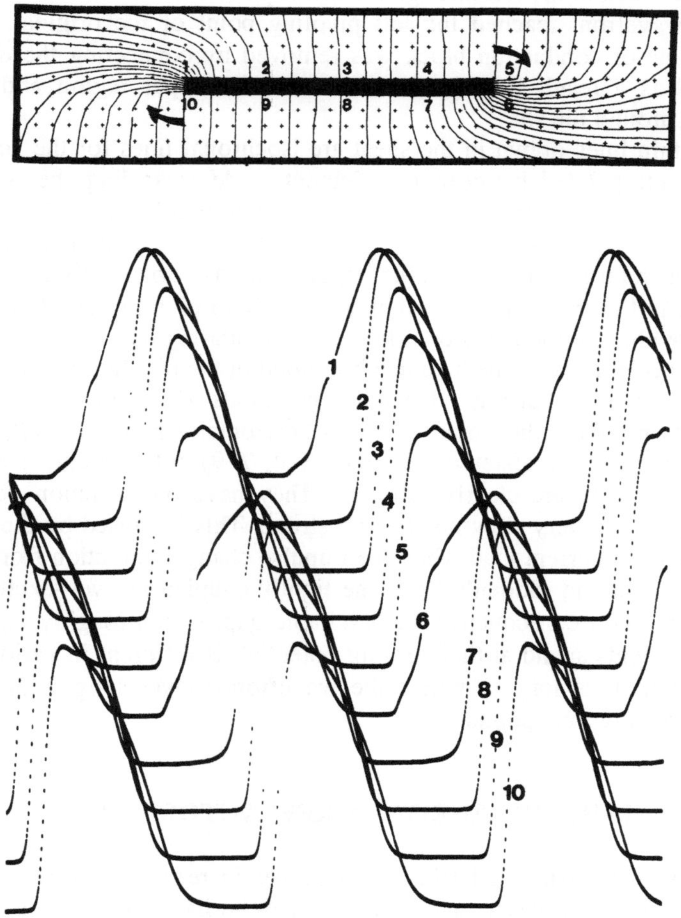

Figure 1: Action potentials of selected elements during circus movement of the activation around a slit in an anisotropic rectangular sheet of excitable elements (above). Isochronal lines (10 ms apart) indicate activation times of the elements. One complete cycle (540 ms) is depicted in the upper panel. Conduction delay at the pivoting points, as apparent by the crowding of isochronal lines, corresponds to broadened action potentials of the elements at the distal side of the pivot (reproduced with permission from [2]).

in a marked difference of the propagation velocity of the impulse in the 2 directions. A horizontal slit was made in the middle of the sheet by deleting the transverse coupling resistances between opposite elements for 20 element pairs (horizontal thick line). The circulating wave was started by temporarily extending the slit at one end to the left edge of the sheet and then stimulating an element in the upper left corner. In this way a wavefront was set up that traveled through the upper half of the sheet to the right, turned around the slit and traveled back to the left through the lower part of the sheet. Just before the returning wavefront reached the left pivoting point of the original slit, the extended line of conduction block was removed by connecting the elements again, and as a result the impulse reentered the upper region and started circulating around the slit.

The thin lines drawn in the sheet are isochronal lines for the activation moments, interpolated between the elements. After scaling the refractory period of the elements to about 330 ms, the isochronal lines are 10 ms apart. One complete cycle (540 ms) of the 'tachycardia' is depicted. The isochronal lines are much farther apart when the impulse travels horizontally than when it propagates in the transverse direction. Around both pivoting points there is a distinct crowding of the isochronal lines; the circulating impulse spends about half of the cycle in the immediate neighborhood of the pivoting points.

In the lower part of Fig. 1 the action potentials of 10 cells adjacent to the slit are shown. When the impulse travels horizontally along the slit, smooth action potentials are registered (elements 2–4, 7–9) but elements 1 and 6 at the pivoting points are clearly different. They have broad action potentials characterized by a very slow notched up stroke. This is caused by the sudden increase of axial current load when the impulse changes direction from vertical to horizontal and suddenly 'sees' the tightly coupled not yet–excited cells downstream. As a consequence, the excitable gap, which roughly equals the time the elements spend at their resting potential, is different in various parts of the circuit; it is longest where the wavefront moves longitudinally and shortest at the pivoting points.

INHOMOGENEITY IN REFRACTORINESS

It is well known that intrinsic differences in refractory period duration exist in the normal ventricle and that they may be increased by interventions such as ischemia or brief periods of enhanced sympathetic activity. It is, however, very difficult, if not impossible, to determine refractory periods at multiple sites in the ventricles during brief, transient episodes of enhanced sympathetic nervous activity using the classical extra–stimulus technique. A method allowing estimation of cardiac refractoriness simultaneously at many sites is therefore highly desirable. Such a method was employed by Lammers and co–

workers [3], who used the average interval between local atrial activations during atrial fibrillation as an index for local refractoriness. The underlying concept was that during multiple wavelet reentry, responsible for fibrillation, cells would be reexcited as soon as they had recovered from refractoriness.

In essence, we used the same approach to determine spatial dispersion in ventricular refractoriness by recording ventricular extracellular electrograms simultaneously at multiple sites during ventricular fibrillation in dogs on total cardiopulmonary bypass, before and after brief periods of stellate ganglion stimulation.

Figure 2 shows the results of one experiment in which VF–intervals and classically determined refractory periods, during a regularly driven rate (basic cycle length 350 ms), were determined at the same site. As can be seen, there is a good correlation between the two parameters. During the paced rhythm, the difference between shortest and longest refractory period is in the order of 30 ms, corresponding to a 14 ms difference in VF–intervals.

Figure 3 shows an example of an experiment where VF intervals at sixteen recording sites are depicted. Electrodes 1 to 6 were on the anterior wall of the left ventricle, electrodes 7 through 14 on the lateral, posterior left ventricular and right ventricular wall, and electrodes 15 and 16 joining with electrode 1. In the middle panel, values are shown before stimulation of the left stellate ganglion; in the lower panel, values directly after left stellate stimulation and in the upper panel, the differences between the two. In this heart, left stellate ganglion stimulation predominantly shortened the VF–interval in the anterior wall, while VF–intervals in the lateral wall and part of the posterior wall were hardly affected.

Out of a total of 11 hearts, a distinct regional effect of either left or right stellate stimulation was found in 6 hearts, whereas in the remaining 5 a more diffuse pattern was found in which the VF–interval was shortened equally in all recording sites. When a quantitative assessment of dispersion in VF–

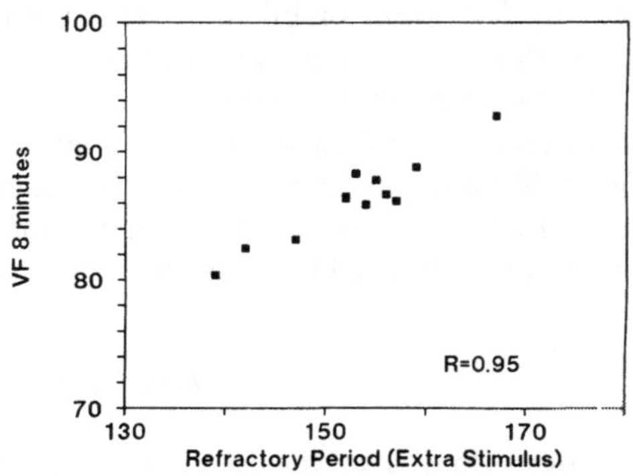

Figure 2: Plot of VF–intervals against classically determined refractory periods at the same eleven ventricular sites (reproduced with permission from [4]).

Figure 3: VF–intervals measured at sixteen electrode sites, positioned in a row and encompassing the circumferences of both ventricles. Middle panel shows values before stellate ganglion stimulation, lower panel those immediately following left stellate ganglion stimulation, upper panel the differences between the two. Note regional effects of left stellate stimulation (reproduced with permission from [4]).

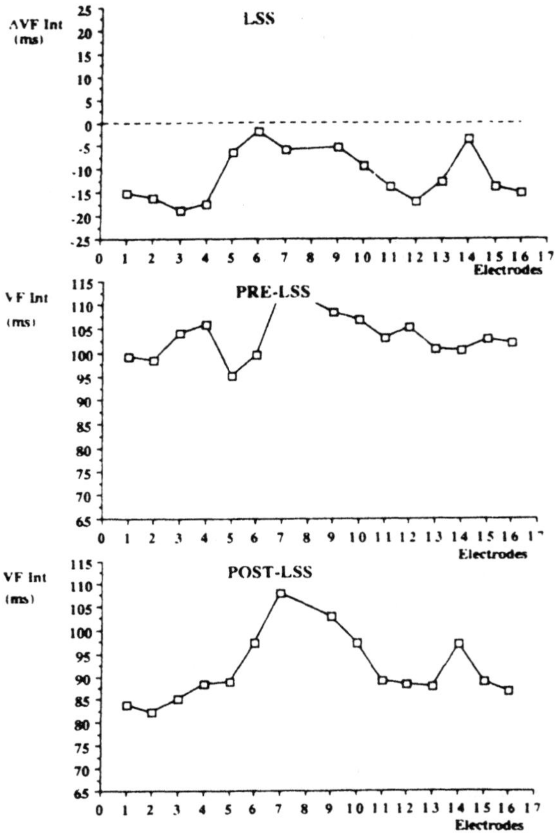

intervals was made, by calculating the variance (i.e. the square root of the standard variation), it was found that a significant increase in dispersion caused by left, right or bilateral stellate stimulation only occurred in the 6 hearts with a distinct regional distribution.

In the intact heart, both intrinsic differences in membrane properties and the consequences of anisotropic conduction will play a role and it will be difficult, if not impossible, to assess the exact contribution of each component. The great advantage of computer simulation is that it allows the study of the influence of each variable on its own.

DISCUSSION

Dr. Jalife: Do you agree that if you could start reentry either in the counter-clockwise or clockwise direction, it means really that one does not need to invoke anisotropy to explain the mechanism of initiation of the arrhythmia? Probably anisotropy modulates the properties of the propagating process once the arrhythmias has been established, but it is possible to induce reentrant activity through unidirectional block, either longitudinally or transversely. That is, the initiation or reentry is probably independent of the fiber orientation. Is that a fair statement to make?

Dr. van Capelle: Yes. It depends very critically on the distribution of resistance. There seems to be a critical resistance value in propagation. The moment that propagation starts to fail is related not only to the discrete mechanism but also to the distributed part of the resistance. I think I saw some kind of draft by Keener, who has developed a beautiful theory (I do not know if it has been published yet), that seems to underline these concepts. This is the way we have to think about it now.

Dr. Gettes: This is beautiful. Is there a critical range of ratios between longitudinal and transverse resistances?

Dr. van Capelle: No, I started off with trying to get the ratio between longitudinal and transverse conduction right, but I had a technical problem with those simulations because of the limited amount of memory. Therefore, the transverse coupling was set so that the transverse direction was just propagated for the normal piece, and that was what we started out with. Incidentally, I had 4 of those resistances leaving the cell. I don't think that is very important if it is one small or 4 large resistances in parallel. Essentially, we had very marginal conduction in the transverse direction and we modulated the longitudinal one so as to get both types of block.

Dr. Joyner: Your issue of the effects of the discrete spacing of cell connections is very important. We have done similar simulations in a one-dimensional case only. Your point of how you arrange the discrete elements makes a great difference. In our simulations, when we use discrete elements, which were actually about the size of one cell, using *delta* X of the 100, or 200 microns, on the basis of whether the cells were arranged longitudinally or transversely in one-dimensional strand, we found that we had to go to astronomical values of coupling resistance to see any of the discretization effects. However, when we considered cells to be in bundles, as you have done here, with 5–6 cells well coupled and then a high resistance gap to another group of 5–6 cells, then we could see discretization effects. Therefore, there was an effective *delta* X which produced bundle-to-bundle conduction. Most of the published models are thoroughly confusing because they did not consider this effect. You have to use totally unrealistic resistance values when you model the system as a homogeneously connected row of isolated cells.

Dr. van Capelle: I could not agree more. It is a very important point and perhaps I did not stress it enough. It is very misleading to represent those elements as cells at the cellular level. We all know that in reentrance tachycardias, if you really look carefully, the scale of the whole thing seems to be so much larger than the scale of the single cell. It must be something like cells lumped together in bundles or something.

REFERENCES

1. Van Capelle FJL: Propagation and reentry in two dimensions, In: *Cardiac Electrophysiology: from Cell to Bedside*, Zipes DP, Jalife J (eds), WB Saunders Co., Philadelphia, 1990, pp 175–182.
2. Van Capelle FJL, Allessie MA: Computer simulation of anisotropic impulse propagation: characteristics of action potentials during reentrant arrhythmias. In: *Cell to Cell Signalling: From Experiments to Theoretical Models*, Goldbeter A (ed), Academic Press, London. 1989, pp 577–588.
3. Lammers WJEP, Allessie MA, Rensma PL, Schalij MJ: The use of fibrillation cycle length to determine spatial dispersion in electrophysiologic properties used to characterize the underlying mechanism of fibrillation. *New Trends in Arrh* 1986;2:109.
4. Janse MJ, Opthof T, Ramdat Misier AR, Vermeulen JT, Frank RGJ, Van Capelle FJL: Sympathetic stimulation causes inhomogeneity in ventricular refractoriness. *New Trends in Arrh* 1990;6:177–183.

CHAPTER 4:

DIFFERENT MECHANISMS FOR REENTRY IN THE EPICARDIAL BORDER ZONE OF HEALING CANINE INFARCTS: THE ROLES OF FUNCTIONAL AND ANATOMICAL CONDUCTION BLOCK

Adam E. Saltman, James Coromilas, Bernd Waldecker, Stephen M. Dillon, and Andrew L. Wit

ABSTRACT

Excitation patterns in reentrant circuits that cause ventricular tachycardia in healing canine myocardial infarcts were mapped. We have found that the circuits can either have a central region of functional or anatomic block. The mechanism causing the central region of block may have important influences on the properties of the reentrant circuits.

INTRODUCTION

Sustained ventricular tachycardia with many characteristics similar to clinical tachycardias can be induced in the canine heart within 5–7 days after complete occlusion of the left anterior descending coronary artery [1, 2]. Complete reentrant circuits that cause the tachycardia have been mapped in this experimental model in a narrow epicardial sheet of surviving muscle called the epicardial border zone [2, 3]. The canine heart is therefore a useful model to investigate mechanisms causing reentry in healing infarcts.

In several of the previously published studies on reentry in the epicardial border zone, including our own, emphasis has been placed on the functional nature of the circuits [2, 3]. The term "functional" is used to describe reentrant circuits that transiently form after an inciting stimulus because of special electrophysiological properties of the participating cardiac muscle fibers. Functional circuits do not have preformed anatomical pathways or a central region of anatomical block around which the impulse circulates [4]. However, now that we have mapped reentrant excitation in a larger number of infarcts, we find that in some instances reentrant circuits may have an anatomical basis, that is, the central region of conduction block around which the impulse circulates may be caused by an anatomical defect in the epicardial border zone. In this communication we briefly describe examples of the two kinds of reentrant circuits, functional and anatomic.

METHODS

The left anterior descending coronary artery was ligated in adult mongrel dogs as previously described [2]. Four days following surgery the dogs were reanesthetized with 10 – 30 mg/kg pentobarbital i.v., as needed to produce a surgical level of anesthesia. The chest was opened through a median sternotomy and the heart suspended in a pericardial cradle, exposing the infarcted anterior left ventricular surface. A 9 cm x 13 cm flexible polymer sheet approximately 0.3 mm thick, containing 300 bipolar electrodes was fixed to the left ventricular epicardium over the infarct with interrupted 2–0 silk sutures around its perimeter. Electrode pairs were 6–9 mm apart. The 300 bipolar electrodes were led into a switch box which allowed selection of either one of two groups of 196 electrodes: One group spaced evenly over the entire surface of the polymer sheet, the other concentrated in a 6 cm by 6 cm square at the center of the electrode array. The signal from each bipolar electrode was preamplified, multiplexed, digitized and recorded on an Ampex model HBR–3000 High Bit Rate analog tape recorder. Lead II and III of the ECG, blood pressure, stimulus pulse tracing and voice log were recorded on the analog FM channels of the tape recorder. The ventricles were stimulated from several different locations during the experimental protocol.

RESULTS

Functional Block and Reentrant Circuits

Figure 1 is constructed from recordings taken from the central group of electrodes in a representative experiment. Following stimulation at a 300 ms pacing cycle length in this experiment through electrodes in the middle of the array (indicated by the pulse symbol at the center of the map), activation spread away from the stimulus site in all directions as indicated by the arrows. The most rapid spread of activation occurred toward the septum and apicolateral free wall of the ventricle. This is the direction of the long axis of the myocardial fiber bundles in the epicardial border zone. The slowest spread of activation occurred toward the base and apex. This is the direction perpendicular to the long axis of the fiber bundles. There is no evidence of conduction block during pacing, nor is an anatomically defined reentrant circuit apparent. Conduction velocities were calculated from the activation map along the long axis and perpendicular to the long axis of the

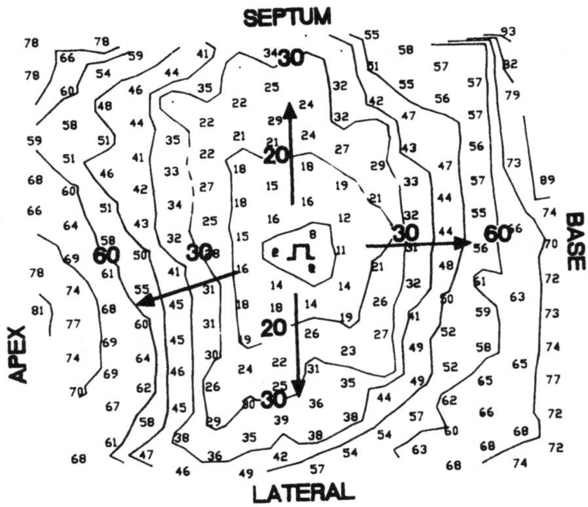

Figure 1: Example of the activation pattern of the epicardial border zone following central stimulation. An activation map constructed from electrogram recordings taken from the central portion of the electrode array is shown. The heart was stimulated at the center of the array at the site indicated by the pulse symbol. The small numbers indicate activation times at each recording site relative to the moment of stimulation. The large numbers designate isochrone times.

myocardial fibers. The conduction velocities were normal at the 300 ms pacing cycle length, 85 cm/s in the direction of the long axis and 35 cm/s in the direction perpendicular to the long axis. It therefore, appears that the transmembrane resting potentials and action potential depolarization velocity in this region may be normal.

Sustained ventricular tachycardia was induced in this experiment by programmed stimulation. An activation map of the reentrant circuit in the epicardial border zone during tachycardia is illustrated in Fig. 2. This activation map was constructed from the electrode array which covered the entire surface of the anterior left ventricular epicardium. The region of this array corresponding to the smaller, more concentrated map shown in Fig. 1 is depicted by the rectangular box placed around the center of the electrode. Activation in the time window shown in the figure began at the 10 ms isochrone (as indicated by the asterisk) toward the lateral margin of the recording array. The excitation wave split into two independent circulating fronts which first traveled transverse to the long axis of the myocardial fibers, one toward the base and the other toward the apex. After approximately

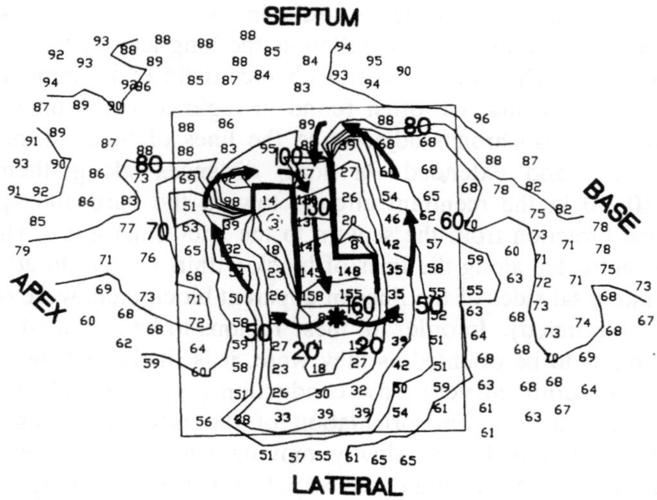

Figure 2: An activation map constructed from a single beat during sustained reentrant tachycardia. The map is constructed from the entire electrode array. Activation in this time window began at the 0 ms isochrone, as indicated by the asterisk. Activation proceeded along two reentrant loops, as illustrated by the arrows. The loop was closed after 160 ms.

40 ms, the basal activation loop turned and rapidly transited the border zone toward the septum. In contrast, the apical wavefront continued to activate the border zone in a transverse fashion toward the apex until approximately 60 ms had elapsed, at which time it finally turned and progressed rapidly toward the septum. After approximately 90 ms had elapsed, the two wavefronts reached the septal edge of the electrode and fused. The common wave thus formed went on to reverse its direction and enter the narrow central common reentrant pathway after about 100 ms. It then traversed the central common pathway rapidly toward the lateral edge of the electrode where it closed the reentrant loop after approximately 160 ms.

Two lines of conduction block, indicated by the dark black lines on the map in Fig. 2, appeared during reentrant tachycardia. Excitation in the circuits circulated around these lines. The lines of block were established because of the large activation time differences on either side of them. Interpolated isochrones in these regions by convention are suggestive of lines of block although there may be alternative interpretations [2]. These lines of block appeared where there had been uniform activation during central stimulation. Therefore, they were not caused by anatomic block but were functional according to the commonly used definition.

The lines of block during tachycardia in the experiment illustrated in Fig. 2, were oriented in the same direction as the long axis of the myocardial fibers (see Fig. 1). This was a general characteristic of functional circuits. Another characteristic that we often found in regions where lines of block developed was that isochrones adjacent to the lines of block were oriented parallel to them and extended for long distances along their length. Isochrones 10–50 in the reentrant circuits toward the base and apex were oriented in the direction from the septum to the lateral margin. There were large regions activated along the lines of block within 10 ms. In addition, on both apical and basal sides of the lines of apparent block there were regions of early activation (circled). In order for these regions to be activated early, and for the isochrones to be oriented as they were, it is likely that these lines of "block" actually permitted very slow conduction (approximately 6 cm/s) to take place across them. The orientation of the isochrones suggests that excitation progressed in broad, slowly moving wavefronts which traveled from the central common reentrant pathway, transverse to the long axis of the fibers toward the apex and toward the base. These characteristics of anisotropic reentry are described in more detail in a previous publication [2].

Anatomic Block and Reentrant Circuits

In some experiments a region of fixed conduction block was found in the epicardial border zone during sinus rhythm or ventricular pacing. An example is shown in Fig. 3. Following stimulation from the center of the electrode array at a cycle length of 300 ms, at the site indicated by the pulse symbol, conduction proceeded away from this site only toward the base without block (as indicated by the arrows), reaching this edge of the electrode after approximately 80 ms. Activation toward the septum progressed evenly at first, but evidence of possible conduction block appeared between the 40 ms and 80 ms isochrones (four interpolated isochrones) as shown by the blocked arrows (T) and the thick black line. Activation did not finally reach the far septal edge of the recording array until almost 90 ms following stimulation. This was a much longer delay than shown in Fig. 1 for septal edge activation (approximately 35 ms). Toward the lateral free wall as well, conduction traveled around a small region of block located immediately lateral to the point of stimulation, as indicated by the thick black line. Toward the apical and apicoseptal edges of the array, conduction progressed for only the first 30–40 ms following stimulation, and then block was encountered. Toward the apicoseptal corner of the activation map, a zone of delayed activation appeared. Excitation arrived there very late, approximately 110–130 ms. after the moment of stimulation, apparently having traveled toward the apex along the septal edge of the recording array. This is indicated by the double–headed

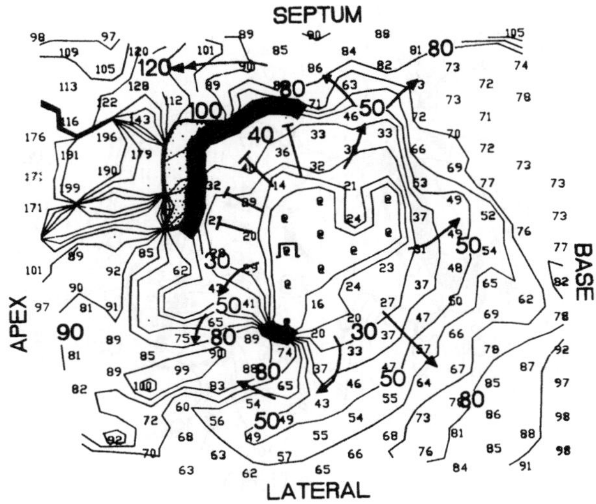

Figure 3: Example of activation pattern following central stimulation at a 280 ms cycle length in an experiment with anatomic block. An activation map constructed from electrogram recordings taken from the central portion of the electrode array is shown. The heart was stimulated at the center of the array at the site indicated by the pulse symbol. The small numbers indicate activation times at each recording site relative to the moment of stimulation. The large numbers designate isochrone times. The region of block is represented by the thick black line and the area shaded by the dots.

arrow. In such areas of fixed block the infarct may extend to the epicardial surface and there may be no surviving epicardial muscle.

An activation map of a single beat during sustained tachycardia in the same experiment as in Fig. 3, is shown in Fig. 4. This activation map was constructed from the group of electrodes which were spaced widely over the entire surface of the epicardial border zone. Activation in the time window in Fig. 4 began within the two 10 ms isochrones, indicated by the asterisks; one located at the apex and the other near the lateral margin. There appeared to be two reentrant excitation waves, each traveling in a circle around its respective line of conduction block (indicated by the thick black lines): the apical excitation wave circulated in a counterclockwise manner, while the basal wave rotated clockwise. Both reentrant circuits employed the central region of the border zone as a rather wide central common pathway. The S-shaped line of conduction block which appeared at the center of the apical reentrant loop in Fig. 4 coincided almost precisely with the line of conduction

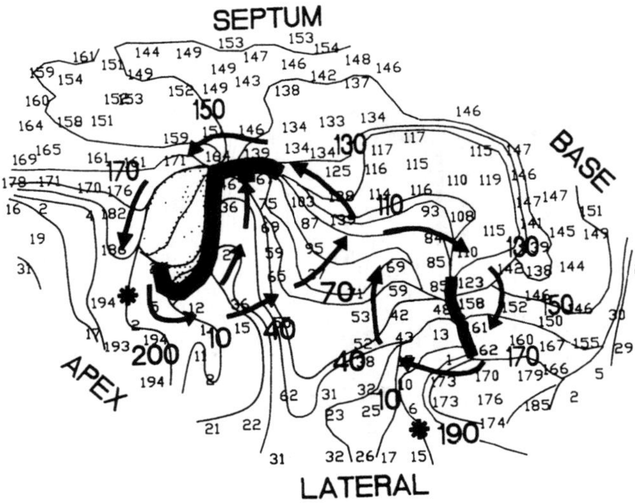

Figure 4: An activation map constructed from a single beat during sustained reentrant tachycardia. The map is constructed from the entire electrode array. Activation in this time window began at the 0 ms isochrone, as indicated by the asterisk. Activation proceeded along two reentrant loops, as illustrated by the arrows. The loop was closed after 180 – 190 ms. The region shaded by the dots is a region of fixed or anatomical conduction block. See text for details.

block in the apicoseptal quadrant of the border zone during the central pacing demonstrated in Fig. 3. Therefore, in this circuit excitation is rotating around a region of fixed anatomic block. The short line of block toward the base, however, appeared only during reentry and was therefore, functional.

The lines of anatomic block in the reentrant circuits in this group of experiments were not oriented only parallel to the long axis of the myocardial fibers as were lines of functional block. This is evident in Fig. 4 which shows that the line of anatomic block in the apical circuit had both longitudinal and transverse segments. These lines of block were thicker than lines of functional block and involved more myocardium. There was no evidence that activation conducted across the block line as in functional, anisotropic reentry. In general, conduction velocities in the region of the epicardial border zone where these circuits were located were depressed to less than half of normal values in both the transverse and longitudinal directions, suggesting the possibility that the transmembrane potentials of the myocardial fibers were depressed.

CONCLUSIONS

The results of our study show that reentrant circuits in healing myocardial infarcts that cause ventricular tachycardia may have different electrophysiological mechanisms. Circuits may be functional and occur in regions where the myocardial fibers do not have severely depressed resting potentials and action potential upstrokes or they may be anatomical and occur in regions with depressed transmembrane potentials. The underlying electrophysiological milieu may influence the properties of the tachycardias. In general, we found that the tachycardias associated with functional circuits had faster rates than those associated with anatomical circuits. Thus, it might be predicted that functional circuits had more of a likelihood to cause hemodynamic deterioration or eventual ventricular fibrillation. In addition, the different kinds of circuits might respond differently to antiarrhythmic drugs. We have found that class IC drugs cannot stop reentrant tachycardias caused by the functional circuits because the drugs cannot block conduction in the circuits [5]. This may be a result of the presence of the relatively normal transmembrane potentials, the upstrokes of which are relatively resistant to block by therapeutic concentrations of drug. We predict that such drugs should block conduction in anatomic circuits because action potentials with depressed upstrokes are more sensitive to sodium channel blockade. We do not, at this time, have the data to prove this prediction.

Acknowledgements

Supported by Program Project Grant HL 30557 and by Grant HL 31393 from the Heart, Lung and Blood Institute of the National Institutes of Health.

DISCUSSION

Dr. Rosen: How would class–3 drugs act on tachycardia in your situation?

Dr. Wit: So far, we have not yet found an antiarrhythmic drug that stops reentrant in the epicardial border zone excitation when the underlying electrophysiological properties indicate normal conduction velocity. I qualify this statement with the fact that we have a limited amount of data. We have done a complete study with flecainide, we have almost finished a study with lidocaine and we have a reasonable amount of data on sotolol as an example of a class–3 drug. We have not been able to stop tachycardia with antiarrhythmic drugs. I would like to call on my modeling colleagues, to develop a model of arrhythmia of this kind and then to predict what a drug has to do in order to stop this type of ventricular tachycardia. Then we could follow with the experimental test.

Dr. Gettes: You use your canine model as a 2D model. What happens below the epicardial surface? Are you really sure that it is really a 2D and not a 3D model?

Dr. Wit: We are pretty certain that activation of the epicardial border zone is two dimensional on a macro scale, as there is no activity below the 5 to 10 surviving cell layers. There is no tissue underneath it. There is no third dimension on gross levels. By 4 or 5 days, all of the myocardial cells below the 6 or 10 layers of cells are disrupted. There is no electrical activity that you can measure with needle electrodes. We have not tried to discern if there is any asynchronous activation within those 5–10 surviving cell layers.

Dr. Corr: The difference between the functional and anatomic block may be simply related to the fact that there is an extension of the fibrosis into the epicardial layer resulting in fixed block. We seem to find it as well.

Dr. Wit: The anatomic block, as you say, results because there is complete extension of the infarct to the epicardial surface as opposed to functional block where there is not. But the properties of the rest of the surviving epicardial border zone seem to be different as well; the conduction velocity in general is slower, even in surviving myocardium and it is much more rate sensitive when there is anatomic block. That might also be a product of the anatomy that we don't quite understand; a thinner epicardial rim may be more depressed in general in terms of transmembrane potentials, and rate sensitivity might be expected to appear. There is also a possibility of more fiber disarray in these particular animals. We do not have that kind of anatomical data yet to support these speculations.

Dr. Corr: You show a lack of influence of shortening of cycle length on functional conduction block. Was that done with a gradual or abrupt change in cycle length?

Dr. Wit: It was done with gradual changes in stimulus cycle length.

Dr. Allessie: Your dogs showing loss of anisotropy in the epicardial border zone can possibly be related to a different anatomical orientation of the fibers. Were those studies done the same number of days after the infarction as the other series, 4–5 days after infarction?

Dr. Wit: Yes. The infarcts were the same age in both groups.

Dr. Sideman: I am trying to relate your model to the whole heart ischemic model. We have studied the phenomenon of creating the reentry loop and we can relate the phenomenon to the endocardial layer, depending on the survival and activity of the Purkinje fiber in the ischemic part of the endocardium. My question is whether we consider the epicardium to be in some sense equivalent to the endocardium as a source of activation focii.

Dr. Wit: That is an excellent question. I have thought about that problem a lot because of the clinical observations that many of these kinds of tachycardias are in fact endocardial in origin, and we have an epicardial model. The way I am thinking at the present time, which is certainly open to revision subsequently, is that we do get reentry very nicely on the epicardium because there is no Purkinje system. Therefore, the properties of the muscle, parallel oriented muscle bundles, separation of muscle bundles, etc. can be brought into play to cause reentry without any short circuiting effect of rapid conduction in a Purkinje system which would prevent all of these properties from being manifested. Maybe what happens sometimes clinically is that when the Purkinje system is destroyed on the subendocardial surface, it leaves a similar layer of surviving muscle fibers on the endocardial surface that might enable the reentrant tachycardias to occur, whereas when the Purkinje system remains with relatively good conduction properties, you cannot get endocardial reentry. That is the way I am thinking about it now. I do not think it makes any difference where the sheet of muscle is, whether it is on the epicardial surface, or it is on the endocardial surface, or even if it is intramural, which it could very well be. It does not necessarily have to be a perfect sheet. This is an animal model which may be more perfect than what one sees in the actual clinical picture, but still enables us to sort out the underlying mechanisms more easily.

REFERENCES

1. El–Sherif N: The figure of 8 model of reentrant excitation in the canine post–infarction heart. In: *Cardiac Electrophysiology and Arrhythmias*, Zipes DP, Jalife J (eds), Grune and Stratton, New York, 1985 pp 363–378.
2. Dillon SM, Allessie MA, Ursell PC, Wit AL: Influence of anisotropic tissue structure on reentrant circuits in the epicardial border zone of subacute infarcts. *Circ Res* 1988;63:182–206.
3. El–Sherif N, Smith NA, Evans K: Canine ventricular arrhythmias in the late myocardial infarction period 8. Epicardial mapping of reentrant circuits. *Circ Res* 1981;49:255–265.
4. Hoffman BF, Rosen MR; Cellular mechanisms for cardiac arrhythmias. *Circ Res* 1981;49:1–15.
5. Coromilas J, Saltman AE, Waldecker B, Dillon SM, Wit AL: Effect of class IC antiarrhythmic drugs on anisotropic reentry in canine infarcts. *Circulation* 1988;78:II–460.

CHAPTER 5:

EFFECTS OF INTERCELLULAR RESISTIVE COUPLING ON ACTION POTENTIAL INITIATION

Ronald W. Joyner and Rosemarie C. Tan

ABSTRACT

The response of an isolated cell to an increased frequency of stimulation is strongly altered by the presence of a coupling resistance (R) to another cell. As the coupling R is gradually decreased, the stimulated cell becomes able to respond successfully to more rapid stimulation and then, at levels of coupling R which allow conduction between the two cells, the coupled pair of cells exhibit arrhythmic interactions not predicted by the intrinsic properties of either cell.

INTRODUCTION

Cardiac activation patterns are produced by multi–dimensional interactions between the spatial distributions of excitable membrane properties and intercellular coupling R. In normal cardiac tissue there are regional and directional differences in coupling R, e.g. high coupling R in the SA and AV node [1], sites of high R between Purkinje fibers and ventricular muscle [2], and also in atrial and ventricular tissue [3, 4]. In ischemic cardiac tissue, there are changes in the spatial distribution of coupling R that lead to slow irregular Action Potential (AP) conduction and "fractionated" activity [5]. We recently showed [6] that after 24 hours of ischemia the ventricular subendocardium had a spatial pattern of small "islands" of excitable tissue which were interconnected by high coupling R. For low frequency stimulation, the adjacent cell groups had nearly synchronous activity; but, as the stimulation frequency was increased, these same groups of cells had highly asynchronous activity, with the activity pattern being different for the different cell groups. This phenomenon is fundamentally different from previously studied phenomena characterized as two coupled spontaneously pacing cells [7] or as an oscillator coupled to a quiescent region [8]. For the phenomenon we observed in ischemic tissue between adjacent cell groups, neither of the cell groups was intrinsically automatic.

The time course of recovery of excitability is known to be closely related to the AP duration (APD) and the APD is itself determined by the frequency of activation and by electrotonic influences [9, 10]. The anatomical complexity of normal and ischemic tissue [11], however, makes tissue experiments difficult to interpret. Antzelevitch et al. [12, 13] studied Purkinje strands with a central compartment containing sucrose or "ischemic" Tyrode's solution. This technique has also been used with SA node–atrial strips [14] and small papillary muscles [15]. Veenstra and DeHaan [7] used spherical aggregates of cultured chick heart cells in which they brought two aggregates in physical contact with each other. Over the course of several hours, electrical coupling between the two aggregates developed. Studies on isolated myocyte cell pairs [16, 17] have been particularly useful in understanding mechanisms for the modulation of gap junction resistances. Although numerous pharmacological agents (e.g. octanol or heptanol) have been shown to reversibly block gap junctional conductance [17], the effects of these agents on the individual cellular properties are not well documented, and in some cases the maximum rate of rise (V_{max}) may be decreased [18].

THEORETICAL METHODS

We used the simultaneous solution of the Beeler and Reuter (BR) equations [19] for each of the two cells. The current flow, I_c (mA) between the cells is assumed to be $(V_1 - V_2)/R_c$, where for each time step V_1 is the membrane potential (mV) of cell 1, V_2 is the membrane potential (mV) of cell 2 and R_c (ohms) is the coupling R. The stimulus protocol used was a sudden change in the basic cycle length (BCL) from 1000 ms to some shorter BCL. In all of our simulations cell 1 received the current stimuli as pulses of current with a duration of 2 ms.

EXPERIMENTAL METHODS

Adult rabbits of either sex weighing 1–1.5 kg were anesthetized using 50 mg/kg Nembutal i.v. and 1000 units heparin. Cells were isolated according to the methods of Bendukidze et al. [20] and were then transferred to an experimental chamber containing (in mM) NaCl (148), KCl (4), $CaCl_2$ (1.8), MgCl2 (0.5), NaH_2PO_4 (0.3), dextrose (5), HEPES (5) at pH 7.4, 36°C. We used the following internal solution (in mM): KCl (120), Na_2CrP (5), MgATP (5), EGTA (11), HEPES (10) with pH adjusted to 7.2 using KOH [21]. The cells were paced at a basic cycle length (BCL) of 500 ms in the current clamp

Figure 1: The upper and lower panels are simulations of the Beeler–Reuter ventricular cell model for membrane potential versus time. The stimulus protocol for each panel was current pulses of 2 ms duration delivered at time zero, at time 1 s, and then a train of stimuli delivered at an interval of either 300 ms (upper panel) or 250 ms (lower panel).

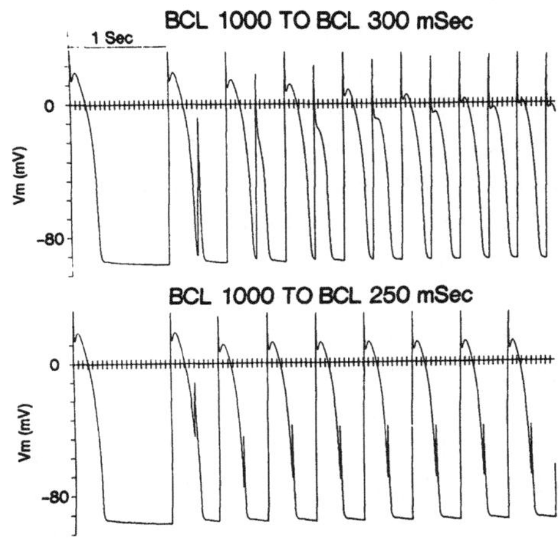

mode with a stimulus (S) duration of 2 ms and a stimulus intensity that provided for 1:1 pacing.

THEORETICAL RESULTS

Figure 1 shows responses of an isolated BR cell model to a sudden change in the BCL. In the upper panel (BCL 1000 to 300 ms) the first stimulus (S) at the shorter cycle length fails to excite the cell. However, the successive occurrences of stimuli at BCL 300 ms lead to an alternating pattern (short APD – long APD) of 1:1 excitation which then resolves over time into a stable 1:1 pattern. The results for a transition from BCL 1000 to BCL 250 are shown in the lower panel. In this case, there is also failure of the first S at the shorter cycle length. The second S at the shorter cycle length succeeds, but the third S fails, leading to a stable pattern of 2:1 activation. From these and other simulations (not shown) we find that for an abrupt change in BCL, the BR model will not allow 1:1 activation for a single isopotential region if the change in BCL is from the standard value of 1000 ms to some value less than or equal to about 250 ms.

Our model consists of two cells, each of which is represented by a BR model system, with each cell having a membrane area of 2.75×10^{-4} cm^2, to provide an input R of 20 MΩ for the BR model, and intercellular resistance, R_c MΩ. All of the stimuli are applied to cell 1. Activation of cell 2 can occur only if an AP generated in cell 1 can conduct across R_c to cell 2. The pacing protocol is a current of 4.0 nA for 2 ms. The asterisks in Fig. 2 identify the stimulus times. In Figure 2, R_c is 275 MΩ (about 14 times the input R of either cell) resulting in an electrotonic response in cell 2 for each AP in

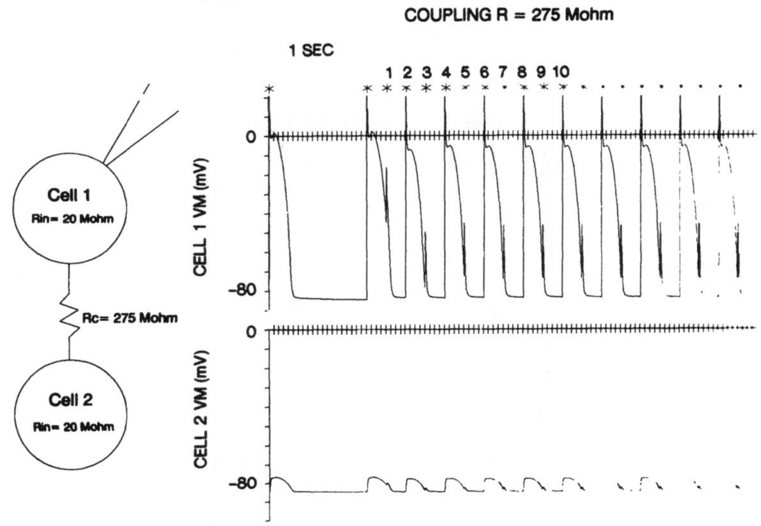

Figure 2: Simulation results for two cells, each represented by the BR model, with the two cells electrically coupled by a R of 275 MΩ. Stimulation pattern is diagramed by the asterisks, with all stimuli applied to cell 1 (upper trace). The lower trace is the solution for cell 2, which at this level of R_c shows passive electrotonic depolarizations associated with each AP in cell 1. The activation pattern in cell 1 at BCL 200 ms is 2:1.

cell 1. The AP durations of cell 1 are shortened by electrotonic coupling, but the phenomenon of 2:1 responses in cell 1 and no AP's in cell 2 has occurred.

If we further decrease the value of R_c to 220 MΩ (Fig. 3) we still produce no AP's in cell 2, but the pattern of AP's in cell 1 is dramatically altered. The first S with a 200 ms interval does not produce an AP. The second S does produce an AP in cell 1. In contrast to the results of Fig. 2 (R_c = 275 MΩ), the third S now does produce an AP in cell 1. This is due to the combined effects of shortening of APD from the premature excitation of the previous S (AP interval of 400 ms compared to the standard 1000 ms) and the shortening of APD produced by the electrotonic influence of cell 2. Cell 1 now displays an alternans pattern of 1:1 responses throughout the rest of the train of stimuli at the 200 ms interval. In terms of the number of AP's generated, it is clear that cell 2, without any AP generation, is exerting a positive effect on cell 1 through shortening of the APD, and thus the refractory period, of cell 1.

Figure 4 shows results for a R_c of 83 MΩ with successful conduction of the AP from cell 1 to cell 2 for the first two stimuli at a 1000 ms interval. In contrast to Fig. 3 (220 MΩ) APD shortening due to the electrical load of

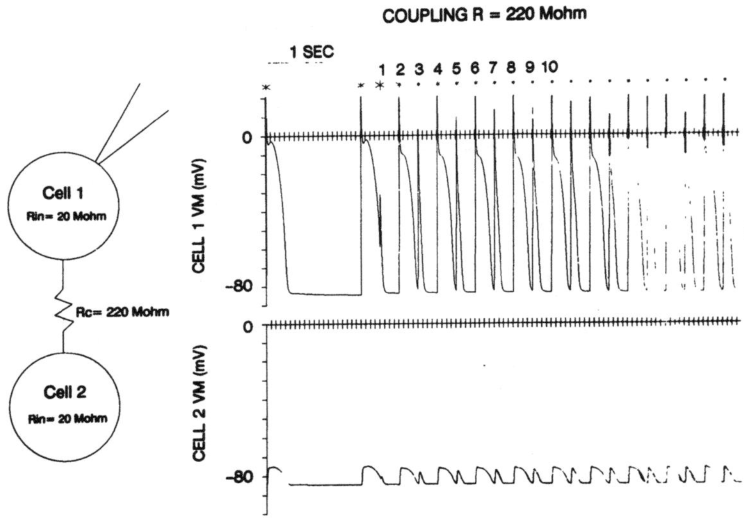

Figure 3: Simulation results for two cells, each represented by the BR model, with the two cells electrically coupled by a R of 220 MΩ. Stimulation pattern is diagramed by the asterisks, with all stimuli applied to cell 1 (upper trace). The lower trace is the solution for cell 2; at this level of R_c shows passive electrotonic depolarizations associated with each AP in cell 1. The activation pattern in cell 1 at BCL 200 is now 1:1 with an alternating APD for the cell 1 AP's.

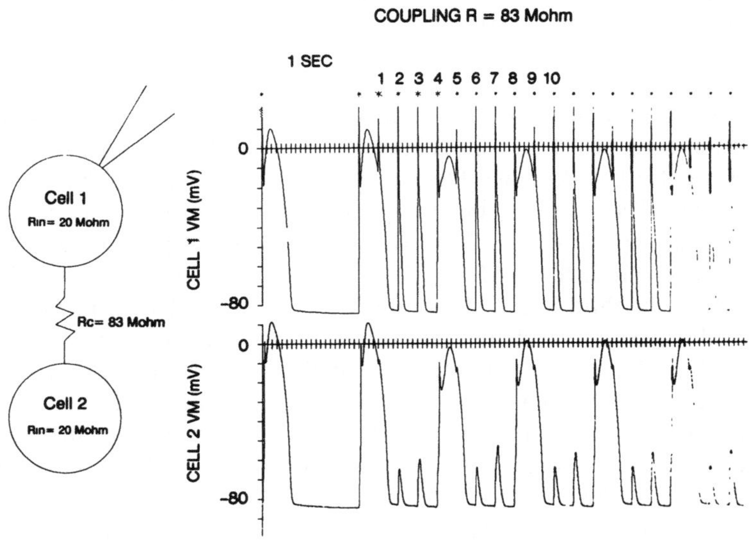

Figure 4: Simulation results for two cells, each represented by the BR model, with the two cells electrically coupled by a R of 83 MΩ. Stimulation pattern is diagramed by the asterisks, with all stimuli applied to cell 1 (upper trace). The lower trace is the solution for cell 2, which at this level of R_c shows AP conduction associated with the first two AP's in cell 1, but at BCL 200 there is a periodic pattern with 4:3 activation in cell 1 and 4:1 activation in cell 2.

cell 2 does not occur for these two stimuli since cell 2 now has an AP in response to these stimuli to cell 1. The first S with a 200 ms interval fails to excite cell 1. The second and third stimuli at the 200 ms interval do produce AP's in cell 1 but these AP's are attenuated to passive responses in cell 2. The fourth S with a 200 ms interval produces an AP in cell 1 and this AP does successful propagate into cell 2. This AP in cell 2 now exerts a significant effect of the AP of cell 1, greatly increasing the APD. This effect causes the fifth S at the 200 ms interval to fail to activate cell 1. As this process repeats, there develops a periodicity with a repeating pattern every four stimuli such that, for every four stimuli there are three AP's in cell 1 and one AP in cell 2. Note that the patterns in cell 1 and two are complementary. Every success in cell 2 is followed by a failure in cell 1. Whenever a periodicity occurred every N stimuli, if M successes occurred in cell 1, there were (N−M) successes in cell 2. When R_c was further reduced to 55 MΩ (not shown) the resulting pattern became 2:1 in both cell 1 and cell 2. This pattern, for cell 1, is the same as the pattern which resulted (Fig. 2) for nearly total uncoupling of the two cells, but now each success in cell 1 is associated with a success in cell 2.

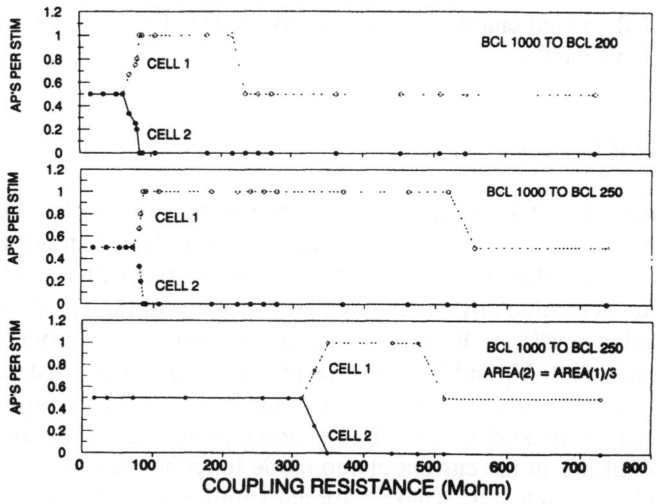

Figure 5: ***Top:*** A summary of simulations in which R_c is systematically varied between cell 1 and cell 2. Results are plotted for the activation patterns as the ratio of number of AP's to number of stimuli in the steady state at BCL 200 ms pacing for cell 1 (open circles) and cell 2 (filled circles). ***Middle:*** A summary as in the Top panel except that the transition is from BCL 1000 to BCL 250 ms pacing for cell 1 (open circles) and cell 2 (filled circles). ***Bottom:*** A summary of simulations as in the Top panel except that the transition is from BCL 1000 to BCL 250 ms and the membrane area of cell 2 is one−third that of cell 1.

The results for the steady state activation pattern at a pacing interval of 200 ms from Figs. 2–4 are summarized in Fig. 5. In the top panel, for each simulation, we plot the ratio of successes per stimuli for cell 1 (open circles) and cell 2 (filled circles). Note that the value of R_c at which the electrotonic load of cell 2 onto cell 1 enables 1:1 activation in cell 1 (240 MΩ) is considerably higher than the input R of either cell (20 MΩ). The value of R_c at which propagation can occur from cell 1 to cell 2 (88 MΩ) is also greater than the input R of either cell.

The values of R_c at which alterations in the activation patterns occur depend on the stimulation frequency and the geometry of the coupled system. In the middle panel of Fig. 5 we show results for simulations in which we made the transition in BCL from 1000 ms to 250 ms. Since the stimulus interval is now longer, not as much APD shortening is required to convert the 2:1 pattern intrinsic to cell 1 in isolation to a 1:1 pattern. This transition now occurs at an R_c of 512 MΩ. If we now alter the geometry of the system such that cell 2 is decreased in membrane area to a value one–third that of cell 1, again using a transition from BCL 1000 to BCL 250 ms, we get the results summarized in the bottom panel of Fig. 5. Comparing the middle and bottom panels of Fig. 5, we see that the value of R_c which enables 1:1 activation of cell 1 is very little affected by a decrease in the membrane area of cell 2, but the value of R_c which enables conduction to occur between cell 1 and cell 2 is increased significantly.

Experimental Results

The experimental system was developed so that two isolated cells could be studied independently and then, by means of an external circuit these same two cells could be electrically connected at any desired value of junctional R (Fig. 6). For two cells, physically coupled as shown in the left panel, there is an intercellular resistance R_c (ohms) and a time varying membrane potential for cell 1 and cell 2, V_1 and V_2 volts, respectively. At every instant in time a current $(V_1-V_2)/R_c$ Amperes leaves cell 1 and enters cell 2 through gap junctions between the two cells. The circuit shown in the right panel of Fig. 6 has a dual preamplifier in the current clamp mode to record the membrane potential V_1 and V_2 of each cell. Two additional amplifiers shown at the top of the diagram take the difference of the two membrane voltages and use an effective gain of $1/R_c$ (Siemens) such that the voltage to current converters continuously supply $(V_1-V_2)/R_c$ Amperes to cell 2 and $(V_2-V_1)/R_c$ Amperes to cell 1. In analogy with the concepts of "voltage" clamp or "current" clamp circuits, this circuit could be thought of as a "coupling" clamp since current is dynamically supplied to each cell to produce the effect of some desired value of coupling R.

Figure 6: Left: Representation of how current flows from cell-to-cell another through their gap junctions. ***Right:*** An experimental system developed to study how two cells which are not physically connected interact with each other when they are electrically coupled with a variable R_c.

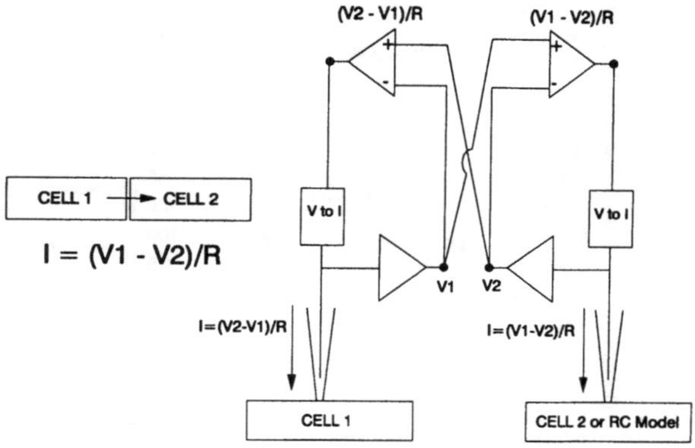

In the present study an RC circuit has been substituted for cell 2, utilizing typical values of membrane input R (20 MΩ) and input capacitance (80 pF) for rabbit ventricular cells and an offset voltage equal to the resting membrane potential of cell 1 so that there was *no* current flow during the resting state. A recording of the AP's from an isolated rabbit ventricular cell and from the passive circuit is shown in Fig. 7 with all stimuli applied to cell 1. The input R for cell 1 was 46 MΩ when depolarized 5 to 7 mV from the resting membrane potential and 30 MΩ when hyperpolarized 5 to 7 mV from rest.

In the left panel of Fig. 7 are the voltage and current traces from cell 1 and cell 2 with the external amplifier gain set to zero (R_c = infinity). The APD of cell 1 is approximately 200 ms with a current threshold of 1.35 mA. Cell 2 has no voltage or current response to the AP initiated in cell 1 when the two cells are uncoupled. The right panel shows the recordings from cell 1 and cell 2 when the gain of the external amplifier is set to provide an effective R_c of 500 MΩ between cell 1 and cell 2. The APD of cell 1 has been significantly shortened from 200 ms to approximately 70 ms in response to electrotonic coupling to cell 2. The current threshold, however, remains unchanged.

The results of varying the value of R_c for this same cell 1 are shown in Fig. 8. One can see the effect on the AP of cell 1 with decreasing values of R_c between cell 1 and the passive RC circuit.

These results over a range of R_c are summarized in Fig. 9 for the cell shown in Figs. 7 and 8. The values of APD, AP Amplitude (APA), and the current threshold (I_{th}) have been normalized to the intrinsic values for cell 1 when it was uncoupled from the passive circuit. The abscissa plots the ratio of the effective R_c to the input R (46 MΩ for small depolarizations) of cell 1. For this cell, a 50% shortening of the APD occurs with an effective R_c of 1000 MΩ. The APA and the current threshold, on the other hand, is only slightly affected even for the lowest R_c tested (less than a 15% change for R_c = 100

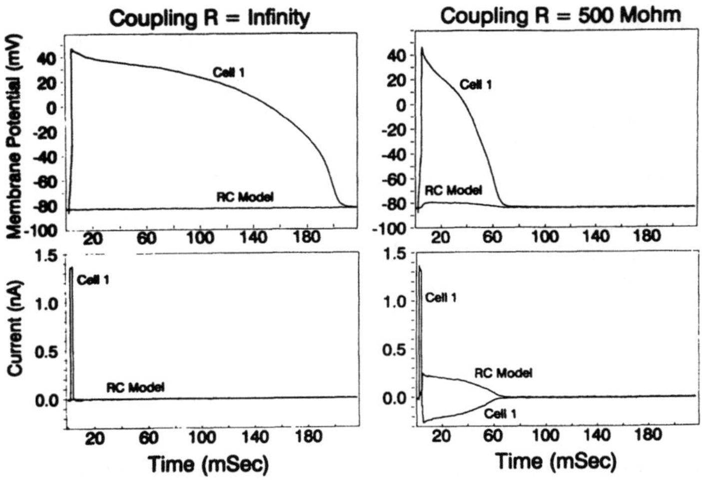

Figure 7: Experimental results for an isolated ventricular cell electrically coupled to a passive RC circuit. ***Left:*** Voltage and current recordings when the two cells are uncoupled. ***Right:*** Voltage and current recordings when the two cells are coupled together with an effective R of 500 MΩ.

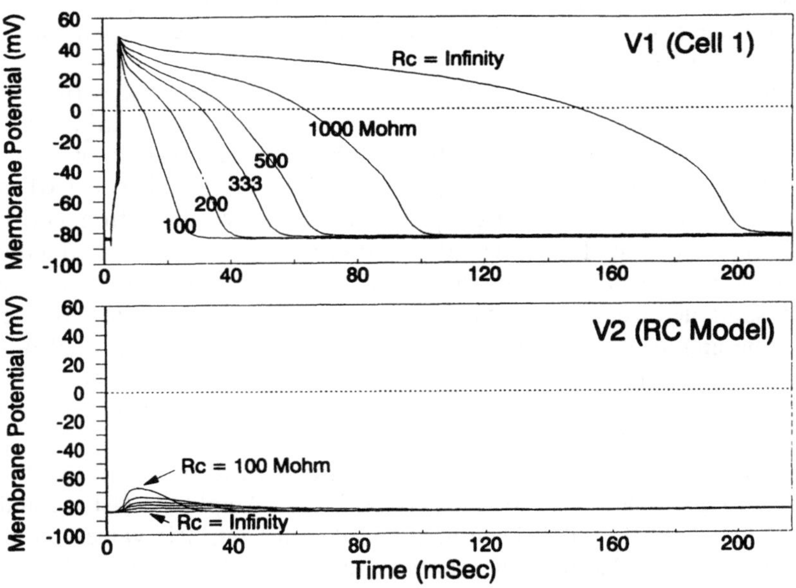

Figure 8: Isolated ventricular cell coupled to a passive RC circuit. ***Top:*** Membrane voltage recordings for cell 1. ***Bottom:*** The passive RC circuit (lower panel) for a wide range of R_c.

MΩ). We have evaluated 6 cells coupled to a passive RC circuit with results consistent with those shown in Fig. 9. As compared to the control uncoupled values, the APD of cell 1 shortened to 52 ± 5% of the control APD for an effective R_c of 1000 MΩ, 36 ± 6% for a R_c of 500 MΩ, 22 ± 4% for a R_c of 250 MΩ, and 13 ± 2% for a R_c of 125 MΩ. The APA was only affected slightly and decreased less than 6% for even the lowest R_c (94 ± 1% of the control APA for a R_c of 100 MΩ). The current thresholds were also very little affected by the R_c, increasing not more than 15% for a R_c as low as 100 MΩ. In all cases, the effects of the imposed R_c were completely reversible when the cell was again uncoupled from the RC circuit by setting the coupling amplifier gain to zero.

Using the same cell shown in Figs. 7–9, cell 1 was paced at BCL 500 ms and then abruptly changed the BCL to 150 ms. In the top panel of Fig. 10, the last two AP's of cell 1 being paced at BCL 500 ms are shown before the BCL is changed to 150 ms, leading to a stable activation pattern of 2:1. If we couple cell 1 to the passive RC circuit, an entirely different response to an abrupt change in pacing frequency can be observed. This time cell 1 has been coupled to the passive RC circuit with an effective R_c of 500 MΩ. In the bottom panel of Fig. 10, we see again the last two AP's of cell 1 while being paced at a BCL of 500 ms. Note the shortening in APD that has occurred just from being coupled to the RC circuit. This time when the BCL is suddenly shortened to 150 ms for cell 1, the stimulus is successful in producing an AP in cell 1. In fact, every stimuli at BCL 150 produces an AP. The ability to

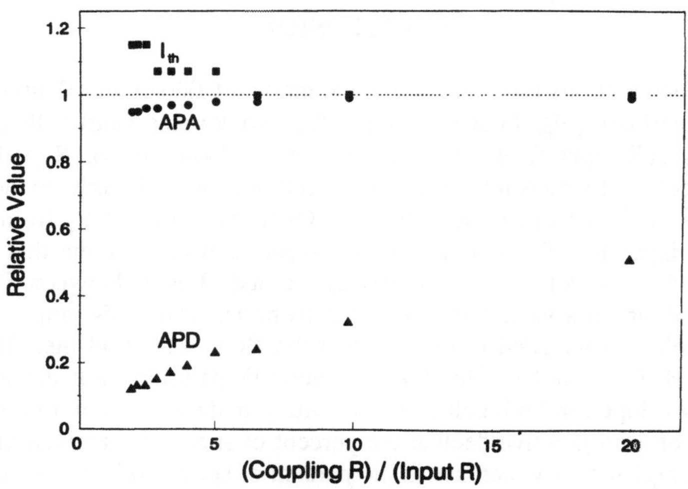

Figure 9: For cell 1 coupled to a passive RC circuit, the vertical axis represents the APD, APA, and I_{th} which have been normalized to their intrinsic uncoupled values. The horizontal axis plots the ratio of R_c to the input R of cell 1.

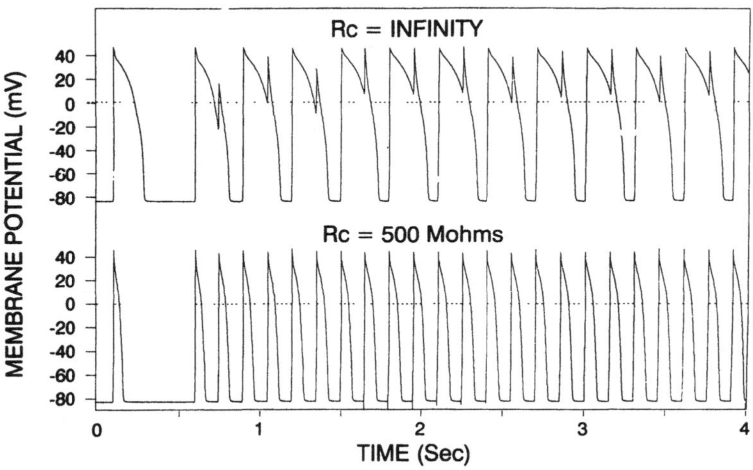

Figure 10: Voltage recordings from cell 1 in response to an abrupt transition in pacing from BCL 500 to BCL 150 ms. In the upper panel cell 1 is uncoupled from cell 2. In the lower panel cell 1 is coupled to the passive RC circuit with an effective R_c of 500 MΩs.

produce a tachycardia in cell 1 which is impossible to produce intrinsically in cell 1 when uncoupled, is accomplished by loading cell 1 with the passive RC circuit.

CONCLUSION

The simulations used the membrane model of Beeler and Reuter [19] for a single heart cell (Fig. 1) or for each of the two coupled heart cells (Figs. 2–5). For a cell input R of 20 MΩ, there was a "window" of R_c values that allowed cell 1 to respond with 1:1 excitation, even though no activation occurred in cell 2, from 83 to 220 MΩ. Over this range of R_c from 4 to 11 times the input R cell 1 could exhibit a significant tachycardia that was not possible when the cell was completely uncoupled. For R_c between 55 and 83 MΩ, we observed a variety of excitation patterns. Over this range, each AP (AP) in cell 1 propagated to cell 2 when the BCL was 1000 ms. However, when the BCL was reduced to 200 ms, a periodic pattern of AP successes and failures developed in both cells. The resulting patterns were not random, but consisted of a progressive decline the percent of successful activations of cell 1 and a complementary increase in the percent of successful activations of cell 2 as R_c decreased from 83 to 55 MΩ. As R_c was further decreased to 55 MΩ and below, every successful activation of cell 1 was matched by a successful activation of cell 2, with each cell now exhibiting the same electrical response

(1:1 activation at BCL 1000; 2:1 activation at BCL 200 ms) as the single uncoupled isolated cell. Figure 6 summarized these effects of R_c and also showed how these effects depended on the frequency of stimulation and the relative sizes of the two cells.

Previous studies on acute myocardial ischemia [20–21] have clearly demonstrated electrotonic influences, among small cardiac regions, which have been proposed to produce ectopic activity, either by re–excitation of normal cells after an AP or by inducing spontaneous activity in non–automatic cells by a flow of "injury current" from depolarized cells to normally polarized cells. The simulations show that the effects of varying R_c in a simple two–cell system can be quite complex. Several features are of particular importance. First, over the "window" of R_c from 83 to 220 MΩ, cell 1 was able to respond to a higher frequency of stimulation than predicted by its intrinsic properties even though no active electrical activity was produced in cell 2. If this phenomenon were occurring between two cardiac regions of the ventricular myocardium, then extracellular recording techniques would not have detected the presence of the second region. Second, the arrhythmic responses over the range of R_c from 55 to 83 MΩ are not due to automaticity (intrinsic or induced), or to delayed after–depolarizations or to micro–reentry. Rather, these arrhythmic responses result from an electrical load which first enables a tachycardia (by shortening the APD of cell 1) and then periodically disables the tachycardia following each successful activation of cell 2, since a successful activation of cell 2 temporarily removes the electrical load of cell 2 on cell 1.

We have previously used isolated rabbit ventricular cells to study the electrical properties of isolated heart cells. We have presented here a technique to artificially provide electrical coupling between two isolated cells in order to study the general phenomenon of how R_c affects the APD, APA, I_{th} and the response to rapid stimulation. For the present experiments we have simplified the experimental design by coupling a real ventricular cell to an RC circuit which approximates the input R and capacitance of a rabbit ventricular cell. Since we are experimentally studying the range of R_c greater than 100 MΩ, the use of a passive "cell 2" is a reasonable choice, since the actual depolarization achieved in "cell 2" is below that which would have produced an active response in an isolated cell.

Acknowledgement

This work was supported in part by NIH grants HL22562, HL41851 and HL27385 and by the Children's Research Center.

DISCUSSION

Dr. Fozzard: One of the factors that you did not mention is the input resistance during the plateau phase of the action potential. I wonder as to what the input resistance ratio is at the resting plateau?

Dr. Joyner: Input resistance of the cells is at least as high or higher during the plateau as it is at rest, but we have not measured this parameter carefully, or its changes with time during the plateau.

Dr. Jalife: Do you think you can simulate reflection?

Dr. Joyner: Simulating reflection is going to be extremely difficult and probably not possible with this experimental two cell model due to the properties of the ventricular cells. To produce reflections we will have to have a delay from cell 1 to cell 2 on the order of the action potential duration. Because of this shortening of the duration due to the electrotonic load, reflection might happen, but the problem is that you must lower the resistance in order to make action potential duration short. When you lower the resistance enough to make conduction succeed, the action potential duration shortening goes away. In order to get action potential conduction to occur with a long delay, you have to have a fairly high resistance and I do not know whether or not there will be enough shortening in the delay to make a reflection. It certainly happened in Purkinje strains, as Charlie Antzelevitch *[12]* and you have shown. I am familiar with Rosanski's work *[15]* with you showing ventricular reflection and we have done some similar experiments as well without seeing ventricular reflection. Certainly, you can not make reflection in a one–dimensional strand in the Beeler–Reuter model *per se*.

Dr. Jalife: Do you think you could do it with the Beeler–Reuter cells?

Dr. Joyner: I do not know. I will certainly look at it.

Dr. Pollack: My question relates to the values of resistance. Isn't the resistance you measured a combination of the resistance of that cell plus all the cells surrounding it? Is it really possible to measure resistance of a single cell as though the cells were in isolation?

Dr. Joyner: The experimental data relates to an isolated heart cell.

Dr. Pollack: What happens when you add the resistance of the single isolated heart cells that you measure in a matrix of cells.

Dr. Joyner: That is totally different. When you have a multi–dimensional syncytium you have current flow out in multi–dimensions and effectively no resistance.

REFERENCES

1. Masson–Pevet M, Bleeker WK, Mackaay AJC, Bouman LN, Houtkooper JM: Sinus node and atrium from the rabbit heart: A quantitative electron microscopic description after electrophysiological localization. *J Mol Cell Cardiol* 1979;11:555–568.
2. Veenstra RD, Joyner RW, Rawling DA: Purkinje and ventricular activation sequences of canine papillary muscle. Effects of quinidine and calcium on the Purkinje–ventricular conduction delay. *Circ Res* 1984;54:500–515.
3. Clerc L: Directional differences of impulse spread in trabecular muscle from mammalian heart. *J Physiol* 1976;255:335–346.
4. Spach MS, Miller WT, Geselewitz DB, Barr RC, Kootsey JM, Johnson EA: The discontinuous nature of propagation in normal canine cardiac muscle. *Circ Res* 1981;48:39–54.
5. Ursell PC, Gardner PI, Albala A, Fenoglio JJ, Wit AL: Structural and electrophysiological changes in the epicardial border zone of canine myocardial infarcts during healing. *Circ Res* 1985;56:436–451.
6. Kienzle MG, Tan RC, Ramza BM, Young ML, Joyner RW: Alterations in endocardial activation of the canine papillary muscle early and late after myocardial infarction. *Circulation* 1987;76(4):860–874.
7. Veenstra RD, DeHaan RL: Electrotonic interactions between aggregates of chick embryo cardiac pacemaker cells. *Am J Physiol* 1986;250:H453–H463.
8. Joyner RW, van Capelle FJL: Propagation through electrically coupled cells. How a small SA node drives a large atrium. *Biophys J* 1986;50:1157–1164.
9. Spear JF, Moore EN: The effect of changes in rate and rhythm on supernormal excitability in the isolated purkinje system of the dog. A possible role in reentrant arrhythmias. *Circulation* 1974;50:1144–1149.
10. Joyner RW: Modulation of Repolarization by Electrotonic Interactions. *Jap Heart J* 1986;27:167–183.
11. Denniss AR, Richards DA, Waywood JA, Yung T, Kam CA, Ross DL, Uther JB: Electrophysiological and anatomic differences between canine hearts with inducible ventricular tachycardia and fibrillation associated with chronic myocardial infarction. *Circ Res* 1989;64:155–66.
12. Antzelevitch C, Jalife J, Moe GK: Characteristics of reflection as a mechanism of reentrant arrhytmias and its relationship to parasystole. *Circulation* 1980; 61:182–191.
13. Antzelevitch C, Bernstein MJ, Feldman HN, Moe GK: Parasystole, reentry, and tachycardia: a canine preparation of cardiac arrhytmias occuring across inexcitable segments of tissue. *Laboratory Investigation* 1983;68:1101–1115.
14. Delmar M, Jalife J, Michaels DC: Effects of changes in excitability and intercellular coupling on synchronization in the rabbit sino–atrial node. *J Physiol* 1986;370:127–150.
15. Rozanski GJ, Jalife J, Moe GK: Reflected reentry in nonhomogenous ventricular muscle as a mechanism of cardiac arrhythmias. *Circulation* 1984;69:163–173.
16. Weingart R, Maurer P: AP transfer in cell pairs isolated from adult rat and guinea pig ventricles. *Circ Res* 1988;63:72–80.
17. Burt JM, Spray DC: Volatile anesthetics block intercellular communication between neonatal rat myocardial cells. *Circ Res* 1989; 65:829–37.
18. Joyner RW, Overholt ED: Effects of octanol on canine subendocardial Purkinje–to–ventricular transmission. *Am J Physiol* 1985;249:H1228–H1231.
19. Beeler GW, Reuter H: Reconstruction of the AP of ventricular myocardial fibres. *J Physiol* 1977;268:177–210.

20. Janse MJ, Van Capelle FJL, Morsink H, Kléber AG, Wilm–Schopman F, Cardinal R, D'Alnoncourt CN, Durrer D: Flow of "injury" current and patterns of excitation during early ventricular arrhythmias in acute regional myocardial ischemia in isolated porcine and canine hearts. *Circ Res* 1980;47:151–165.
21. Janse MJ, Van Capelle FJL: Electrotonic interactions across an inexcitable region as a cause of ectopic activity in acute regional myocardial ischemia. *Circ Res* 1982;50:527–537.

CHAPTER 6:

REENTRY OF CARDIAC EXCITATION: A SIMULATION STUDY

Yoram Rudy and Weilun Quan

ABSTRACT

A computer model of a ring–shaped cardiac fiber is used to examine responses of propagation to premature stimuli under different degrees of cell–to–cell coupling and different degrees of membrane excitability. Non sustained and sustained reentry are induced by a properly timed premature stimulus. The likelihood of inducing reentry is proportional to the degree of cellular uncoupling, demonstrating the importance of cellular uncoupling in the genesis of reentry. Alternans in action potential duration are also demonstrated during circus movement and are caused by the alternating kinetic properties of the slow ionic currents.

INTRODUCTION

Reentry is a major cause of arrhythmias following myocardial ischemia or infarction. Recent experiments demonstrated high degree of cellular uncoupling in cardiac tissue under ischemic conditions [1, 2]. There is increasing experimental evidence that abnormality of cell–to–cell coupling is likely to be involved in slow conduction, conduction block, and reentry. Our previous simulations [3, 4] demonstrated that conduction disturbances which lead to reentry, such as slow conduction, decremental conduction and conduction block, can be caused by high degree of cellular uncoupling. In general, propagation in cardiac tissue is determined by the source–sink relations. The interplay between the degree of intercellular coupling and the state of membrane excitability determines success or failure of propagation. In this chapter, we examine effects of changes in the degree of cellular coupling and in membrane excitability on the formation of unidirectional block and reentry. We simulate these effects using a discontinuous ring–shaped model of myocardial tissue that represents a closed pathway of reentry.

Characteristics of the reentrant action potentials and their relations to the kinetics of membrane ionic channels are also investigated in the same model. Details can be found in a recent publication [5].

METHODS

The model used in this study is a representation of a closed pathway of conduction in reentrant rhythm (Fig. 1). This model is an extension of our discontinuous 1–dimensional model [3, 4]. The fiber is composed of up to 1500 cells of realistic dimensions. Cells are connected to each other by a resistive network representing the gap junction. The electrical activity of the membrane is represented by the Beeler and Reuter model [6] with the Ebihara and Johnson [7] fast sodium current.

RESULTS

Initiation of Reentry

The action potential elicited by a basic stimulus at the top cell (cell 1) propagates down both branches of the ring and collides at the bottom cell (Fig. 1). By applying a properly timed premature stimulus at a point in the left branch during the refractory period of a propagating action potential we were able to elicit a premature action potential. It propagated a short distance, decrementally, in the antegrade direction before block occurred while, in the retrograde direction, the premature action potential propagated around the ring fiber and returned to its point of initiation. Thus, reentry was induced.

We define the "vulnerable window" (TW) as the time interval in the refractory period of a propagating action potential during which unidirectional

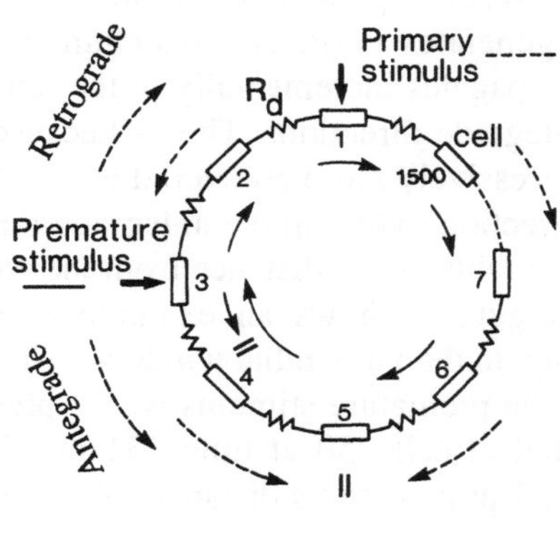

Figure 1: Ring–shaped model of reentry. R_d, gap junction resistance (from [5], by permission of the American Heart Association, Inc.).

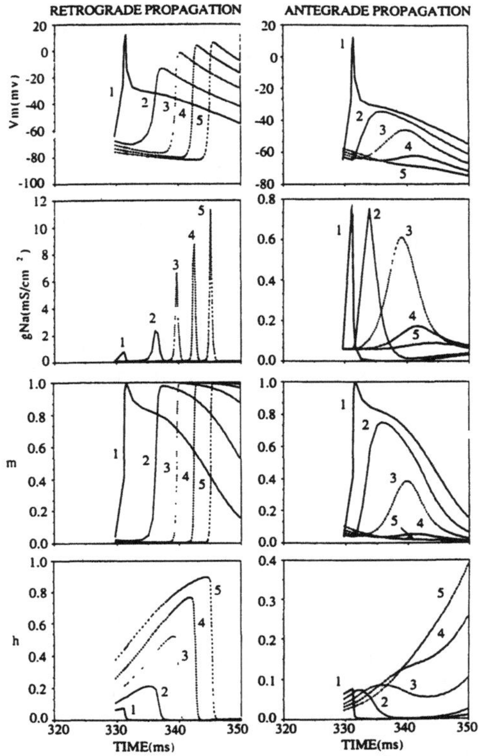

Figure 2: Antegrade decremental propagation (right) and retrograde incremental propagation (left) induced by a premature stimulus in vulnerable window. Vm - membrane potential (mV), gNa - sodium channel conductance (milliSiemens/cm²), m - activation gate and h - inactvation gate. The parameters are displayed at increasing distances (curves 1 to 5) from the site of premature stimulation in both directions. Note different gNa scales in left and right panels (from [5], by permission of the American Heart Association, Inc.).

block and reentry can be induced. The vulnerable window can also be represented as a distance in the space domain (SW). Outside this window, it is impossible to induce unidirectional block and reentry; an action potential induced by a premature stimulus either propagates or blocks in both directions. When a premature stimulus is applied inside the window, the membrane generates a critical fast sodium current, giving rise to an action potential that propagates incrementally in the retrograde direction and decrementally in the antegrade direction. This is because in the retrograde direction the tissue is progressively more recovered as the distance from the window increases in this direction, while in the antegrade direction the membrane is progressively less excitable as the distance from the window increases.

Figure 2 shows an example of propagation induced by a premature stimulus in the vulnerable window. The basic stimulus was applied at cell 1 at t=0. The premature stimulus was applied in the wake of a propagating action potential at cell 250 at time 331 ms. In the retrograde direction, the action potential propagated a distance of 40 cells before reaching the region of fully

Figure 3: Spatial distribution of membrane excitability in neighborhood of vulnerable window during refractory period. Solid curves represent parameters; squares represent first spatial derivatives. For notation see Fig. 2 (from [5], by permission of the American Heart Association, Inc.).

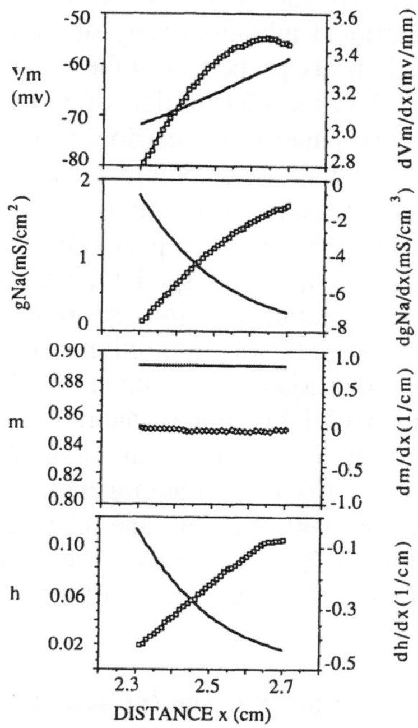

excitable membrane. The conduction velocity was only 19.5 cm/s from cell 250 to cell 240. As the distance increased, the conduction velocity gradually increased, reaching 38.2 cm/s from cell 220 to cell 210. In the antegrade direction, propagation gradually diminished as a result of decreased membrane excitability in this direction. The action potentials gradually decreased in amplitude and in conduction velocity (17.5 cm/s from cell 250 to cell 245 and 9.7 cm/s from cell 245 to cell 240). Note the graded nature of electrical excitation induced in the vulnerable window. In the retrograde direction gNa recovers slowly from curve 1 (0.78 mS/cm^2) to curve 2 (2.4 mS/cm^2), reflecting the time course of recovery of the inactivation parameter h. Note that the activation parameter m is practically fully recovered and does not determine the behavior of gNa. In the antegrade direction, there is a sharp decrease in gNa from curve 3 (0.62 mS/cm^2) to curve 4 (0.18 mS/cm^2). This decremental behavior reflects the sharp decrease in the activation parameter m. Note that the inactivation parameter h recovers more slowly.

It is clear from the above discussion that the inducibility of unidirectional block and reentry is related to the spatial inhomogeneity (asymmetry) of excitability at the vulnerable window. It should be emphasized that the spatial inhomogeneity is introduced by a propagating action potential and is not an intrinsic property of the membrane, hence the term functional inhomogeneity of excitability. Figure 3 shows the distribution of excitability properties in the neighborhood of the vulnerable window. The top panel is the distribution of membrane potential along the fiber (V_m). The second panel is the distribution of the maximum fast sodium channel conductance (gNa). The sodium activation gate m and the inactivation gate h, which were measured at the time gNa reached its maximum, are plotted in the third and fourth panels, respectively.

To quantify spatial inhomogeneity of excitability, we computed first derivatives of V_m, gNa, m, and h and plotted the results in Fig. 3. We found that dV_m/dx was approximately proportional to $-dgNa/dx$ and $-dh/dx$ in the neighborhood of the vulnerable window. Therefore, dV_m/dx is a measure of the spatial functional inhomogeneity in membrane excitability. In terms of underlying sodium channel kinetics, this functional inhomogeneity results mainly from the inhomogeneity in the inactivation gate h, which is a major determinant of membrane refractoriness during propagation. Note that in contrast to h, m is fully recovered and displays no spatial inhomogeneity ($dm/dx=0$) in the vicinity of the window.

Vulnerability to the Induction of Reentry

As explained in the previous section, the size of the vulnerable window in the time domain (TW) is a measure of the vulnerability of the tissue to the induction of unidirectional block and reentry. We examined the effects of cellular uncoupling and of membrane excitability on the tissue vulnerability. Vulnerability (TW) as a function of gap junction resistance is shown in Fig. 4A (curve 1). As the gap junction resistance increases, vulnerability increases as well, with accompanying decrease in propagation velocity (curve 2). For normal cellular coupling (R_d=2 ohm cm^2), the vulnerability is about 0.5 ms. Very precise timing (TW<0.5 ms) of a premature stimulus is required for induction of unidirectional block and reentry, but for a high degree of cellular uncoupling (R_d = 200 ohm cm^2), the vulnerability is increased to 30 ms. For such a wide window, unidirectional block and reentry can be easily induced. The simulations demonstrate, therefore, that as the degree of cellular uncoupling is increased, velocity of propagation decreases (due to long delays at gap junctions) and vulnerability to reentry increases. In the previous section we showed that the vulnerability is determined by the degree of functional inhomogeneity of excitability at the vulnerable window. As regional propagation delay increases, due to long delays at gap junctions, the degree of functional inhomogeneity in the state of the membrane increases as well. As demonstrated in Fig. 3, this increased asymmetry at the vulnerable window reflects an increase in the spatial gradient of the sodium inactivation gate h.

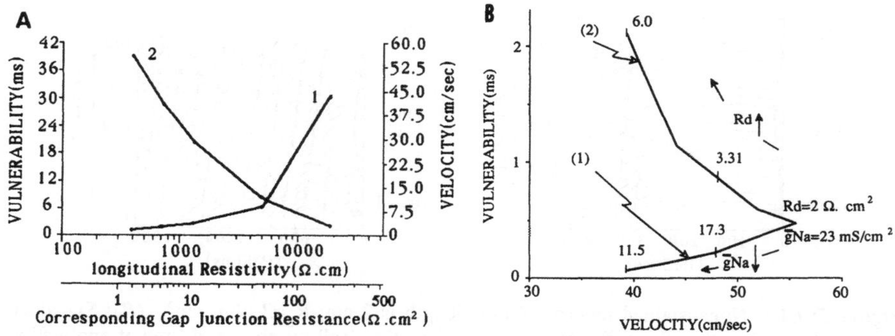

Figure 4: ***(A):*** Changes in vulnerability due to changes in gap junction resistance R_d. Curve 1 shows an increase in vulnerability when R_d increases. Curve 2 shows velocity of propagation as a function of R_d. ***(B):*** Vulnerability at different propagation velocities due to changes in $\bar{g}Na$ (1) and R_d (2). Values of $\bar{g}Na$ and R_d are depicted along curves. R_d, gap junction resistance; $\bar{g}Na$, maximum sodium conductance (from [5], by permission of the American Heart Association, Inc.).

In contrast to the effect of cellular uncoupling described above, uniform reduction in sodium channel conductance (*intrinsic* membrane excitability) resulted in slow propagation with *decreased* vulnerability to the induction of reentry (Fig. 4B). This is because a uniform reduction in membrane excitability produces a shift of the vulnerable window to a more recovered portion of the repolarization phase of the action potential. This shift to a less steep portion of the action potential brings about reduction in spatial gradients and reduces spatial inhomogeneities in the state of the membrane at the vicinity of the window. The result is lower vulnerability to the induction of unidirectional block and reentry.

Characteristics of Reentrant Action Potentials

Reentry induced by a premature stimulus could be sustained or nonsustained. Figure 5A shows an example of nonsustained reentry; the gap junction resistance is 2 ohm cm². For high degree of cellular uncoupling (gap junction resistance of 50 ohm cm²) sustained reentry is obtained, Fig. 5B. The increased uncoupling brings about a fivefold decrease in propagation velocity and reentry is sustained. The duration of the reentrant action potential is initially very short and then increases during a transient period until stable propagation with a duration of 120 ms (less than half the 270 ms duration of the normal action potential) is obtained.

Another characteristic of the reentrant action potential in our simulations are beat–to–beat alternans in action potential duration (APD). To elucidate the

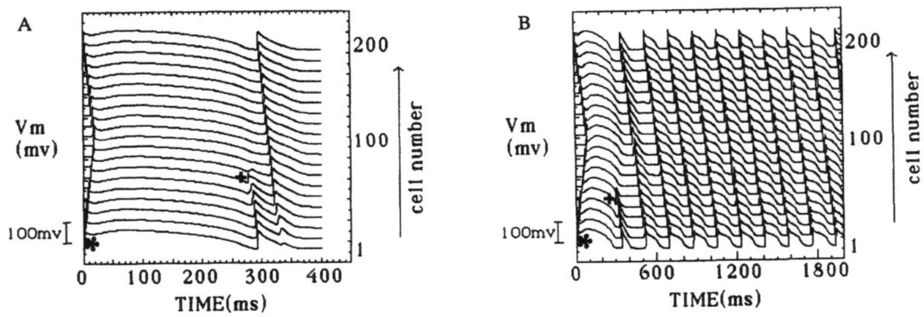

Figure 5: (A): Nonsustained reentry. Circle length is 2 cm; R_d=2 ohm cm². ***(B):*** Sustained reentry. Circle length is 2 cm; R_d=50 ohm cm². Location (cell number) and time of application of initial stimuli are depicted by star; location and time of application of premature stimuli are depicted by plus sign (from [5], by permission of the American Heart Association, Inc.).

underlying mechanism responsible for this long–short pattern of alternans, we computed the various membrane ionic currents and the corresponding gate variables (not shown). The results identified beat–to–beat alternans in the kinetic properties of the slow inward current (I_s) due to a large–small alternating pattern of the channel gating variables (d and f) and its conductance (g_s). Similar dynamic changes were displayed by I_k and its activation variable X. The alternating kinetics of both channels contribute to the beat–to–beat alternans in APD.

DISCUSSION

Reentry induced by premature stimulation is often attributed to nonuniform dispersion of refractoriness and the presence of large disparities in refractory periods. This hypothesis implies that spatial variations in membrane properties must be present for initiation of reentry. The simulations described in this chapter demonstrate that reentry could be induced by a properly timed and properly placed single premature stimulus even in a homogeneous fiber with uniform membrane properties. Propagation itself creates a certain degree of functional inhomogeneity that provides the necessary conditions for unidirectional block and reentry.

The window of vulnerability defined by our simulations is a measure of the inducibility of reentry and reflects the degree of functional inhomogeneity in the state of the membrane. The simulations demonstrate the opposite effects of cellular uncoupling and of reduced membrane excitability on the vulnerability to reentry. An increase in the degree of cellular uncoupling brings about propagation delays which result in a higher degree of asymmetry of excitability at the vulnerable window. The result of the increased asymme–

try is greater vulnerability to unidirectional block and reentry. In contrast, a uniform reduction in membrane excitability shifts the vulnerable window to a more recovered portion of the repolarization phase of V_m. This shift in the "operating point" results from the fact that with depressed membrane excitability, excitation can occur only at more recovered portions of the action potential. Without the shift, a stimulus of equal strength will be blocked in both directions under the conditions of reduced excitability, and unidirectional block (and therefore reentry) cannot be induced. The shift to a less steep portion of the V_m curve reduces spatial gradients of membrane excitability at the window and vulnerability is reduced. Note that both, increased gap junction resistance and reduced membrane excitability result in slow conduction. However, increased uncoupling increases vulnerability while decreased excitability brings about reduction in vulnerability. It is clear, therefore, that slow conduction alone is not sufficient to increase inducibility of reentry, and that the degree of functional inhomogeneity (rather than propagation velocity) is the important determinant of the tissue vulnerability. It should be emphasized that these results apply to uniform changes in membrane excitability and cellular coupling throughout the tissue. In fact, nonuniform reduction of membrane excitability due to nonuniform distribution and action of drugs will augment spatial inhomogeneities, rendering the tissue more vulnerable to the induction of reentry. Moreover, the induction of sustained reentry depends not only on the inducibility of unidirectional block but also on the ability of the tissue to sustain circus movement. This implies that drugs that reduce the wavelength (by reducing propagation velocity and/or refractory period) increase the probability that reentry will be sustained.

The reentrant action potentials are often characterized by a short duration. We have shown that both slow currents I_s and I_k are responsible for the shortening of APD. This relates to the fact that the time constants of the activation and inactivation variables of the slow currents are longer than the interval between successive depolarizations of a cell during reentry. We have also shown that the interaction between the head and tail of the reentrant action potential produces beat-to-beat alternans in action potential duration that reflect alternating kinetics of the slow channels. Recently [8], similar oscillations in action potential duration were observed experimentally in an *in vitro* ring preparation during circus movement.

Acknowledgements

This study has been supported by grant HL33343 from the National Institutes of Health and a Postdoctoral Fellowship (to Dr. Quan) from the American Heart Association, North East Ohio Affiliate, Inc. Computer time was provided by Pittsburgh Supercomputing Center grant PSCA 29 (NSF).

DISCUSSION

Dr. Allessie: In biological experiments, you can often induce reentry by a single stimulus. Are you able to induce reentry in your model without applying premature stimuli?

Dr. Rudy: We can induce reentry in the ring–shaped model without applying a delayed premature stimulus, provided that inhomogeneities are introduced in the reentry pathway. The simulations that are described here are limited to a completely homogeneous pathway in which *functional* inhomogeneity is introduced by the propagation of the basic wavefront. In the presence of intrinsic, anatomical, inhomogeneities, reentry could be induced by a single stimulus. We have simulated inhomogeneities in refractoriness, membrane excitability, fiber cross–section, and in the degree of cellular coupling. Results and details of these simulations were published *[5]*. It is interesting to note that abrupt changes in load are likely to be encountered by the circulating wavefront at pivot points. A similar situation is represented by sites of changes in fiber cross–section in our ring model, and the simulations demonstrate that these sites are vulnerable to the development of unidirectional block.

Dr. Allessie: How important is asymmetry of the propagating wave?

Dr. Rudy: The purpose of the model simulations was to study the effects of various active and passive factors on the inducibility of reentry, and to quantify these effects in terms of a "window of vulnerability." We were particularly interested in the effects of different degrees of cell–to–cell coupling and of membrane excitability.

Dr. van Capelle: I assume that the size of the vulnerable window will depend very much on stimulus strength. A strong stimulus, will result in a larger window than a small one. If you change the resistances, then the threshold will also change. When you defined the width of the window, did you always use the same current or did you change it so that it would always be comparable, say, to twice threshold?

Dr. Rudy: We have characterized the effects of stimulation strength on the vulnerable window *[5 (Fig. 6)]*. In this study we were interested in the isolated effects of tissue properties (e.g. membrane excitability, cellular coupling) on the induction of reentry. We, therefore, used a constant stimulus of 500 $\mu A/cm^2$.

Dr. Joyner: Let me try to answer Dr. Allessie's point. There is more than one way to make this inhomogeneous in order to create unidirectional block. One of the ways is certainly going to be due to spatial inhomogeneity in the refractory period, which is making these resistances spatially inhomogeneous. If you make one side of the loop have a basically longer refractory period than the other side (and we know that happens in spatially homogeneous tissue) then you will get your reentry with the premature stimulus applied at the same location as the regular one. In the real world, if you put them in the same place you will never have exactly the same functional refractory period on both sides of the loop.

Dr. Rudy: This is absolutely true. We have simulated reentry in inhomogeneous rings including inhomogeneities in refractory period, in membrane excitability, in fiber cross-section and in gap junction resistance. All of these inhomogeneities provide conditions that are necessary for unidirectional block. I have shown here that the anatomical inhomogeneities along the reentry pathway are not necessary and that reentry can be induced by premature stimulation, even in a completely homogeneous tissue.

REFERENCES

1. Kléber AG, Riegger CB, Janse MJ: Electrical uncoupling and increase of extracellular resistance after induction of ischemia in isolated, arterially perfused rabbit papillary muscle. *Circ Res* 1987;51:271–279.
2. Yan G, Cascio W, Kléber AG: Evidence for inhomogeneous cellular uncoupling in ventricular muscle during ischemia (abstract). *Circulation* 1988 ;78(4):II-639.
3. Rudy Y, Quan W: A model study of the effects of the discrete cellular structure on electrical propagation in cardiac tissue. *Circ Res* 1987;61:815–823.
4. Rudy Y, Quan W: The effects of the discrete cellular structure on propagation of excitation in cardiac tissue: a model study. In: *Cell Interaction and Gap Junctions (Vol. II)*, Sperelakis N, Cole W (eds), CRC Press, Boca - Raton, Florida, 1989, pp 123–141.
5. Quan W, Rudy Y: Unidirectional block and reentry of cardiac excitation – a model study. *Circ Res* 1990;66:367–382.
6. Beeler GW, Reuter H: Reconstruction of the action potential of ventricular myocardial fibers. *J Physiol London* 1977;286:177–210.
7. Ebihara L, Johnson EA:: Fast sodium current in cardiac muscle. A quantitative description. *Biophys J* 1980;32:779–790.
8. Frame L, Simson M: Oscillations of conduction, action potential duration, and refractoriness. *Circulation* 1988;78:1277–1287.

CHAPTER 7:

CONDUCTION VELOCITY VARIABILITY IN THE ISCHEMIC MYOCARDIUM AS THE CAUSE OF REENTRANT ARRHYTHMIAS

Shlomo A. Ben–Haim and Yoram Palti

ABSTRACT

Using a model of a homogeneous two–dimensional (2D) matrix of conducting cells, we derive the relationship between the relative standard deviation (RSD) of conduction velocity and the probability of occurrence of a locally ordered dispersion of velocities, shown to have the potential to result in circular propagation. This probability is found to be insignificant when the RSD was lower than 10%, but increased dramatically for RSD's greater than 10%.

INTRODUCTION

Reentry arrhythmias are the major cause of death from cardiovascular disease. A number of models or mechanisms have been proposed to explain the generation of reentry arrhythmias in the ischemic or damaged heart [1–6]. Nevertheless, none of these models can explain the formation of reentry loops with no modifications of the basic electrophysiological characteristics of the myocardium [5, 6]. Here, we introduce a concept that the generation of reentry arrhythmias may be due to the increased variance of the conduction properties of the cardiac tissue, rather than to modification of these properties.

Using a model of a homogeneous 2D matrix of excitation–conducting cells, we have derived the relationship between the RSD and the probability of occurrence of a locally ordered dispersion of velocities, shown to have the potential to result in circular propagation. This probability was found to be insignificant when the RSD was lower than 10%, but increased dramatically for RSD's greater than 10%. On the basis of experimental RSD, the calculated probability for vortex formation is one in 10,000 normal heart beats and one in two ischemic heart beats. The model provides new insight into the mechanism of reentry arrhythmias.

METHODS

Consider a homogeneous conducting layer (Fig. 1), in which the activation front of any pathway can propagate in one of three directions: one forward and two opposite lateral directions. Let the refractory period prevent any backward conduction and the pathway velocities' order be random (no covariance assumption). We will apply the analysis to a set of k adjacent pathways.

Each pathway has a probability P of having an average conduction time Y larger or smaller by α percent than the average tissue conduction time μ:

$$P(Y) \Big|_{\{Y \geq \mu(1+\alpha)\, \cup\, Y \leq \mu(1-\alpha)\}} = (1 - \mathrm{erf})\,[\{(\alpha)\sqrt{n}\} / \{\sqrt{\sigma_r}\}] \quad (1)$$

where n is the number of conducting cells in a pathway and σ_r is the relative standard deviation, RSD (standard deviation/average), assuming a Gaussian distribution of μ_i between the cells. The difference between the velocity of the fastest and slowest members of the set is the "maximal velocity difference" (MVD).

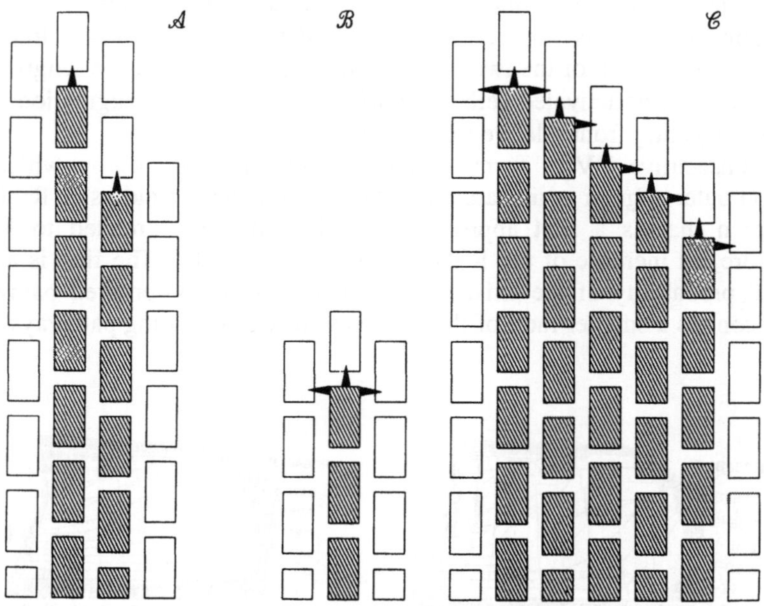

Figure 1: (a) A simplified model for conducting myocardium. Two pathways are drawn for illustrative purposes, a fast one and a slow one. (b) Three possible propagation directions of an activated cell are reduced into two possible directions when organized in an ordered velocity pathway set (c).

RESULTS AND DISCUSSION

We calculated numerically the probability of occurrence of five and seven adjacent, ordered, velocity gradient pathways as a function of MVD and the RSD (σ_r). This was done using the Monte Carlo algorithm and a Gaussian random number generator (RNNOM, IMSL). The results are illustrated in Fig. 2. Figure 3 is a cross–section of Fig. 2b and shows that for an RSD of 0.08, the P values were nonsignificant for MVD higher than 10%, while for an RSD of 0.26, significant P values were noted for every MVD tested.

In the ordered velocity situation there are only two possible propagation directions, lateral and forward, for all except the fastest pathways. This is due to the fact that the faster lateral pathway always blocks one of the lateral propagation directions for the slower pathway.

In an isotropic myocardial model the probabilities of a forward and lateral propagation are equal. Therefore, the average propagation direction is 45° to the previous propagation direction. We now define a "basic vortex element" (BVE) (Fig. 4) as the combination of an occurrence of an ordered velocity arrangement with the oblique (45°) propagation direction.

A propagating depolarization front from the endocardium towards the epicardium can form a vortex movement by propagation with four consecutive BVE's (180° rotation), followed by a shortcut through the endocardial conducting system, as was recently measured [7]. A BVE has a trapezoid shape, in which each of the two lateral borders is made up of the length of an integer number of activated cells. The MVD determines the angulation of the BVE propagating front relative to its base.

Consecutive BVE's form a vortex movement (Fig. 4), in which the polygon outer angular points are all part of a circle, with a radius of R. It can be shown that as a first approximation, R is inversely related to MVD. Therefore, an increase of the RSD will have two effects. The first is an increased probability of the existence of adjacent velocity–ordered pathways. The second is a smaller radii of the evoked vortex propagating pathways. The

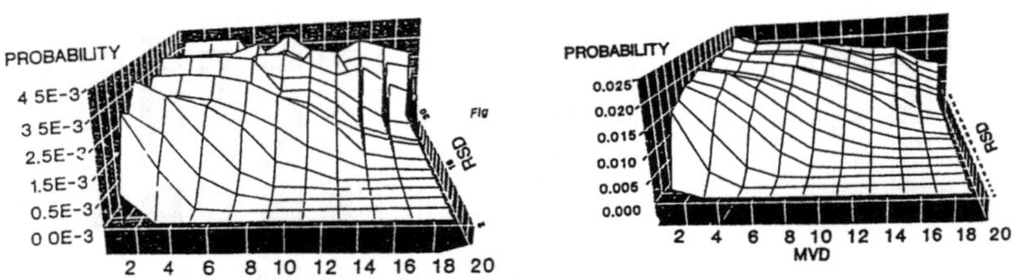

Figure 2: The probability of occurrence of (a) seven or (b) five adjacent pathways, as a function of the myocardial RSD and the MVD.

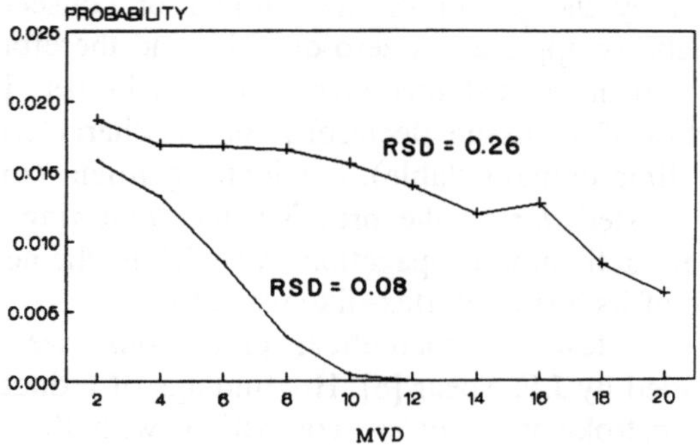

Figure 3: Two cross-sections of Fig. 2b at RSD of 0.08 (——) and RSD of 0.26 (- - -).

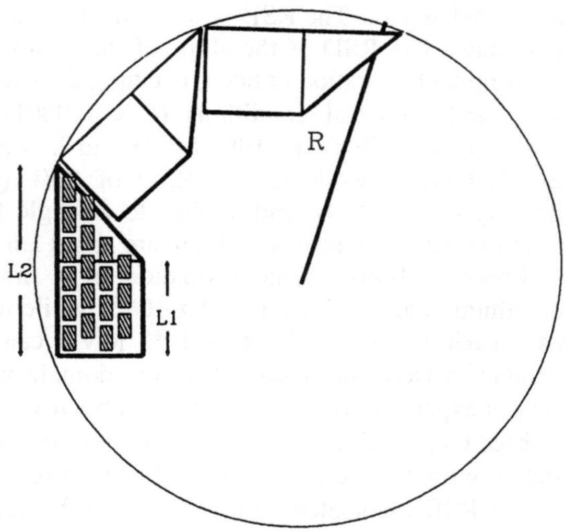

Figure 4: Vortex movement. A graphical description of the BVE, composed of a set of velocity-ordered pathways. L_1 and L_2 denote the distance of activation propagation of the two lateral pathways of the set. Three consecutive BVE's show the initiation of a vortex movement of a wave front.

effect of a small change in the value of RSD is amplified several orders of magnitude (10^5 to 10^{10}) by its highly nonlinear relation to the MVD (Fig. 2).

Upon completion of a full vortex ring, reentrant activation may take place. The condition to be satisfied for this phenomenon is that the vortex propagation time is shorter than the dominant pacemaker interval and longer than the refractory period of the first cells of the ring. This reentrant activation will be recognized as a premature activation. The fate of such an activation depends on the kinetics of the tissue electrophysiologic characteristics. Whenever the tissue electrophysiologic characteristics change faster than the heart rate (time domain unstable), the probability of a repeated vortex movement through the same route equals the previous probability raised by the power of the number of consecutive reentrant beats. This probability approaches zero quickly, and therefore cannot be considered as a mechanism for sustained reentrant arrhythmias. However, in the case that the kinetics of the tissue electrophysiologic characteristics is slower than the heart rate (time domain stable), a velocity gradient combination similar to the one that existed during the previous reentrant ring will be encountered by the present activation propagation wave. Thus, the new wave front will follow the route of its former vortex–like wave front.

To test our theoretical predictions, we use the experimental data provided by J.F. Spear [8]. His findings of reduced action potential amplitude and upstroke velocity are compatible with the commonly reported changes occurring during ischemia. Calculation of the intra–experimental variance, not reported previously, shows that while the average electrophysiological characteristics decrease during ischemia, the absolute standard deviation remains unchanged. The RSD is dramatically increased as compared with the normal state. The RSD of the slope of the action potential (dV/dt_{max}) is the most prominent alteration noticed, reaching 26% during ischemia as compared to 12% under normal conditions ($P < 0.0001$). The probability of the formation of a BVE (with 10% MVD and $k = 5$) at RSD of 12% (normal tissue) is 0.0026, while for an RSD of 26% (ischemic tissue) the BVE probability equals 0.014 and applies to a single fifth of adjacent myocytes, since in the whole myocardium there are about $1.6\ 10^7$ adjacent myocyte fits.

The probability of the formation of a first vortex ring in a normal myocardium is approximately 7.2×10^{-4}, while in the ischemic myocardium it may reach $5.9 \times 10^{-1}1$. If a 10% MVD can be associated with vortex movement in ischemic tissue, in a time domain unstable myocardium, a bi–geminy is expected. In a time domain stable myocardium the appearance of at least three consecutive ventricular premature beats (VPBs), i.e. a short VT, is expected within a few minutes. Furthermore, for any given MVD the increased RSD associated with the acute ischemia increases by 4– to 5–fold the probability of vortex ring formation.

SUMMARY

The model developed deals with the probability of generation of a single reentry cycle (equivalent to a VPB) and a sustained reentry (equivalent to VT). The predictions of the model are compatible with the clinical experience, i.e., highly variable electrophysiologic parameters found in severely ill patients, as well as the high prevalence of reentry arrhythmias encountered during ischemia. The proposed model provides us with new insight into normal and abnormal cardiac functions that may be of utmost importance in understanding, diagnosing, evaluating, and treating reentry arrhythmias and possibly numerous other biologic phenomena.

Acknowledgement

The authors wish to thank Professor J.F. Spear for providing the experimental data.

DISCUSSION

Dr. Rudy: Did you use the average of conduction velocity or conduction time?

Dr. Ben Haim: Conduction velocity.

Dr. Jalife: If I understand you correctly, the greater the curvature the higher the probability to have reentry, but there must be a limit to that curvature.

Dr. Ben Haim: It is true that, according to our model predictions, the greater the curvature the higher the probability for its occurrence. However since the myocardium has finite dimensions, it can encompass only reentry propagation pathways with a radius smaller than that of the myocardial thickness. Another important prediction made by the model is that reentry loops going from the endocardium to the epicardium have higher probability than smaller reentry loops.

Dr. Jalife: What does your model tell us about the stability of the vortex?

Dr. Ben Haim: We have two kinds of stabilities. We have the time stable myocardium, in which the characteristics of the tissue will stay the same for several beats; this means that if the refractory period is shorter than the overall vortex–conduction time, ventricular tachycardia will ensue and the vortex movement will continue. The second case is the time unstable myocardium which will cause the reentrant movement to activate the myocardium prematurely, but the next beat will propagate normally. This case corresponds to a PVB.

REFERENCES

1. Ursell PC, Gardner PI, Albala PJ, Fenoglio JJ, Wit AL: Time course of structural and electrophysiologic changes in the epicardial border zone of canine myocardial infarcts during infarct healing. *Circ Res* 1983;56:11–23.
2. Gardner PI, Ursell PC, Fenoglio JJ, Wit AL: Electrophysiologic and anatomic basis for fractionated electrograms recorded from healed myocardial infarcts. *Circulation* 1985;72:596–611.
3. Plonsey R, Rudy Y: Electrocardiogram sources in 2–dimensional anisotropic activation model. *Med Biol Eng Comput* 1980;18:87–94.
4. Plonsey R, Barr RC: Current flow patterns in two–dimensional anisotropic bisyncytia with normal and extreme conductivities. *Biophys J* 1984;45:557–571.
5. Spach MS, Miller WT, Geselowitz DB, Barr RC, Kootsey JM, Johnson EA: The discontinuous nature of propagation in normal canine cardiac muscle: Evidence for recurrent discontinuities of intracellular resistance that affect the membrane currents. *Circ Res* 1981;48:39–54.

6. Spach MS, Dobler PC: Relating extracellular potentials and their derivatives to anisotropic propagation at a microscopic level in human cardiac muscle: Evidence for electrical uncoupling of side–to–side fiber connections with increasing age. *Circ Res* 1986;58:356–371.
7. Ogawa S, Miyazaki T, Sakai T, Hosokawa M, Sakurai K, Nakamura Y: Epicardial mapping during induction of nonsustained polymorphic ventricular tachycardia in a 7–day–old canine myocardial infarction model. *Am Heart J* 1987;114:34–41.
8. Spear JF, Michelson EL, Moore EN: Reduced space constant in slowly conducting regions of chronically infarcted canine myocardium. *Circ Res* 1983;53:176–185.

Propagation of the Cardiac Impulse

CHAPTER 8:

DETERMINANTS OF ACTION POTENTIAL TRANSFER FROM CELL TO CELL

Robert Weingart

ABSTRACT

Cell pairs isolated from adult rat and guinea pig ventricles were used to study the transfer of action potentials from cell to cell and to assess the electrical properties of cardiac gap junctions. An attempt was made to correlate the impulse transfer with the conductance of the gap junction membrane and the number of operational gap junction channels.

INTRODUCTION

Impulse propagation in cardiac tissue occurs by means of local circuit currents. Local circuits represent closed loops which involve four functional elements: i) channels carrying the excitatory inward current, ii) intracellular longitudinal resistance, r_i (cytoplasmic resistance, r_c, in series with the nexus or gap junction resistance, r_n) iii) membrane capacitance, and iv) extracellular longitudinal resistance, r_o. Together with the topology of the cell arrangements, these elements determine the speed with which an action potential propagates. Elements i), ii), and iv) are of great interest because they are known to undergo changes under certain pathological conditions. So far, impulse propagation was examined in intact hearts, preparations dissected from different areas of the heart, or synthetic strands of myocytes (for references, see [1]).

Cell pairs offer the possibility to study impulse transmission at the cellular level. Combining current–clamp and voltage–clamp experiments, this preparation allows to explore the transfer of action potentials and determine the state of cell–to–cell coupling simultaneously [1]. This article describes experiments performed on pairs of adult myocytes. The results obtained contribute to a better understanding of the relative importance of passive electrical properties on impulse propagation.

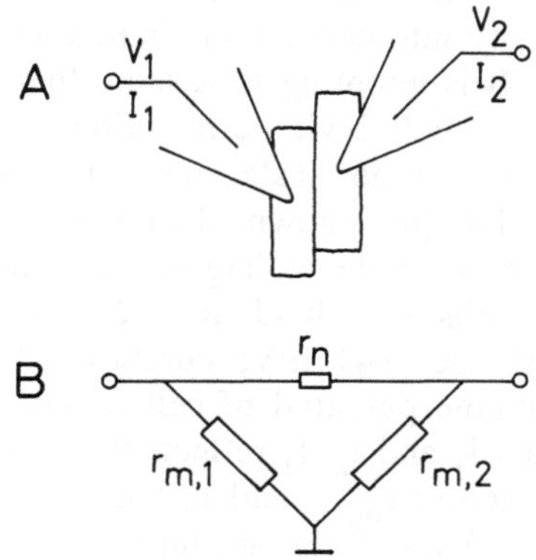

Figure 1: Experimental arrangement. ***(A)*** Cell pair with attached patch pipettes enabling the measurement of membrane voltages (V_1, V_2) and currents (I_1, I_2). ***(B)*** Equivalent circuit used to analyze the resistive elements involved: $r_{m,1}$, $r_{m,2}$, resistances of the non–junctional membranes of cell 1 and cell 2; r_n, resistance of the gap junction membrane located between the cells (reprinted by permission of the American Heart Association [1]).

METHODS

Cell were isolated from adult rat and guinea pig ventricles by means of an enzymatic treatment. Pairs of cells were selected to perform electrical measurements. For this purpose, each cell of a cell pair was attached to a patch pipette so as to enable whole–cell, tight–seal recording (Fig. 1A). The pipettes were connected to separate amplifiers that allowed operation in the current–clamp and voltage–clamp mode. The current–clamp mode was used to inject rectangular current pulses and record action potentials. The voltage–clamp mode was employed to determine the conductance (reciprocal resistance) of the gap junctional membrane (g_n) and/or gap junctional channel (γ_n). In essence, it enabled the potential of the non–junctional membrane of each cell (V_1, V_2) to be controlled and thereby the voltage gradient across the gap junction (V_n) to be set, while monitoring the associated current flow through each pipette (I_1, I_2). Thus, when a voltage pulse was administered to cell 1 (V_1), while cell 2 was kept at the holding potential, V_1 reflected the V_n and I_2 corresponded to the current flowing across the gap junction, I_n (Fig. 1B). From these quantities, g_n or γ_n (I_n/V_n) was determined. All experiments were performed at room temperature (20 – 23°C). A detailed description of the method is given elsewhere [2].

RESULTS AND DISCUSSION

Action Potential Transfer

Figure 1 illustrates the electrical properties of an isolated cell pair under control conditions. Panel A shows a set of four standard signals obtained in the current–clamp mode. Injection of a depolarizing current pulse of threshold intensity into cell 1 (I_1) elicited an action potential in cell 1 (V_1) and cell 2 (V_2). It is tempting to assume that the action potential was transferred from cell 1 to cell 2 via current flow across the gap junction. When displayed at expanded time scale, no delay was detectable between the two action potentials (not shown). Panel B depicts the appropriate signals obtained after switching to the voltage–clamp mode. Initially, the membrane potential of both cells was held at –52 mV. Thereafter, a depolarizing clamp pulse (amplitude: +41 mV; duration: 200 ms) was applied to cell 1, while the membrane potential of cell 2 was maintained. This gave rise to the current signals I_1 and I_2. I_1 reflects the sum of two current components, $I_{m,1}$ (current flow across $r_{m,1}$), and I_n (current flow across r_n); I_2 corresponds to I_n. The ratio I_2/V_1, defining g_n, turned out to be 245 nS. At the very beginning of the experiment, g_n exhibited a value of 110 nS. The signals displayed in Fig. 2 were recorded 12 min later, when g_n had increased spontaneously. Under control condition, cell pairs usually had g_n ranging from 200 – 1000 nS [2, 3].

Figure 3 illustrates the behavior of a cell pair with impaired impulse propagation. Panel A shows current–clamp records at slow time scale. Application of small current pulses to cell 1 (I_1) provoked subthreshold depolarizations in both cells (V_1, V_2). The amplitudes of the voltage changes differed considerably, implying a substantial voltage drop across r_n. Application of larger stimuli evoked action potentials in both cells in most trials. However, most interestingly, on two occasions cell 2 failed to fire an

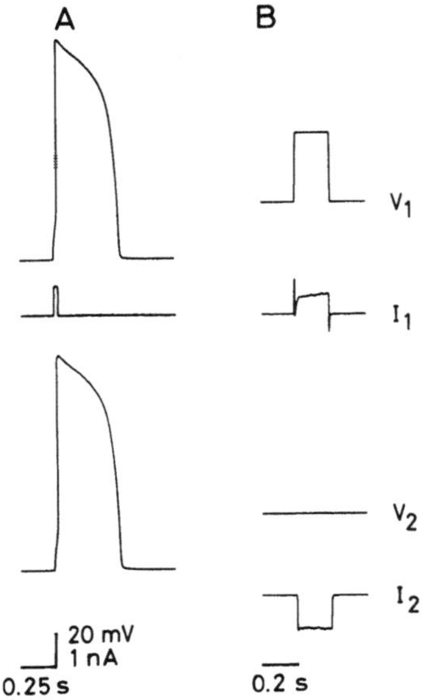

Figure 2: Electrical properties of a cell pair exhibiting normal intercellular coupling.
(A) Records obtained in the current–clamp mode. Current injection into cell 1 (I_1) evoked an action potential in cell 1 (V_1) and cell 2 (V_2). Vm = –85 mV.
(B) Signals recorded in the double voltage–clamp mode. Establishment of a transjunctional voltage gradient (V_1) gave rise to the junctional current flow (I_2). g_n = 245 nS (by permission of the American Heart Association, Inc. [1]).

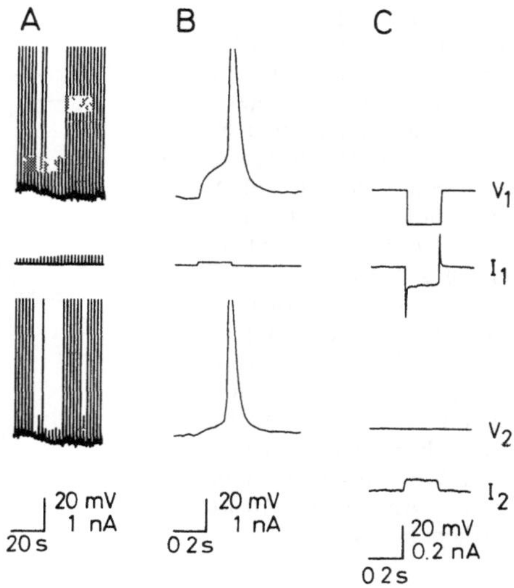

Figure 3: Transfer of action potentials in a partially uncoupled cell pair.
(A) Subthreshold current injection into cell 1 (I_1) produced electrotonic responses of different size in cell 1 (V_1) and cell 2 (V_2). Suprathreshold current injections elicited action potentials in cell 1 and, with two exceptions, in cell 2.
(B) Successful transfer of an action potential from cell 1 to cell 2, displayed at faster time scale.
(C) Determination of g_n by means of the double voltage–clamp approach. A voltage gradient was applied across the gap junction (V_1) which gave rise to a transjunctional current flow (I_2). g_n= 3.2 nS (by permission of the American Heart Association, Inc. [1]).

action potential thus giving rise to an impulse block. Panel B shows a successful action potential transfer at faster time scale. Current injection into cell 1 (I_1) was associated with a large electrotonic response in cell 1 (V_1) and a small one in cell 2 (V_2). The former eventually reached threshold, the latter remained subihreshold. Nevertheless, eventually both cells fired an action potential. Obviously, in cell 1 the action potential results from *direct* stimulation, whereas in cell 2 it is a consequence of *indirect* stimulation via current flow through the gap junction. Figure 3C depicts voltage–clamp records obtained under this condition. Analysis of the transjunctional voltage gradient (V_1) and the accompanying junctional current (I_2) revealed a g_n of 3.2 nS.

Figure 4 shows selected action potentials, displayed at fast time scale. Injection of a short stimulus produced a direct depolarization of cell 1 (V_1) and, via current flow across the gap junctional, an indirect depolarization of cell 2 (V_2). V_1 was large enough to reach threshold, while V_2 was too small. Consequently, an action potential was elicited in cell 1, but not in cell 2 as yet. The upstroke of the action potential provided an extra voltage gradient across the gap junction. The subsequent intercellular current flow depolarized cell 2 towards the threshold and, after a delay of 24 ms, released an action potential. Conversely, V_2 also influenced V_1. The early plateau phase of V_1 shows a depression attributable to a repolarizing current provided by cell 2. This cross–talk nicely illustrates the electrical interaction between the two cells.

Utilizing the experimental approach outlined in Figs. 2 to 4, we attempted to establish a correlation between the state of impulse transfer and g_n. For this purpose cell pairs were used exhibiting spontaneous uncoupling or

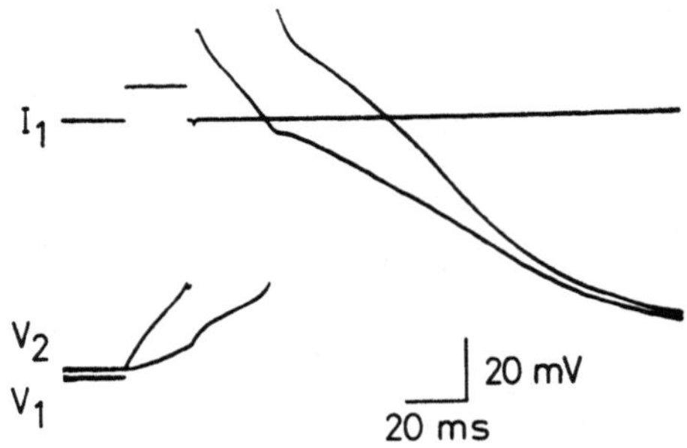

Figure 4: Tracings of action potential transfer in a partially uncoupled cell pair. Current injection into cell 1 (I_1) gave rise to electrotonic responses of different amplitudes in cell 1 (V_1) and cell 2 (V_2). In cell 1 the threshold was reached towards the end of the stimulus eliciting an action potential. The upstroke of this action potential provided the current to further depolarize cell 2. Within a delay of 24 ms, cell 2 fired an action potential, too. To emphasize the temporal relation, signal V_2 was offset by 3 mV (by permission of the American Heart Association, Inc. [1]).

induced uncoupling caused by temporary exposure to cardiac glycosides, or low $[Na^+]_o$ [3]. Successful impulse transfer was found in cell pairs with values of g_n ranging from 200 to 3.75 nS. In these preparations, the action potential developed virtually simultaneously in both cells. Occasional failures in action potential transfer were observed in cell pairs whose g_n was 6.45 to 2.65 nS. In these cases, impulse transfer occurred with a delay of up to 24 ms. In intact cardiac tissue, such a delay would correspond to a conduction velocity of 0.5 m/s, i.e. 1/100 of that observed under physiological conditions. Impulse transfer was completely blocked when g_n was smaller than 1.3 nS.

Single Channel Conductance

Exposure of cell pairs to long–chain alcohols (e.g. heptanol, octanol), or general anesthetics (halothane) leads to reversible electrical uncoupling [4]. We made use of this pharmacological approach to search for current signals arising from putative single gap junction channels. Inspection of I_n signals at high gain, recorded immediately before complete uncoupling or early during recovery from uncoupling, revealed discrete current signals, attributable to random opening and closing of single gap junction channels [5]. Figure 5 illustrates an example of a single channel event recorded in the presence of 3 mM heptanol. In Fig. 5A the upper trace represents the transjunctional voltage gradient applied, V_n (+50 mV; 1 s), and the lower trace the associated junctional current, I_n. The current signal exhibits a discrete current step reflecting the opening and closing of a single channel. Figure 5B shows the amplitude histogram of the current record using a sampling interval of 0.2 pA.

Figure 5: Electrical activity of single gap junction channel in the presence of 3 mM heptanol. *(A)* Voltage gradient across the junctional membrane (V_n) and associated transjunctional current flow (I_n). Dashed lines indicate the discrete current level obtained from the analysis outlined in B.
(B) Amplitude histogram of current record, plotting the normalized number of occurrences versus amplitude of the current. Bin width: 0.2 pA. g_n = 78 pS.

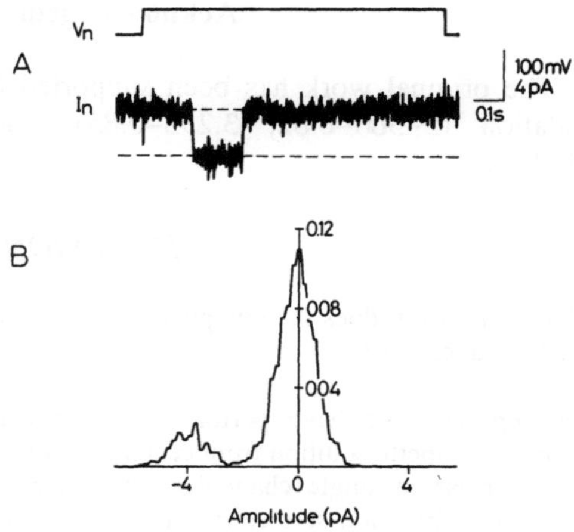

The amplitude of the peaks corresponds to the normalized time spent at a particular current level, and the distance between the peaks indicates the size of the prevailing current level. The large peak represents the background fluctuations and the small peak the open channel activity. The analysis yielded a unitary current level of 3.9 pA and a single channel conductance, γ_n, of 78 pS.

This experimental approach was then used to determine the electrical properties of gap junction channels. It turned out that γ_n is not affected by V_n and V_m, the voltage across the junctional and non-junctional membrane, respectively. Pooling all data, our experiments revealed a mean single channel conductance, γ_n, of 35 – 50 pS, depending on the charge carriers used [5, 6].

CONCLUSION

Studies on action potential transfer together with the data on g_n, the conductance of gap junction membrane, and γ_n, the conductance of gap junction channel, allow an estimate of the number of gap junction channels involved during the impulse transfer from cell to cell. Experiments examining the electrical properties of single gap junction channels revealed a unitary conductance of 35 – 50 pS. This suggests that 75 – 110 channels have to be open simultaneously in order to allow a safe impulse transfer. (In the presence of 25 – 35 channels there is no action potential transfer at all.) Assuming an open channel probability of 0.16 [7], the number of channels involved would increase by 84%.

Acknowledgement

The original work has been supported by the Swiss National Science Foundation (5–360–0.82; 3.253–0.85), and the Swiss Foundation of Cardiology.

DISCUSSION

Dr. Fozzard: How does the composition of the pipette solution influence the single channel conductance?

Dr. Weingart: We have performed preliminary experiments in this direction. Changing the pipette solution from cesium aspartate to potassium chloride produced a sizable increase in single channel conductance (a doubling). This finding was a surprise since diffusion studies revealed that these channels accept particles up to a size of 1,000 dalton, thus giving the impression of large pores. Of course, there is a difference in the mobility. However, it does not explain the large change in conductance.

Dr. Strauss: Have you looked at the mean open time of the channels under the two conditions?

Dr. Weingart: With our experimental approach, the patch pipettes are sitting on non–junctional membranes and not on gap junctional membrane. In other words, we do not have direct access to the gap junction membrane. Thousands of channels are present between two adjacent cells, hence we cannot perform kinetic studies.

Dr. Rudy: Do you have any idea if and how heptanol affects your results?

Dr. Weingart: It is hard to tell from our experiments if heptanol influences the single channel conductance, because we need heptanol to visualize the single channel signals. However, Habo Jongsma has tried to investigate the *de novo* formation of gap junctions pushing two myocytes together. Utilizing this approach, he got similar values for the single channel conductance. This suggests that the influence of heptanol on the single channel conductance is small.

Dr. Marban: There has been some controversy about the relative roles of physiological pH and pCa in controlling gap junctions; I know you have done a lot of work in this area. What is your feeling about the relative roles that they are playing on the gap junction conductance?

Dr. Weingart: As you said, electrical uncoupling can be achieved by interventions aimed at elevating intracellular calcium or protons. In case of calcium, a large increase is required. Therefore, this mechanism may not play a role under physiological conditions. In case of protons, however, moderate changes are sufficient to modify gap junctions. Hence, it is likely that protons are involved in regulating gap junctions under physiological conditions.

Dr. Joyner: What happens when you put aspartate into one cell and chloride into the other cell, would you get a uni–directional difference in conductance in channel conductance?

Dr. Weingart: We have carried out experiments with mixed solutions, adding small cations and large anions to one pipette and large cations and small anions to the other one. If we do this, we observe time–dependent conductance changes, a phenomenon which looks like rectification.

REFERENCES

1. Weingart R, Maurer P: Action potential transfer in cell pairs isolated from adult rat and guinea pig ventricles. *Circ Res* 1988;63:72–80.
2. Weingart R: Electrical properties of the nexal membrane studied in rat ventricular cell pairs. *J Physiol (Lond)* 1986;370:267–284.
3. Maurer P, Weingart R: Cell pairs isolated from adult guinea pig and rat hearts: Effects of $[Ca^{2+}]_i$ on nexal membrane resistance. *Pflügers Arch* 1987; 409:394–402.
4. Niggle E, Rüdisüli A, Maurer P, Weingart R: Effects of general anesthetics on current flow across membranes in guinea pig myocytes. *Am J Physiol* 1989;256:C273–C281.
5. Rüdisüli A, Weingart R: Electrical properties of gap junction channels in guinea–pig ventricular cell pairs revealed by exposure to heptanol. *Pflügers Arch* 1989;415:12–21.
6. Rüdisüli A, Weingart R: Electrical properties of gap junction channels in guinea–pig ventricular cell pairs. *Experientia* 1989;45:A47.
7. Veenstra RD, De Haan RL: Measurement of single channel currents from cardiac gap junctions. *Science* 1986;233:972–974.

CHAPTER 9:

THE EFFECT OF THE FINITE EXTRACELLULAR RESISTANCE ON IMPULSE CONDUCTION IN THE VENTRICULAR MYOCARDIUM

André G. Kléber and Gan–Xin Yan

ABSTRACT

In densely packed ventricular muscle, the extracellular volume fraction amounts to 20–25% of total volume. The methods used to measure the electrical resistance of these narrow extracellular clefts are discussed. Evidence is provided from discussion of published work that the resistance to current flow in the extracellular compartment is approximately of the same magnitude as the resistance in the intracellular compartment. The extracellular electrical resistance therefore exerts an important influence on impulse propagation within the ventricular wall.

INTRODUCTION

The concept of heart muscle forming a functional syncytium with individual cells being connected by gap junctions has become widely accepted over the past decades [1]. During propagation of the electrical impulse, the local current generated at the head of the wavefront of excitation flows along a pathway which not only encounters the resistances within and between the cells, but also the resistances of the extracellular narrow clefts. The mere finding of the extracellular fraction of cross–sectional area of ventricular muscle being considerably lower than the intracellular fraction [2] already may suggest that, for the case of densely packed ventricular muscle, the extracellular electrical resistance may form a considerable obstacle to local current flow, and therefore to impulse propagation. Moreover, for the case where densely packed myocardium comes into contact with an infinite volume conductor (e.g. at the endocardial boundary), an effect of the non–uniform extracellular resistance on conduction velocity is predicted [3].

The experimental studies performed to assess the magnitude of the extracellular resistance (especially in comparison to the intracellular resist–

ance) have provided results which differ considerably according to the methods used [4–6]. This article makes an attempt to discuss the different experimental approaches to the measurement of extracellular resistance and the different results available from literature with the question in mind of whether and to what extent the extracellular resistance in ventricular muscle may influence propagation velocity.

THE METHODOLOGICAL APPROACH

The basic method to assess the extracellular resistance in cardiac muscle has been given by the classical work of Weidmann [4], and is explained in Fig. 1. The major advancement with this method consisted in the possibility to separate the lumped resistance of the extracellular compartment from the lumped resistance of the intracellular compartment. This was achieved by using the physiological process of propagation in the longitudinal direction of the fibers.

If all the extracellular resistive elements in a given cross–sectional area of a cylindrical muscle fiber are lumped into a single resistor, r_o (Ω/cm), and all the intracellular resistive elements are lumped into a single intracellular resistance, r_i (Ω/cm) the whole fiber can be represented by a simple core conductor. In such a simplified electrical equivalent circuit, the amplitude of the transmembrane action potential, ΔV_m, which is equivalent to the driving force for propagation, is divided by the local circuit currents into an extracellular (ΔV_o) and intracellular component ($\Delta V_i = \Delta V_m - \Delta V_o$), because the local currents flow through r_o and r_i in series:

$$r_o/r_i = |\Delta V_o| \;/\; |\Delta V_m| - |\Delta V_o| \qquad (1)$$

For the simplified case of this core conductor therefore the ratio r_o/r_i can be calculated from the measurements of ΔV_m (action potential amplitude) and ΔV_o (the amplitude of the extracellular wavefront, which is equivalent to the amplitude of the extracellular bipolar electrogram).

The second measurement consists in the application of an extracellular subthreshold current pulse between two electrodes placed at either end of the fiber. Close to the site of current injection, the current partially flows from the extracellular to the intracellular space until, at remote sites (> 3 length constants from the site of current injection) the passive resistive elements balance current flow, and consequently, transmembrane current becomes zero. Within this so–called extrapolar region, the measurement of the extracellular voltage gradient, $V_o/\Delta x$, and the strength of the applied current, I, yield a value of a resistor per unit length, r_t (Ω/cm), which is composed of r_o and r_i in parallel:

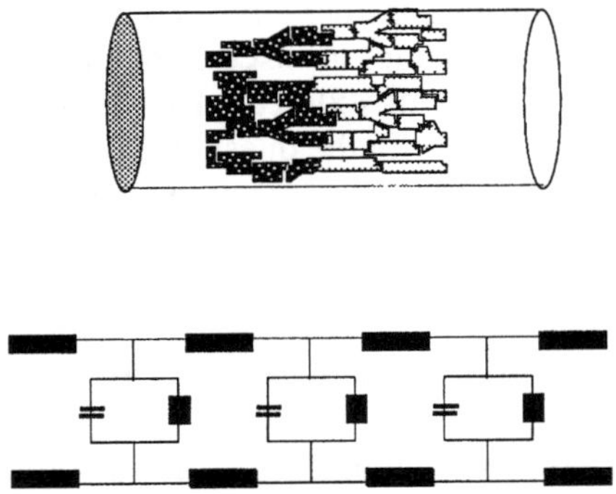

Figure 1: Top: Schematic representation of the cellular network in a cylindrical muscle fiber. The dark cellular elements represent excited cells delivering local current at the border of the wavefront of propagation. *Bottom:* Simplified equivalent electrical circuit used for data analysis. (continuous core conductor model). The resistors of the extracellular space are represented by a single longitudinal resistor, r_o; the resistive elements of the intracellular compartments are represented by a single intracellular resistor, r_i. Capacitive and resistive membrane elements are also shown.

$$(V_o/\Delta x)\,(1/I) = r_t = r_o r_i / r_o + r_i \qquad (2)$$

From the measurements of r_o/r_i and r_t, the values of r_o and r_i can be calculated. For methodological details, see [4–6].

All the experimental investigations which used this method are based on the assumption that the complex electrical network of cardiac muscles behaves like the simple core conductor shown in Fig. 1. This assumption implicates that within a given cross–sectional area all extracellular and intra–cellular resistive elements are in average of the same magnitude (homogenei–ty). Consequently, all the extracellular sites (and, correspondingly, all the intracellular sites) within any cross–sectional area should be equipotential at any given time during excitation or flow of subthreshold current.

EXPERIMENTAL RESULTS AND DISCUSSION

The Magnitude of the Extracellular Resistance

The original method described by Weidmann used isolated cylindrical muscular trabeculae from sheep heart which were soaked in silicon oil. The surrounding oil acted as an electrical insulator and thus forced both the local electrical currents during excitation and (externally applied) subthreshold currents to flow through the tissue's proper extracellular and intracellular spaces. In this original description however, a technical compromise was made, because the fibers were metabolically supplied by a thin layer of Tyrode's solution (thickness 25 μm) which acted to a certain degree as a shunt resistor (see below). The two modifications of Weidmann's method handled the problem of the extracellular shunt around the cylindrical preparation differently. Buchanan et al. [5] used a thin layer of superfusate (100 μm thickness) whereas in the other modification [6] the surrounding fluid layer was completely avoided, because the cylinders were arterially perfused and kept in a artificial gaseous atmosphere (which acted as the electrical insulator). All three methods allow to measure, albeit to a different qualitative degree, changes in cell–to–cell coupling, i.e. in the longitudinal intracellular resistor, r_i. For the measurement of the extracellular longitudinal resistance, r_o, however, the results obtained differ considerably among the three methods mentioned above: the ratio r_o/r_i was shown to be 1/10 with a surrounding fluid layer of 100 μm [5], 1/3 with a surrounding fluid layer of 20 μm [4] and 1/1 with absence of surrounding fluid [6]. For the potential role of the narrow extracellular clefts as a resistive obstacle to local current flow, and consequently to propagation, this ratio r_o/r_i is crucial: In the first case, (r_o/r_i = 1/10) the extracellular resistance would be of minor importance and 90% of the resistance to local current flow during propagation would be concentrated in the intracellular compartment ("cytoplasmic" resistance, junctional resistance). In the last case (r_o/r_i = 1/1) the resistances of the intracellular and extracellular compartments would be of equal importance.

Several observations from literature and from our laboratory, which indicate that the resistance of the extracellular clefts in densely packed ventricular tissue is approximately of the same magnitude as the resistance of the intracellular compartment ($r_o/r_i = 1$), are of interest:

(i) With a ratio $r_o/r_i = 1$, the driving force for propagation (action potential amplitude of approximately 100 mV) is divided to provide an extracellular wavefront amplitude of approximately 50 mV. A very similar value (46 mV) was found on the epicardial surface of dog hearts in the longitudinal fiber direction [7]. This indicates that for ventricular tissue this ratio is relatively independent of the measuring site (endocardial papillary muscle, 50 mV, versus epicardial surface, 46 mV), although it may be modified theoretically by local differences in tissue architecture. A further argument for the high value of the extracellular resistance is given by the measurements of injury potentials. In myocardial injury, currents are driven through the injured membrane sites by the gradients in membrane potential between injured and non–injured regions. Similar to the situation described for normal propagation above, these currents are divided by the intracellular and extracellular resistors. In the cases where the extracellular and intracellular resistors form the only obstacle to current flow, with no

interference by an additional seal resistor of a suction electrode, amplitudes of injury potentials of 50 mV in mechanical injury [8] and ischemic injury [9] have been reported.

(ii) The absolute values of extracellular resistivity, calculated from the measured values of r_o and an assumed value for the extracellular/intracellular volume fraction [10], were found to be 47Ωcm[5] with a ratio r_o/r_i of 1/3 and 63Ωcm[6] with a ratio r_o/r_i of 1/1. Since the first value is somewhat lower than the resistivity of Tyrode's solution, and the latter is, as expected, between the resistivity values of Tyrode's solution (52Ωcm) and blood (84Ωcm), the calculation of the first value is likely to have been influenced by an underestimation of the ratio r_o/r_i.

(iii) The effect of a small shunt resistance (a fluid layer of 20 μm thickness) on the amplitude of the extracellular bipolar electrogram, ΔV_o, the ratio r_o/r_i, and on conduction velocity, θ, is shown in Fig. 2. The left panel shows a normal bipolar electrogram, obtained after stimulation of the papillary muscle at the tip. The right panel shows the bipolar electrogram after putting a droplet of Tyrode's solution on the papillary muscle. This small amount of Tyrode's solution formed a liquid film of approximately 20 – 30 μm thickness around the muscle. The thickness was measured through the binocular microscope with a CCD camera. Comparison of both panels shows that the presence of a 20 – 30 μm fluid layer, i.e. of a small electrical shunt, reduced the amplitude of thc extracellular electrogram from 49 mV to 38 mV, produced a small increase of conduction velocity and reduced the upstroke velocity of the bipolar electrogram.

All the above arguments taken together suggest that in ventricular muscle the electrical resistance of the extracellular space is of approximately the same value as the resistance of the intracellular space. Therefore, r_o has to be considered when interpreting the importance of changes in cell-to-cell coupling (changes of r_i) for changes in impulse propagation. This is because

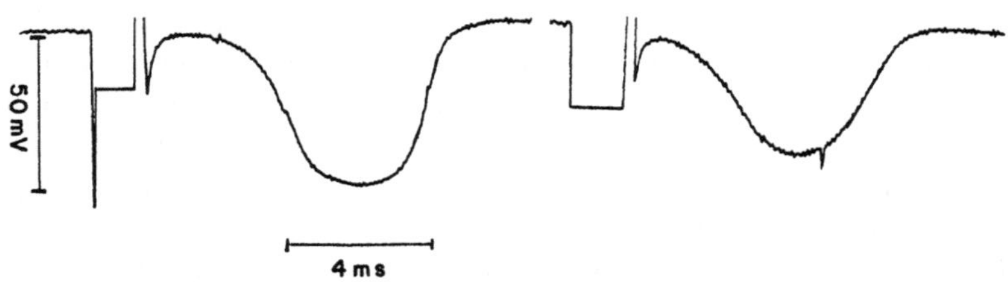

Figure 2: Extracellular bipolar electrograms measured on the surface of an arterially-perfused rabbit papillary muscle surrounded by a artificial gaseous atmosphere, interelectrode distance 2.14 mm. ***Left:*** Amplitude = 49 mV, $(\Delta V_o/\Delta t)_{max}$ = 57 V/s. θ = 56 cm/s. ***Right:*** Amplitude = 38 mV, $(\Delta V_o/\Delta t)_{max}$ = 24 V/s. θ = 60 cm/s. Calibration marks: 50 mV (vertical) and 4 ms (horizontal).

the large series resistance r_o will attenuate the effect of any changes in r_i. Moreover, changes of r_o are predicted to modify impulse conduction to the same extent than changes in r_i as shown in the next paragraph.

Changes of Extracellular Longitudinal Resistance and Conduction Velocity

The importance of r_o as an obstacle to local current flow during propagation can be tested by comparing the changes of r_o due to an experimental intervention to the changes of the conduction velocity, θ. For the intracellular resistance r_i (within a limited range of r_i) it has been shown that the changes of θ can be closely predicted from simple linear cable theory. According to Tasaki and Hagiwara [11], the calculated conduction velocity θ^* is given by:

$$\theta^* = K / \sqrt{r_o + r_i} \qquad (3)$$

The effects of changes in r_o and r_i on θ and comparison with θ^* are shown for several experimental interventions in Table 1. Acute interruption of coronary flow causes a drop of perfusion pressure within approximately 15s. As an concomitant event, the collapse of the intravascular space produces a increase of r_o (and a yet unexplained increase of r_i). The resulting decrease of conduction velocity was closely predicted by the cable theory [6, 12].

These changes were most likely independent of ischemic metabolism, because they occurred regardless of whether the gaseous atmosphere surrounding the papillary muscle was oxygenated or not. A decrease of r_o and a concomitant increase of θ have been produced by interventions which cause vasodilation (Table 1). Both the application of Verapamil and a 50% reduction of extracellular Ca^{++} caused a small decrease of r_o and a concomitant increase of θ, which was close to the theoretically predicted value [13]. Although the changes of r_o and θ per se are not dramatic with these interventions and may not have a large physiological impact, they can only be

Table 1: Effect of changes in extracellular resistance, r_o and intracellular resistance, r_i on conduction velocity θ (%). Comparison with calculated velocity θ^* (%) predicted from linear cable theory.

	r_i	r_o	θ	θ^*
Arrest of perfusion	18	35	−13	−14
Verapamil (0.5 μM)	3	−15	4	3
$[Ca^{++}]_o$ = 0.6 mM	−5	−26	6	9

explained by θ depending on r_o in a way predicted by linear cable theory, i.e. with a r_o/r_i ratio of approximately unity. Consequently, these experiments demonstrate the important role of r_o as an obstacle to local current flow. In a core conductor, a zero extracellular resistance, e.g. a 3-fold increase of r_i (cell-to-cell uncoupling) would decrease θ by 42%. In a core conductor with $r_o/r_i = 1$, a 3-fold increase of r_i would decrease θ only by 29%.

The Effect of a Boundary Between Packed Myocardial Tissue and a Volume Conductor on Conduction Velocity

If the electrical resistance of the restricted extracellular space were to have an decelerating effect on the conduction velocity, then the contact of myocardial tissue with a volume conductor of zero resistance should increase θ at the surface of the muscle. This problem was elegantly treated by Suenson [3] and is illustrated in Fig. 3.

Suenson used Weidmann's "silicon oil" method to measure the influence of a variable shunt resistor around the cylindrical fiber on θ. With a

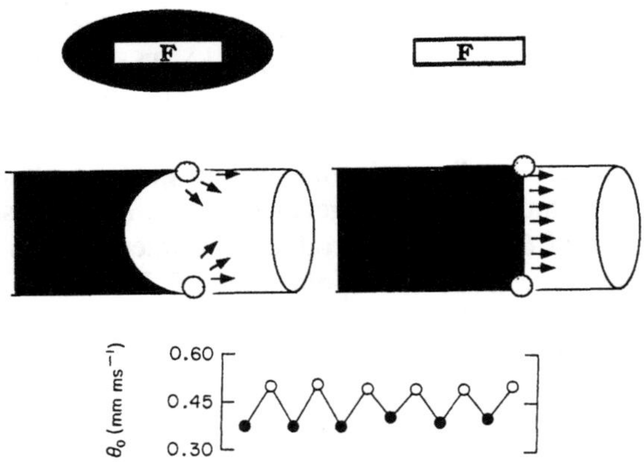

Figure 3: Top: Schematic representation of a superfused cylindrical muscle fiber soaked in a large volume conductor *(left)* or superfused with a small layer (20 μm thickness) of Tyrode's solution. ***Mid:*** Expected shapes of excitation wavefronts moving from left to right. The large volume conductor produces a bent wavefront, with the surface cells leading ahead of the core cells. In the ideal core conductor *(right)*, the wavefront is located in a radial plane. Arrows indicate flow of local current at the wavefront. ***Bottom:*** Measurement of conduction velocity θ at the fiber surface, in the presence (open symbols) and absence (filled symbols) of a volume conductor. Reproduced from [3] with permission.

large shunt resistor, the wavefront at the surface of the fiber is predicted to move ahead of the inner wavefront, and the steady state velocity will be higher than in absence of the volume conductor. Since the cells at the surface wavefront have to deliver local current not only in the longitudinal direction, but also in the radial direction, the amount of current available for charging the local membrane capacity becomes less, and therefore the upstroke velocity of both the transmembrane action potential and the extracellular electrogram is reduced. This can be seen in Fig. 2 where the application of only a 20–30 μm Tyrode's layer around a papillary muscle kept in a gaseous atmosphere reduced the upstroke velocity of the extracellular electrogram by 50% and increased θ by approximately 10%. The effect of a surface shunt resistance as recorded by Suenson is shown on the bottom panel of Fig. 3. As mentioned by Suenson, the effect depended on the geometry of the fibers, and (as expected) was markedly expressed for very thin muscle cylinders soaked in the large volume conductor. For the largest fibers tested (diameter 1 mm), the ratio of the surface velocity in presence of a large extracellular shunt to the velocity in presence of a small extracellular shunt (20 μm fluid layer) was 1.1. To what extent the shunting effect of cavity blood might have a physiological accelerating effect on surface velocity of subendocardial fibers at certain sites remains to be determined.

Acknowledgements

Original work was supported by the Swiss National Science Foundation and by the Swiss Foundation of Cardiology.

DISCUSSION

Dr. Gettes: Do you figure that the intact heart situation as the extracellular space collapses is equal from endocardium to epicardium or do you think there is preferential effect which may influence the changes in conduction earlier on?

Dr. Kléber: There is no method available which would allow to investigate this problem.

Dr. Noble: Were you implying that the blood in the ventricular cavity might affect the electrical activity of the endocardial myocytes?

Dr. Kléber: Simply by virtue of having a low extracellular resistor at this site, an increase of conduction velocity is expected to occur. The electrical activity of the myocytes *per se* will not be changed.

Dr. Noble: But is there not a layer of normal myocytes at the surface which might insulate the myocytes from the effect of the surrounding fluid?

Dr. Bassingthwaighte: I wonder how you treat your preparation. The question is whether or not there is an endothelial layer surrounding the bundle which would give the separation between an interstitial extracellular fluid layer around the myocytes. This would preserve that extracellular layer of fluid around the muscle cells. If you have a bathing medium that has no albumin in it, then the endothelial barrier would tend to be leaky between the ISF and the bathing medium. Albumin lowers the paracellular conductance through the clefts between the endothelial cells. The endothelial cells surrounding the bundle would be effectively destroyed by rubbing the fiber bundle in mounting it.

Dr. Kléber: The preparation is kept in a humidified gaseous atmosphere and perfused through its proper arterial tree. Thus, there is no "bathing medium" in the proper sense. The perfusion medium contains red blood cells, albumin and dextran. We consider the possibility of fluid penetrating through an endothelial barrier as rather small. We cannot exclude a very small layer of fluid at the surface, however. If such a fluid layer was important, the real r_o/r_i ratio would be even > 1.

REFERENCES

1. Weingart R, Rüdisüli A, Maurer P: Cell–to–cell communication. In:*Cardiac Electrophysiology, From Cell to Bedside*, Zipes DP, Jalife J (eds). Saunders WB, Orlando, FL, 1990, pp 122–127.
2. Tomanek RJ, Searls JC, Lachenbruch PA: Quantitative changes in the capillary bed during developing, peak and stabilized cardiac hyperthrophy in the spontaneously hypertensive rat. *Circ Res* 1982;51:295–304.
3. Suenson M: Interaction between ventricular cells during the early part of excitation in the ferret heart. *Acta Physiol Scand*, 1985;125:81–90.
4. Weidmann S: Electrical constants of trabecular muscle from mammalian heart. *J Physiol (Lond)* 1970;2101041–1054.
5. Buchanan JW, Oshita S, Fujino T, Gettes LS: A method for measurement of internal longitudinal resistance in papillary muscle. *Am J Physiol* 1986;251: H210–H217.
6. Kléber AG, and Riegger CB: Electrical constants of arterially perfused rabbit papillary muscle. *J Physiol (Lond)* 1987;385:307–324.
7. Roberts DE, Hersh LT, Scher AM: Influence of cardiac fiber orientation on wavefront voltage, conduction velocity, and tissue resistivity in the dog. *Circ Res* 1979;44:701–712.
8. Eyster JAE, Gibson WE: Electrical characteristics of injuries to heart muscle. *Am J Physiol* 1947;150:573–579.
9. Kléber AG, Janse MJ, van Capelle FJL, and Durrer D: Mechanism and time course of S–T and T–Q segment changes during acute regional myocardial ischemia in the pig heart determined by extracellular and intracellular recordings. *Circ Res* 1978;42:603–613.
10. Polimeni PI: Extracellular space and ionic distribution in rat ventricle. *Am J Physiol* 1974;227:676–683.
11. Tasaki I, Hagiwara S: Capacity of muscle fiber membrane. *Am J Physiol* 1957; 188:423–429.

12. Kléber AG, Riegger CB, Janse MJ: Electrical uncoupling and increase of extracellular resistance after induction of ischemia in isolated, arterially perfused rabbit papillary muscle. *Circ Res* 1987;61:271–279.
13. Cascio WE, Yan GX, Kléber AG: Passive electrical properties, mechanical activity and extracellular potassium in arterially perfused and ischemic rabbit ventricular muscle: effects of calcium entry blockade or hypocalcemia. *Circ Res* 1990;66:1462–1473.

CHAPTER 10:

THE EFFECTS OF POTASSIUM ON ANISOTROPIC CONDUCTION

Lucas V.A. Boersma, Josep Brugada, Charles J.H.J. Kirchhof and Maurits A. Allessie

ABSTRACT

A cryo–technique was used to create a thin layer of perfused anisotropic epicardium in Langendorff perfused rabbit hearts. The heart was stimulated with a frequency of 0.5 Hz and the extracellular concentration of K^+ was gradually increased. The resulting effects on anisotropic conduction were studied by using a high resolution mapping system. In this setting, no preferential effect of K^+ on the ratio of anisotropy or on the occurrence of local conduction block could be detected.

INTRODUCTION

In the chronic phase after myocardial infarction, cardiac arrhythmias can occur because of disturbances in impulse conduction. Most of the times, reentry is the underlying mechanism of ventricular tachycardia or fibrillation. Since the study of Sano et al. [1], who has shown that conduction velocity was higher in a direction parallel to the fiber orientation than in a direction perpendicular to it, it has become clear that anisotropy might be a factor of importance in the genesis and the mechanisms of reentrant tachyarrhythmias. In a study performed by Dillon et al. [2], reentrant circuits were found in the thin epicardial layer overlying the area of infarction, exhibiting pronounced anisotropic properties. Slow transverse conduction (perpendicular to the fiber orientation) in some parts of the epicardium was found to facilitate the occurrence of these functional arrhythmias. Similar results were found in studies on ventricular tachycardia in uniform anisotropic epicardium of the rabbit, performed by Schalij [3] and Allessie et al. [4].

In one of their studies on anisotropic conduction, Spach et al. [5] stated that the safety factor for conduction in anisotropic tissue is lower for longitudinal conduction than for transverse conduction. As a result, impulse propagation will become decremental and will preferentially block in the direction with the highest speed of conduction. They showed that premature stimulation could induce longitudinal conduction block, followed by a short series of reentrant beats. Other interventions have been used in several studies to depress impulse conduction. When conduction was impaired by raising the extracellular concentration of K^+, Tsuboi et al. [6] showed that conduction velocities in all directions were equally depressed by K^+, but that conduction block occurred more readily in longitudinal than in transverse direction. However, Delmar et al. [7] reported that by increasing K^+, conduction velocity was more depressed and conduction block occurred preferentially in a transverse direction. During progressive uncoupling of cardiac cells by heptanol, Delmar et al. [8] and Balke et al. [9] showed that transverse conduction was more susceptible to conduction block.

In the light of these different findings, we studied the effect of increased extracellular concentration of K^+ on anisotropic conduction in Langendorff perfused rabbit hearts. Using a cryo–technique we created a two dimensional (2D) uniform anisotropic layer on the epicardial surface of the left ventricle (LV). The epicardial activation pattern was studied with a high resolution mapping system (240 recording sites). The goal of the study was to find out whether during slow pacing (2000 ms interval), inactivation of the fast sodium channels by increased extracellular K^+, would yield detectable different effects on conduction velocity and the occurrence of local conduction block in longitudinal or transverse direction.

METHODS

Flemish Giant rabbits (3.5–4.5 kg) were used in this study. After heparinization (1000 I.U.) the rabbits were killed by cervical dislocation. The hearts were removed after a midsternal incision and connected to a Langendorff perfusion system. The Tyrode used in the perfusion system had the following mM composition: NaCl 130, KCl 6.0, $CaCl_2$ 2.2, $MgCl_2$ 0.6, $NaHCO_3$ 20.1, $Na2HPO_4$ 1.2 and glucose 12.0. The pressure was kept constant at 50 mmHg resulting in a flow of about 35 ml/min. The temperature was 37°C and pH was 7.35. The Tyrode was saturated by carbogen a mixture of 5% of CO_2 and 95% of O_2. A freezing method was used to create a thin epicardial layer of ventricular myocardium. A cryoprobe was inserted in the LV through an incision in the left auricle. The heart was placed in a tissue bath containing Tyrode at 30°C and the coronary circulation was temporarily interrupted. The cryoprobe was then filled with liquid nitrogen (–192°C) for 7 minutes. After this period the coronary circulation was restored and the heart was emerged from the tissue bath.

As previously described [4] this procedure results in complete destruction of the endocardial and intramural layers of the LV, leaving only a thin epicardial layer of about 1.0 mm intact. Conduction velocity and refractory periods of the remaining viable myocardium were not affected by the cryo– technique. However conduction in this 2D epicardial layer transverse to the fiber orientation, could proceed over a longer distance because it was no longer bypassed by fast conducting wavefronts in deeper layers.

To correlate the electrophysiological findings with the structure of the myocardium, histological studies were performed after the experiment. Sections parallel to the surface of the heart were made and stained according to the method described by Gordon and Sweets [10]. The activation pattern in the epicardial layer was studied by a high resolution mapping system. A rectangular mapping electrode (2.1 cm^2) containing 256 individual leads (inter–electrode distance 1 mm) was placed on the left ventricular wall. The signals were amplified (bandwidth 1–400 Hz), A–D converted (8 bits) and stored on videotape for later analysis. On–line analysis could be performed by an X–Y display showing any combination of up to 16 individual electrograms and a computer program which produced color coded activation maps. The steepest part of the intrinsic deflection was taken as the moment of local activation.

RESULTS

During control (extracellular concentration of K^+ 6.0 mmol) the heart was stimulated at 0.5 Hz at the center of the mapping electrode. By using such long pacing intervals the possibility was excluded that spatial dispersion of refractory periods could influence the conduction properties. Epicardial stimulation resulted in regular activation patterns with an ellipsoid shape, with wavefronts propagating from the center toward the edges of the mapping electrode (Fig. 1). Isochrone lines were drawn at 10 ms intervals. As can be seen the distance traveled between successive isochrones is much greater for impulse conduction to the left and the right than in perpendicular directions. By comparing this activation pattern to the structure of the myocardium, the electrical axis of fast conduction correlated well with the anatomical fiber direction. The slow conducting wavefronts propagated transverse to the orientation of the fibers.

Conduction velocities were determined as the distance traveled normal to the isochrones per unit time. Four values of conduction velocity for impulse propagation were measured, two parallel and two perpendicular to the fiber orientation, in opposite directions from the site of stimulation. In longitudinal direction the wavefronts were usually very narrow, whereas broad wavefronts existed in transverse direction. In the experiment given in Fig. 2, during 6 mmol of K^+, conduction velocities in opposite longitudinal directions were 71

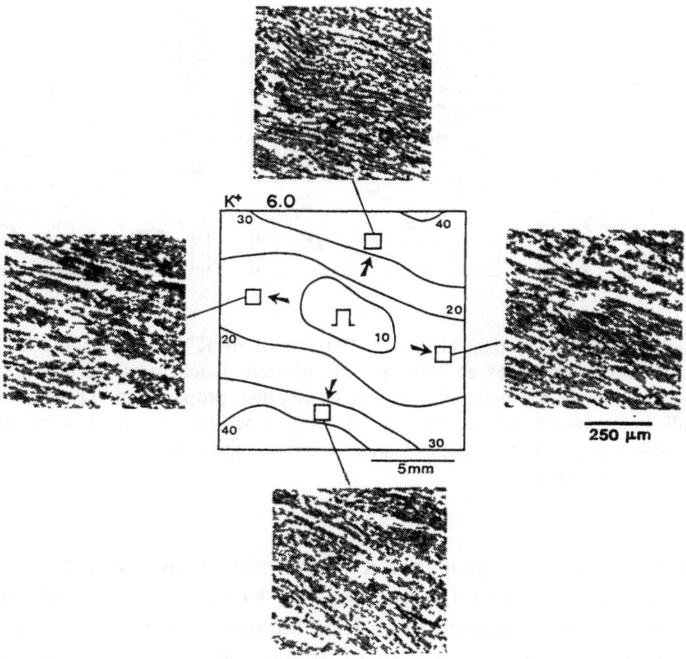

Figure 1: Isochrone map at 6.0 mmol of K^+ during pacing at 2000 ms interval at the center of the map. Isochrones are drawn at 10 ms intervals. Arrows indicate the direction of propagation. The four panels surrounding the map show the local fiber orientation of the epicardium.

and 52 cm/s whereas conduction velocity transverse to the fiber orientation was 29 and 16 cm/s. This shows that there are large spatial differences in conduction velocity both longitudinal and transverse to the fiber orientation. Despite these regional inhomogeneities, the ellipsoid pattern of the isochrones clearly demonstrated the existence of uniform anisotropy.

When K^+ was increased to 11 mmol conduction was clearly depressed. The isochrones maintained their ellipsoid shape, indicating that conduction was still clearly anisotropic. Conduction velocities were measured to be 56 and 39 cm/s in longitudinal directions. In transverse direction conduction upward was 17 cm/s. In a downward direction, transverse conduction velocity could no longer be measured accurately, because of marked local slowing of conduction in the middle lower part of the preparation. As a result the wavefront turned around both sides of this area of slow conduction and propagated oblique to the fiber orientation.

At 13 mmol of K^+ (just before total inexcitability occurred at 14.0 mmol) large arcs of conduction block developed in the lower part of the preparation. In the upper part conduction was slowed down even further but

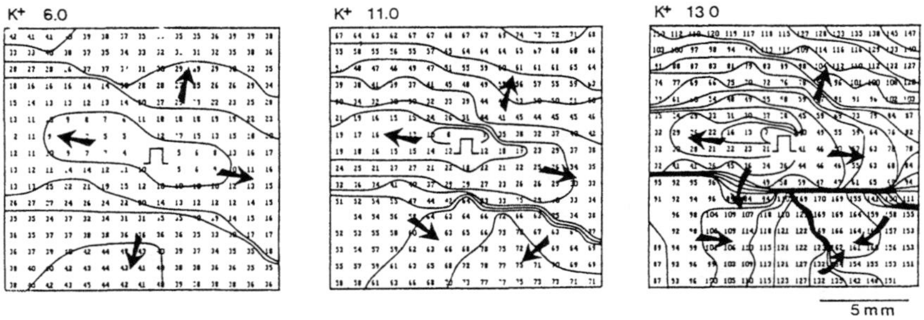

Figure 2: Activation maps at 6.0, 11.0, and 13.0 mmol of K^+ during pacing at 2000 ms interval. Local activation times at 240 sites are plotted. Isochrones are drawn at 10 ms intervals. Arrows indicate the general direction of impulse propagation. Local crowding of isochrones indicates areas of slow conduction. A delay in activation time of more than 40 ms between two adjacent electrodes (1 mm) was taken to represent conduction block and is indicated by heavy black lines.

the activation wave still propagated uniformly. Longitudinal conduction towards the left was 24 cm/s (34% compared to control). Towards the right the impulse started to propagate after a long latency with a velocity of 18 cm/s (35% of control). In upward transverse direction conduction was now 7 cm/s (24% of control). In the lower part of the preparation, two large transverse lines of block disturbed the activation wave propagating downwards. A small area between these two lines of block was still able to conduct at very low speed. Most of the tissue behind the lines of block, was now activated by wavefronts turning around the arcs of block outside the area of the mapping electrode. In addition to the long arcs of transverse block the lower part of the preparation also developed a line of longitudinal block.

When all experiments are taken into account, raising the extracellular concentration of K^+ did not show a consistent preferential effect on longitudinal conduction. The average ratio for anisotropy (θ_L/θ_T) showed a statistically insignificant change from 2.6 to 3.2 ($p = 0.26$). In our opinion this does not necessarily demonstrate that such preferential depression does not exist. However the significance of such a preferential depression of longitudinal conduction would be rather small and is fully masked by the more important preexisting regional heterogeneities in conduction properties of the myocardium. Conduction block may occur both in transverse and in longitudinal direction in different experiments or in different parts of the ventricle in the same experiment. Although transverse block was observed more frequently, this can not be taken as evidence that transverse conduction has a lower safety factor for conduction. Because the wavefronts in transverse direction are broad, the chance that conduction block occurs is also higher.

We believe that increasing the extracellular concentration of K^+, unmasks areas which have an intrinsic lower safety factor for conduction not dependent on fiber orientation. Our results seem to be in good agreement with the latest findings of Spach et al. [11], who also found that when the active membrane properties are depressed in uniform anisotropic myocardium, conduction block occurred both in longitudinal and in transverse direction. We conclude that in a two–dimensional sheet of anisotropic myocardium, an increase in the extracellular concentration of K^+ has no demonstrable preferential effects on conduction in longitudinal or in transverse direction.

DISCUSSION

Dr. Joyner: In looking at your isochrones and your definition of where a block occurs, it seemed a little bit ambiguous to me. You can define a block as occurring at the last point in which you could record electrical activity. If I am looking at the maps properly, as you come around this loop, the cells are oriented longitudinally, where a block occurred, and yet the isochrones are transverse at that point. Even though the fibers may have been oriented longitudinally with respect to the loop, the isochrones have actually turned and you are propagating transversely at that point. To say that propagation is occurring transversely or longitudinally in a region is very different; it depends on whether we are talking about the anatomy of the fibers or the alignment of the isochrones. It looks to me that you could argue one side or the other because the isochrones have become transverse to the fibers. So whether it is transverse or longitudinal depends on how you want to call it.

Dr. Allessie: We used two criteria for block. First, a delay of more than 40 ms should exist between two recording sites at a distance of 1 mm. Second, there should be a change in direction of propagation distal to the line of block. Our definition of longitudinal and transverse conduction relates to the anatomy of the fibers and not to the alignment of the isochrones. Since the direction of propagation is at right angles to the isochrones, during longitudinal conduction the isochrones are orientated perpendicular to the fiber axis and during transverse conduction the isochrones are parallel to the fiber orientation.

Dr. Janse: Your results with the lowest possible conduction velocities in longitudinal and transverse directions in the two–dimensional preparation fit rather well with our own results in ischemic myocardium, where conduction velocity in the longitudinal direction does not get below some 20 to 25 cm/s before a block occurs. Do you get similar values in the ring preparation?

Dr. Allessie: During sustained circus movement tachycardia in the ventricular ring preparation, depression of the fast sodium channels by potassium could produce marked local conduction delays in some segment of the ring while the tachycardia still continued. Locally the effective conduction velocity could be much slower than 20 cm/s before block occurred. However, it is very hard to know what is happening on a microscale. If the pathway of conduction becomes tortuous or proceeds in a zig–

zag manner, the actual conduction velocity may be considerably higher than the apparent velocity.

Dr. Jalife: Recently, Dr. Carmen Delgado did a study in my laboratory *[12]* on longitudinal vs. transverse propagation in sheep ventricular muscle. In her hands, during high potassium superfusion, those preparations consistently and significantly showed transverse block at a time when longitudinal propagation was maintained. Moreover, simulations done in collaboration with Bruce Steinhaus suggested that the safety factor for propagation, defined as the ratio of available charge over charge needed to activate downstream tissue, is greater in the longitudinal than in the transverse direction. This was confirmed by Carmen Delgado, Mario Delmar and Dante Chialvo *[12]* who showed in L–shaped ventricular preparations that even though, at relatively long cycle lengths, the threshold for excitation is lower when measured during propagation in the transverse direction, during premature stimulation, longitudinal propagation may occur in the presence of transverse block.

Dr. Allessie: In our perfused 2D–epicardial sheets of ventricular myocardium, paced at a long cycle length of 2000 ms, we could not detect any systematic differential effect on conduction when the potassium concentration of the perfusate was progressively increased. Sites of block could develop anywhere in the epicardium both during longitudinal and transverse conduction. Our conclusion was that even if the safety factor is different for longitudinal and transverse conduction, quantitatively such a difference is not very large and is overruled by other spatial inhomogeneities in the margin of safety for conduction. In the present study we did not investigate the role of premature activation on anisotropic conduction and conduction block.

Dr. Bassingthwaighte: The question is how clearly we should rely on the Streeter viewpoint, that there is a smooth shifting of the angle of the fiber direction from the epicardium to the endocardium. The question has several aspects. One is that the heart can be regarded as a folded toroid, one fold upon the other, and so there should be a junctional region between these, to take the simplest view. Streeter's model does not quite allow that. Secondly, if you look at the myocardium as perfused by capillaries which are parallel to the muscle fibers, you see that the branching vascular network feeds nicely into muscle bundles. If you inject some non–watery substance via a needle inserted into myocardial tissue, it splits the heart into separate bundles which are perfused by separate arteries. So these things do not line up nicely, but there are groups which are separated clearly from each other. The third thing is that the embryological origin is different in the subendocardial region from the subepicardial region, another reason for fiber direction difference. So my question is, you have laid out a model that is dependent on nice smooth relationships between fiber directions, but in the larger sense this can not be true. The Streeter idea needs to be reexamined in order to put it into the 3D context and relate it to ischemic situations in which you create the environment for it. How will you go about that?

Dr. Allessie: What is the thickness of those muscle bundles?

Dr. Bassingthwaighte: The groups are about 20 capillaries thick, that is to say 200–400 μ thick. There is a layer of loose aureolar connective tissue in between them, about 20 μ thick. I think the loose connective tissue layer allows bundles to slide on each other.

Dr. Allessie: The histological examination of our preparation was done from sections cut parallel to the epicardial surface. They showed a regular parallel orientation of the muscle fibers, separated by thin strands of reticulum tissue. There were spatial irregularities in the density of reticulum which certainly might create local differences in electrical coupling of the myocardial fibers. Since we did not attempt a 3D reconstruction of our preparation I can give you no information on the exact 3D arrangement of the ventricle. Since our epicardial preparation was only 1 mm thick, we think that the shifting of the fiber angle from epicardium to endocardium, as described by Streeter, did not have an important effect on epicardial excitation.

REFERENCES

1. Sano T, Takayama N, Shimamoto T: Directional difference of conduction velocity in the cardiac ventricular syncytium studied by microelectrodes. *Circ Res* 1959;7:262–268.
2. Dillon SM, Allessie MA, Ursell PC, Witt AL: Influence of anisotropic tissue structure on reentrant circuits in the epicardial border zone of subacute canine infarct. *Circ Res* 1988;63:182–206.
3. Schalij MJ: Anisotropic conduction and ventricular arrhythmias. *PhD Thesis*, University of Limburg, 1988.
4. Allessie MA, Schalij MJ, Kirchhof CJHJ, Boersma L, Huybers M, Hollen J: Experimental electrophysiology and arrhythmogenicity. Anisotropy and ventricular tachycardia. *Eur Heart J* 1989;10 suppl. E:2–8.
5. Spach MS, Miller WT III, Geselowitz DB, Barr RC, Kootsey JM, Johnson EA: The discontinuous nature of propagation in normal canine cardiac muscle. Evidence for recurrent discontinuities of intracellular resistance that effect the membrane currents. *Circ Res* 1981;48:39–54.
6. Tsuboi N, Kodama I, Toyama J, Yamada K: Anisotropic conduction properties of canine ventricular muscles. *Jpn Circ J* 1985;49:487–498.
7. Delmar M, Delgado C, Chialvo D, Michaels DC, Jalife J: On the problem of anisotropic propagation in ventricular muscle. In: *Lethal arrhythmias resulting from myocardial ischemia and infarction*. Rosen MR, Palti Y, Kluwer Academic Publishers, Boston/Dordrecht/London. 1989:181–197.
8. Delmar M, Michaels DC, Johnson T, Jalife J: Effects of increasing intercellular resistance on transverse and longitudinal propagation in sheep epicardial muscle. *Circ Res* 1987;60:780–785.
9. Balke WC, Lesh MD, Spear JF, Kadish A, Levine JH, Moore EN: Effects of cellular uncoupling in anisotropic canine ventricular myocardium. *Circ Res* 1988;63:879–892.
10. Gordon H, Sweets HH: Simple method for the silver impregnating of reticulum. *Amer J Pathol* 1936;12:545–550.
11. Spach MS, Dolber PC, Heidlage JF: Influence of the passive anisotropic properties on directional differences in propagation following modification of the sodium conductance in human atrial muscle. A model of reentry based on anisotropic discontinuous propagation. *Circ Res* 1988;62:811–832.
12. Delgado C, Steinhaus B, Welmar M, Chialvo DR, Jalife J: Directional difference in excitability and margin of safely for propagation in sheep ventricular epicardial muscle. *Circ Res* 1990;67:97–110.

CHAPTER 11:

EFFECT OF RYANODINE AND OTHER MODULATORS OF INTRACELLULAR CALCIUM ON IMPULSE PROPAGATION DURING SIMULATED ISCHEMIA

Leonard S. Gettes and Toru Maruyama

ABSTRACT

Changes in impulse propagation with acute and simulated ischemia are briefly discussed in relation to the maximum rate of rise of the action potential upstroke, the extracellular – intracellular resistances and the potassium concentrations.

INTRODUCTION

An understanding of clinically relevant changes in impulse propagation demands an understanding of the active and passive membrane properties which regulate impulse propagation in the intact heart, their modification by the various disease states and drugs which alter impulse propagation. The changes in impulse propagation which occur in the setting of acute myocardial ischemia offer one such example. It has been known for many years that shortly after the acute interruption of blood flow in the coronary artery, propagation of the cardiac impulse into the ischemic area becomes dramatically slowed and that this slowing of impulse propagation is one of the factors that leads to reentrant ventricular arrhythmias and to sudden cardiac death [1].

RESULTS AND DISCUSSION

The changes in impulse propagation following acute coronary ligation parallel the changes in extracellular potassium which occur [2] and the inhomogeneities in the changes in extracellular potassium are matched by inhomogeneities in the changes in impulse propagation. However, the changes in conduction velocity associated with acute ischemia cannot be explained entirely by the changes in extracellular potassium. Several investigators have shown that the changes in conduction velocity are greater than those predicted by the changes of extracellular potassium associated with acute ischemia [3, 4, 5]. These potassium independent effects on conduction can be attributed to a decrease in resting membrane potential (RMP) beyond that mediated by the changes in extracellular and intracellular potassium and by non–resting membrane potential effects i.e. changes in conduction velocity due to factors other than the change in RMP. These include changes in internal and external longitudinal resistance, possible changes in intracellular sodium which will influence the sodium inward current, and changes in the relationships between the RMP, the sodium inward current, $\dot{V}_{max}$ of the action potential upstroke, and conduction velocity.

Kléber [6] compared the RMP changes associated with acute ischemia in the guinea pig heart to those anticipated by changes in the potassium equilibrium potential as calculated from the observed extracellular potassium changes. He demonstrated that although the fit was remarkably good, it was not perfect. The observed changes in RMP were approximately 5 mV greater than those predicted by the change in equilibrium potential until the range of 55 – 60 mV. This additional effect on RMP could be explained by the fall in pH and PO_2 which occur. For instance, within the first 30 minutes of acute ischemia, following the acute interruption of coronary flow, extracellular pH falls from a control of about 7.4 to levels of approximately 6.0. Such changes in pH have been observed to cause slight changes in RMP. In addition, PO_2 decreases to zero shortly after the onset of acute ischemia and also causes slight changes in RMP. These additional effects on RMP would be anticipated to cause further changes in $\dot{V}_{max}$ and conduction velocity.

Kagiyama, et al. [7] showed that decreasing pH at any level of extracellular potassium resulted in a progressive decrease in conduction velocity. However, the change in conduction velocity was greater than anticipated by the change in $\dot{V}_{max}$. This greater than anticipated change might be caused by an increase in extracellular resistance (r_o) and/or intracellular (r_i) resistance, i.e. cellular uncoupling. Extracellular resistance increases rapidly following acute interruption of coronary flow [8]. An increase in r_i occurs after approximately 10 minutes of no flow or simulated ischemia [8, 9] and has been largely attributed to the development of intracellular acidosis and to an increase in intracellular calcium [10].

In the superfused guinea pig papillary muscle model, exposure to solutions modified to simulate ischemia, i.e. K = 9 mM, pH = 0.5, PO_2 = 20–30, glucose = 0, results in a fall in $\dot{V}_{max}$ of the action potential upstroke, an increase in r_i and a decrease in conduction velocity [9]. The decrease in $\dot{V}_{max}$ is greater than can be anticipated by the change in RMP alone [11]. The cause of this unanticipated decrease in $\dot{V}_{max}$ has not been elucidated. We have found that factors which influence intracellular calcium can modulate this change in $\dot{V}_{max}$ and the rise in intracellular longitudinal resistance, r_i, without influencing the change in RMP [11]. Ryanodine, 1 x 10^{-6}M, decreases the non–resting membrane potential component of the $\dot{V}_{max}$, bringing it closer to that anticipated by the change in RMP, while Bay–K 8644, 1 x 10^{-7}M accentuates it. Similar changes following Bay–K 8644 have been observed in the isolated perfused rabbit heart by Kabell [12]. In addition, he found that verapamil affected $\dot{V}_{max}$ in a manner similar to the effects of ryanodine. The non–resting membrane potential component of the change in $\dot{V}_{max}$ may reflect a change in the relationship between RMP and $\dot{V}_{max}$ such as has been demonstrated to occur when extracellular calcium is increased [13] or extracellular pH is decreased [7]. An increase in extracellular sodium would also provide an explanation for a non–resting membrane potential dependent decrease in $\dot{V}_{max}$. Whether intracellular sodium increases within 20 minutes of the onset of no flow ischemia is controversial, with some investigators showing no change [4, 14] and others showing an increase [15, 16, 17]. Nonetheless, the ability of ryanodine and verapamil to lessen the decrease in $\dot{V}_{max}$, the rise in r_i and the decrease in conduction velocity, and that of Bay K–8644 to exaggerate these changes, suggests that the changes in intracellular calcium, known to occur in acute ischemia, may exert a primary effect on both the active and passive membrane properties which are responsible for the slowing in conduction occurring in this setting.

The changes in impulse propagation associated with acute ischemia are frequency–dependent; they become more pronounced as driving rate is increased [18]. This frequency–dependent or rate–dependent component of the conduction slowing cannot be attributed to a rate–dependent increase in extracellular potassium [3, 17, 19]. The more likely explanation for the rate–dependent component of the conduction delay is the rate–dependence of the decrease in $\dot{V}_{max}$, and the rate dependence of the increase internal longitudinal resistance [9]. The change in both parameters becomes more pronounced as dependency rate is increased. It is likely that these rate–dependent effects are due to rate–dependent increases in intracellular sodium and intracellular calcium [20, 21].

CONCLUSION

Changes in impulse propagation associated with acute and simulated ischemia, can be explained very well by the decreases in the maximum rate of rise of the action potential upstroke and the increases in extracellular and intracellular resistance. The changes in $\dot{V}_{max}$ reflect the changes in sodium inward current as observed in isolated myocytes [22] and are due not only to the change in RMP induced by changes in K, pH and PO_2, but also to RMP independent factors which may be triggered by changes in intracellular calcium. Such changes include changes in the relationships between RMP, sodium inward current, and changes in intracellular sodium. The immediate rise in extracellular resistance and the later rise in intracellular resistance also contributes to the slowing of conduction. Changes in intracellular calcium and in intracellular pH contribute to the changes in intracellular resistance [1]. Thus, the clinically observed changes correlate well with, and are likely explained by, changes observed in isolated myocytes, cell pairs, and single fibers. Yet care must be taken in extrapolating results obtained in such isolated systems to the clinical setting without careful testing in rigorous models.

Acknowledgement

This research was supported by NHLBI grants P01 HL27430 and R37 HL 38885.

DISCUSSION

Dr. Corr: Could you lower your oxygen tension below 20 mmHg? Would you then get an exact rate of change that was equal to the response above 20 mmHg?

Dr. Gettes: We did not try to do that. When we lower PO_2 below 20 mmHg in this setting we see a much more rapid and dramatic influence, which is usually irreversible.

Dr. Corr: At a level of PO_2 of 20 mmHg, you are certainly going to have oxidation of fatty acids.

Dr. Gettes: I would not be at all surprised.

Dr. Strauss: The only comment I would make is that the proposed pathophysiologic cascade you have presented would be highly dependent on the model that you use in your experiments, because so many of the critical elements of your system are pH-sensitive. Sodium-calcium exchange is pH-sensitive, as is the calcium release channel. Hence, an ischemia where you see acidosis as a by-product of high energy phosphate breakdown is likely to yield quite different results than in a model that has an acid load imposed simultaneously with a reduction in pO_2. In the latter case you would expect less energy consumption in the face of an equivalent reduction in pO_2. Thus, a lot of the elements of the model would be very dependent upon what the key stresses are and the rate at which the changes in ions occur within the cell.

Dr. Marban: Are you, in your preparation, supplying calcium in the superfusate during the ischemic period? If so, is there a difference from no-flow ischemia?

Dr. Gettes: I think there is a difference. We do supply calcium. We carry it to about 1.8 mmM, which is what we have in our bath. It is one of the many differences I think between our simulation and others.

REFERENCES

1. Gettes LS: Effect of ischemia on cardiac electrophysiology, in Fozzard HA, Haber E, Jennings RB, Katz AM, Morgan HE (eds): *The Heart and Cardiovascular System. Scientific Foundations*. Raven Press, NY, in press, 1990.
2. Coronel R, Fiolet JWT, Wilms-Schopman FJG, Schaapherder AFM, Johnson TA, Gettes LS, Janse MJ: Distribution of extracellular potassium and its relation to electrophysiologic changes during acute myocardial ischemia in the isolated perfused porcine heart. *Circulation* 1988;77:1125–1138.
3. Hill JL, Gettes LS: Effect of acute coronary artery occlusion on local myocardial extracellular K^+ activity in swine. *Circulation* 1980;61:768–778.
4. Kléber AG, Janse MJ, Wilms-Schopman FJG, Wilde AAM, Coronel R: Changes in conduction velocity during acute ischemia in ventricular myocardium of the isolated porcine heart. *Circulation* 1986;73:189–198.
5. Morena H, Janse MJ, Fiolet JWT, Krieger WJG, Crijns H, Durrer D: Comparison of the effects of regional ischemia, hypoxia, hyperkalemia, and acidosis on intracellular and extracellular potentials and metabolism in the isolated porcine heart. *Circ Res* 1980;46:634–646.
6. Kléber AG: Resting membrane potential, extracellular potassium activity and intracellular sodium activity during acute global ischemia in isolated perfused guinea pig hearts. *Circ Res* 1983;52:442–450.
7. Kagiyama Y, Hill JL, Gettes LS: Interaction of acidosis and increased extracellular potassium on action potential characteristics and conduction in guinea pig ventricular muscle. *Circ Res* 1982;51:614–623.
8. Kléber AG, Riegger CB, Janse MJ: Electrical uncoupling and increase of extracellular resistance after induction of ischemia in isolated, arterially perfused rabbit papillary muscle. *Circ Res* 1987;61:271–279.
9. Hiramatsu Y, Buchanan JW Jr, Knisley SB, Koch GG, Kropp S, Gettes LS: Influence of rate-dependent cellular uncoupling on conduction change during simulated ischemia in guinea pig papillary muscles: effect of verapamil. *Circ Res* 1989;65:95–102.
10. Steenbergen C, Murphy E, Levy L, London RE: Elevation in cytosolic free calcium concentration early in myocardial ischemia in perfused rat heart. *Circ Res* 1987;60:700–707.
11. Maruyama T, Knisley S, Buchanan J, Gettes L: ryanodine suppresses conduction slowing, decrease of V^{max}, and cellular uncoupling induced by simulated ischemia. *Fed Proc* 1990;4:A681 (abstr).
12. Kabell G: Modulation of conduction slowing in ischemic rabbit myocardium by calcium-channel activation and blockade. *Circulation* 1988;77:1385–1394.
13. Gettes LS, Reuter H: Slow recovery from inactivation of inward currents in mammalian myocardial fibres. *J Physiol (Lond)* 1974;240:703–724.
14. Wilde AAM, Kléber AG: The combined effects of hypoxia, high K^+, and acidosis on the intracellular sodium activity and resting potential in guinea pig papillary muscle. *Circ Res* 1986;58:249–256.
15. Neubauer S, Balschi JA, Springer CS, Smith TW, Ingwall JS: Intracellular Na^+ accumulation in hypoxic vs. ischemic rat heart: evidence for Na^+–H^+ exchange. *Circulation* 1987;76 (Suppl IV):IV–56 (abstr).

16. Pike MM, Kitakase M, Marban E: Increase in intracellular free sodium concentration during ischemia revealed by ^{23}Na NMR in perfused ferret hearts. *Circulation* 1988;78:II–151–II–151 (abstr).
17. Tani M, Neely JR: Role in intracellular Na^+ in Ca^{2+} overload and depressed recovery of ventricular function of reperfused ischemic rat hearts: possible involvement of H^+–Na^+ and Na^+–Ca^{2+} exchange. *Circ Res* 1989;65:1045–1056.
18. Hope RR, Williams DO, El–Sherif N, Lazzara R, Scherlag BJ: The efficacy of antiarrhythmic agents during acute myocardial ischemia and the role of heart rate. *Circulation* 1974;50:507–514.
19. Harper JR, Engle CL, McCarthy JJ, Martin DG, Johnson TA, Gettes LS: Rate–dependent conduction slowing during occlusion is not due to a rate–dependent rise in potassium. *Circulation* 1989;80 (Suppl II):500 (abstr).
20. Cohen CJ, Fozzard HA, Sheu SS: Increase in intracellular sodium ion activity during stimulation in mammalian cardiac muscle. *Circ Res* 1982;50:651–662.
21. Lado MG, Sheu SS, Fozzard HA: Changes in intracellular Ca^{2+} activity with stimulation in sheep cardiac Purkinje strands. *Am J Physiol* 1982;243:133–137.
22. Sheets MF, Hanck DA, Fozzard HA: Nonlinear relation between $\dot{V}_{max}$ and I_{Na} in canine cardiac Purkinje cells. *Circ Res* 1988;63:386–398.

CHAPTER 12:

LOW DIMENSIONAL CHAOS AND THE TRANSITION FROM RHYTHMIC TO ARRHYTHMIC BEHAVIOR IN CARDIAC TISSUE

José Jalife and Dante R. Chialvo

ABSTRACT

Nonlinear mathematical tools are used to analyze the transition from rhythmic to arrhythmic behavior in isolated cardiac Purkinje fibers. Pacing the fibers with stimuli of relatively high amplitude at decreasing cycle lengths reveals period doubling bifurcations en route to highly irregular activity. The seemingly random behavior corresponds to low dimensional chaos in that it follows a simple deterministic law and has sensitive dependence on initial conditions. These results may be useful in understanding heart rate dependent patterns of cardiac excitation and mechanisms of some cardiac arrhythmias.

INTRODUCTION

The heart is a highly dimensional dynamical system that may undergo transitions from regular to irregular electrical activity analogous to the transition from laminar to turbulent flow in fluid dynamics. Until recently [1–4], descriptions of such transitions in cardiology had been somewhat superficial and no formal approaches had been proposed to determine their mechanisms. Here we briefly summarize some recent work from our laboratory in which we used a dynamical systems' approach to analyze the rate–dependent transitions from rhythmic to arrhythmic response patterns of isolated non–pacemaking cardiac Purkinje fibers during repetitive electrical stimulation. The most appealing aspect of these studies is that the highly complex behavior of these fibers can be accurately predicted by a relatively simple analytical model on the basis of a few relevant parameters, without any regard to ionic or molecular mechanisms. In fact, the overall approach makes use of some universal mathematical rules which are derived from the analysis of a wide variety of physical, chemical and biological systems [5].

Monotonic Recovery and the Farey Tree

In spite of their different nature, many nonlinear systems with finite recovery time behave predictably in much the same way [2, 5]. For example, in the case of a cardiac Purkinje fiber superfused with 7 mM KCl Tyrode solution, monotonic recovery of excitability may give rise to predictable steady state activation patterns [4, 7], in response to repetitive stimulation of relatively low amplitude at decreasing basic cycle lengths (BCL's). As a matter of fact, the sequence of n:m (stimulus:response) phase–locking patterns, which occur as the stimulus cycle length is systematically altered, conforms to the rules prescribed by the so–called Farey Tree [2, 6]. Indeed, as

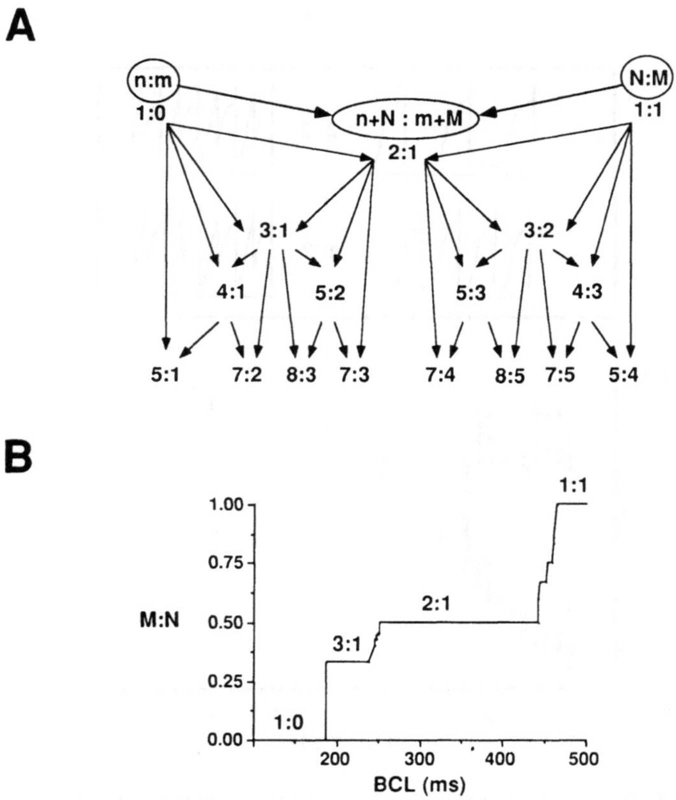

Figure 1: ***(A)*** Farey rule used to predict consecutive phase locking patterns in the case of monotonic recovery of relevant parameters and low stimulus amplitudes. ***(B)*** A plot of activation ratio as a function of basic cycle length (BCL) for the same case as in Panel A.

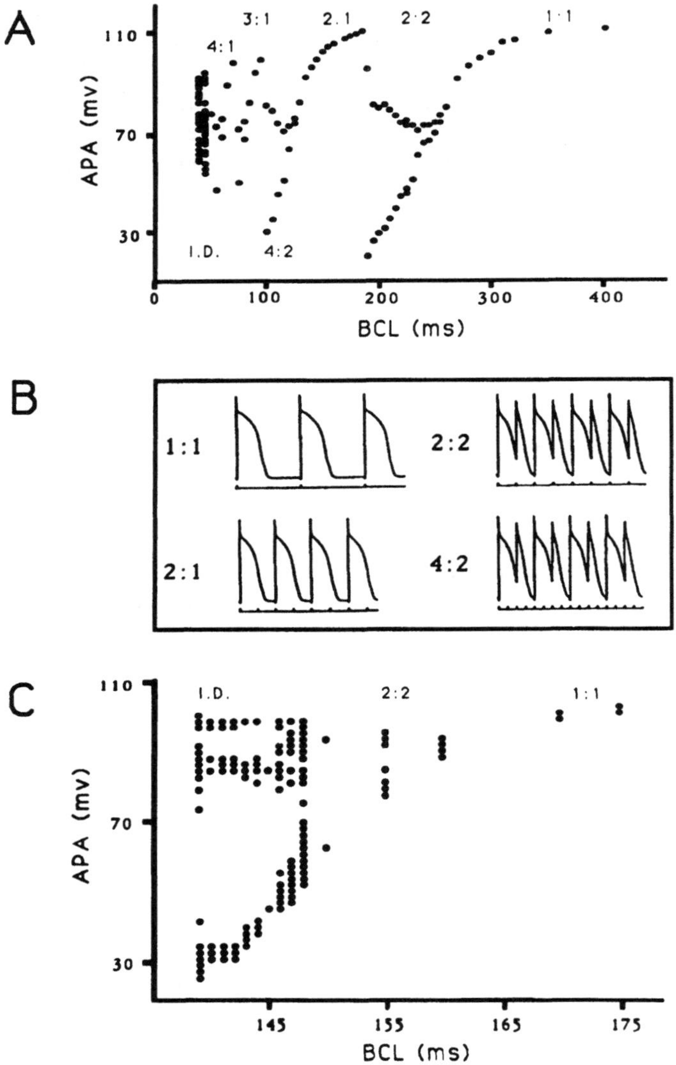

Figure 2: (A) Bifurcation diagram of action potential amplitude (APA) versus BCL for intermediate level of stimulation amplitudes. ***(B)*** Examples of period doubling bifurcations. *(C)* Bifurcation diagram of APA's versus BCL for relatively high level of stimulation amplitudes (modified from ref. [15], by permission).

shown in Fig. 1A, if one defines "n" as the number of stimuli applied in a given periodic sequence of "m" successful activations, for any two n:m and N:M neighboring phase–locking patterns, there is always an intermediate n+N:m+M phase–locking pattern. Therefore, in this case when the activation ratio (i.e. M/N) is plotted as a function of the basic driving cycle length, the resultant plot is a staircase (Fig. 1B) which, in theory, is infinitely detailed [2, 8–10].

Complex Rate–dependent Transitions and Chaotic Behavior

The previously described rule is no longer applicable when the stimulation amplitude is high enough to elicit premature (i.e. short) action potentials. In that case, the nonlinear restitution of the action potential duration now comes into the picture to give rise to more complex rate dependent behavior. Moreover, the presence of a supernormal phase of excitability [11] itself plays an important role in the appearance of complex periodic sequences other than those included in the Farey tree [7, 12–14]. In fact, pacing the fibers with repetitive pulses of relatively high amplitude at decreasing BCL's may reveal period doubling bifurcations en route to highly irregular activity. This is illustrated in Fig. 2A which presents a bifurcation diagram showing the relationship between action potential amplitude (APA) and BCL in a short (- 3 mm) Purkinje fiber driven repetitively with pulses of 2 ms in duration at 1.5 x diastolic threshold applied through a suction pipette. Period doubling bifurcations were demonstrable as alternans in APA giving rise to various n:m locking patterns. As the BCL was decreased in small steps, 1:1 was replaced by 2:2; 2:1 by 4:2; 3:1 by 6:2, etc., until irregular dynamics (ID) were established at very brief BCL's [15].

The analog recordings in Fig. 2B were obtained from a different experiment to illustrate the actual appearance of the action potentials for the respective stimulus:response patterns. Top traces are the transmembrane potential recordings and bottom traces show the stimulus timing. The 1:1 trace was recorded at a BCL of 500 ms, 2:2, 2:1 and 4:2 were recorded at BCL's of 145, 140 and 75 ms, respectively.

In Fig. 2C we present another bifurcation diagram obtained from the same fiber after the stimulus amplitude was increased to 2.6 x threshold. Under these conditions, 2:2 locking was followed by an additional period doubling bifurcation which yielded a pattern of 4:4, and then by irregular activity at a BCL of 138 ms. The structure of the bifurcation diagram in this case was similar to that observed during the period doubling route to chaos in other deterministic systems [5].

The Return Map in the Chaotic Regime

To establish whether a deterministic law governs the irregular patterns obtained with pulses of high amplitude applied at high frequencies, data may also be presented as a return map in which the value of APA after stimulus n+1 is plotted versus that for stimulus n. This kind of plot shows directly the repetition period (if any) of the pattern. For example, if each stimulus is followed by a response of constant APA, then only a single point will be seen on the identity line (i.e. APAn = APAn+1); the return map and the activity can be said to be of period 1; in other words, the n:m pattern is 1:1. When the APA of responses after successive stimuli alternates, the pattern is 2:2 and two dots will appear in the return map at either side of the identity line. During irregular dynamics when the evolution may be chaotic, the period of the system is undefined, but the return map shows a simple function (i.e., an attractor) which may have a strange shape (hence the name "strange attractor") and is evidence of the deterministic law for the system [5].

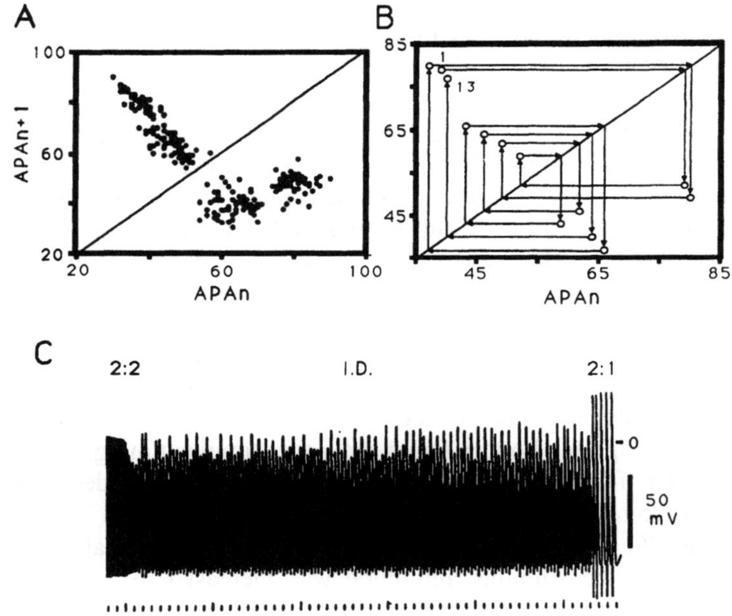

Figure 3: Low dimensional chaos in a Purkinje fiber. *(A)* A return map of APAn versus APAn+1 during the aperiodic sequence. *(B)* Return map for same experiments during 13 sequential responses (open symbols). Arrows show the trajectories of graphical solutions obtained by extrapolation through the identity line. *(C)* Analog recording during period doubling cascades. Modified from ref. [15] by permission.

Thus, we constructed return maps of irregular sequences of action potential amplitude in an effort to provide a rigorous proof for the presence of chaotic dynamics. In Fig. 3A, the lower record corresponds to a long sequence of transmembrane potentials obtained during pacing at a BCL of 180 ms, flanked by brief records during pacing at BCL of 185 ms (2:2 locking, left) and 175 ms (2:1 locking, right). The APA data at BCL = 180 ms are plotted in the return map of Fig. 3A. Successive data points were confined to specific regions of the map, rather than being distributed randomly. In addition, the dynamics are described by a function which has a minimum and consists of two segments: a long descending segment that intersects the identity line and a shorter ascending segment. Most importantly, the steepness of the descending segment (slope = −1.441) demonstrates that during the irregular dynamic regime the system has a sensitive dependence on initial conditions in that a small difference in the APA of two consecutive action potentials leads to very great ones after several beats. This is illustrated by the return map of Fig. 3B, obtained by following 13 sequential action potential amplitude values in the aperiodic portion of the record shown in panel A. The arrows indicate the sequence of the map and the open symbols the respective APA's starting at 1 and ending at 13. Clearly, as shown by the order of appearance of consecutive data points in the map, the small differences of subsequent APA's increases. This indicates that the irregular activity in these preparations indeed corresponds to low dimensional chaos [15].

The Analytical Model of Cardiac Excitation

To study systematically the dynamics observed in our experiments, we have devised a relatively simple analytical model [13, 15] whose main features are illustrated in Fig. 4A. Such features rest primarily on two major arguments: 1) There is a nonlinear recovery in membrane properties including excitability (1/threshold), time to full repolarization (TFR = action potential duration (APD) + latency to excitation (L)) and action potential amplitude (APA), during the diastolic interval; and 2) There are reciprocal interactions between diastolic interval (i.e., recovery time) and the degree of recovery of such properties. The inset of Fig. 4A illustrates these features. S_n represents the first stimulus following a sudden change in BCL, after which constant BCL is maintained. For beat n+1, threshold (Th) action potential duration (APD) action potential amplitude (APA) and latency (L) are functions (Z, G, Q and F, respectively) of the diastolic interval after beat n (DI_n) so that

$$DI_n = BCL - L_n - APD_n \qquad (1)$$

$$Th_{n+1} = Z(DI_n) \qquad (2)$$

$$APD_{n+1} = G(DI_n) \quad (3)$$
$$APA_{n+1} = Q(DI_n) \quad (4)$$
$$L_{n+1} = F(DI_n) \quad (5)$$

DI_n is approximated by Eq. (1) and the individual functions are obtained experimentally [13]. An iterative procedure is used to solve this set of equations in a computer. For a given BCL and stimulus magnitude, the calculated diastolic interval in each step determines the next latency, action potential and

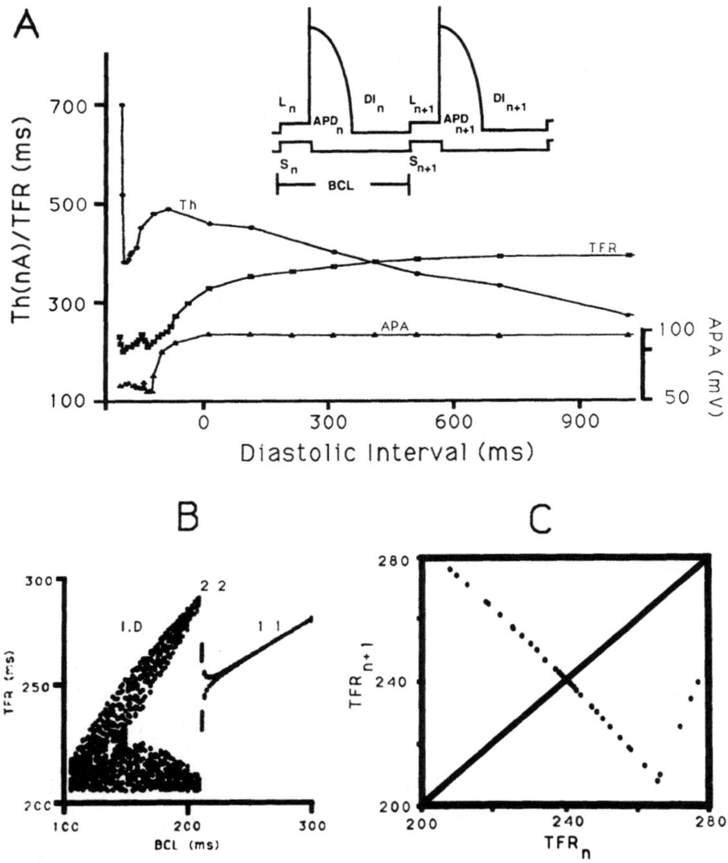

Figure 4: Difference–equation model of cardiac excitation. ***(A)*** Threshold (Th) time to full repolarization (TFR = L + APD) and action potential amplitude (APA) functions used as parameters in the model. ***Inset:*** features of the model. ***(B)*** Bifurcation diagram obtained by iteration of the model equation. ***(C)*** Return map of TFR vs TFR + 1 during chaotic regime shows an attractor with a minimum and slope of < –1 (modified from ref. [15] by permission).

threshold. The values thus obtained are now used to establish the next diastolic interval, and so on. The overall procedure is repeated iteratively until a steady state response is obtained. The predicted bifurcation sequence for TFR versus BCL in an experiment similar to that in Figs. 2C and 3 is presented in Fig. 4B. In both experimental and model results, chaos occurred after an initial period doubling bifurcation. In addition, the theoretically derived return map during chaotic activity was also similar to that obtained experimentally. This is shown in Fig. 4C which helps to illustrate the mechanism responsible for the particular low dimensional map obtained both with the model and in the experiment. Firstly, the slope (< -1) of the descending segments in the APA (Fig. 3A) and TFR (Fig. 4C) maps was the result of the presence of a steep fast component of the TFR function at diastolic intervals between -100 and 0 ms; secondly, the ascending branch in either map reflected prolongation of TFR at the shortest diastolic intervals; finally, the presence of supernormal excitability facilitated activation at very short diastolic intervals where recovery of APA and TFR were notched (Fig. 4A). The dynamical consequences of return maps with both a steep slope and a minimum value are well established [2, 5]; the existence of a minimum in the map guarantees that higher period doubling will occur, and the steep negative slope is sufficient for the appearance of sensitive dependence on initial conditions and, consequently, for the establishment of chaotic dynamics.

CONCLUSIONS

The electrical activity of cardiac tissues may undergo transitions from periodic to highly aperiodic electrical activity. We have studied the global rate–dependent behavior of isolated nonpacemaking cardiac Purkinje fibers through a systematic exploration of their response to single as well as repetitive electrical stimuli. In addition, we have used a difference–equation model to analyze in detail the behavior observed experimentally. Pacing the fibers at decreasing cycle lengths revealed period doubling bifurcations associated with changes in stimulus:response locking in such a way that 1:1 was replaced by 2:2; 2:1 by 4:2; 3:1 by 6:2; etc., until irregular dynamics were manifest at brief cycle lengths. At relatively higher stimulus amplitudes, decremental pacing led to a cascade which followed the sequence: 1:1 → 2:2 → 4:4 → ID. Analysis of the irregular dynamics using appropriate mathematical tools revealed that the behavior corresponded to low dimensional chaos in that it followed a deterministic law and was highly sensitive to changes in the initial conditions. Moreover, iterative procedures using the difference–equation model demonstrated that the critical features leading from regular to chaotic activity include a non–monotonic recovery of excitability during diastole and a steeply sloped action potential duration restitution curve at early diastolic intervals. The demonstration of period doubling and of chaotic dynamics in isolated cardiac tissues shows that it is possible to use analytical techniques to determine the bases of cardiac rhythm and conduction disturbances. Application of nonlinear dynamics techniques to the study of the transition from rhythmic to arrhythmic behavior in the heart may be useful in advancing our understanding of the mechanisms of sudden cardiac death.

REFERENCES

1. Glass L, Guevara MR, Shrier A: Universal bifurcations and the classification of cardiac arrhythmias. *Ann NY Acad Sci* 1987;504:168–178.
2. Glass L, Mackey M: *From Clocks to Chaos. The Rhythms of Life.* New Jersey, Princeton Univ Press, 1988.
3. Guevara MR, Glass L, Shrier A: Phase locking, period doubling bifurcations, and irregular dynamics in periodically stimulated cardiac cells. *Science* 1981; 214:1350–1352.
4. Chialvo DR, Jalife J: Non–linear dynamics of cardiac excitation and impulse propagation. *Nature* 1987;330:749–752.
5. Berge P, Pomeau Y, Vidal C: *Order Within Chaos. Towards a Deterministic Approach to Turbulence.* New York, John Wiley & Sons, 1984, pp 1–329.
6. Keener JP: On cardiac arrhythmias: AV conduction block. *J Math Biol* 1981;12:215–225.
7. Chialvo DR, Jalife J: On the nonlinear equilibrium of the heart. Locking phenomena and chaos in cardiac Purkinje fibers. In: Zipes DP, Jalife J (eds), *Cardiac Electrophysiology and Arrhythmias: From Cell to Bedside.* WB Saunders Co, 1990, pp 201–214.
8. Schrier A, Dubarsky H, Rosengarten M, Guevara MR, Nattel S, Glass: Prediction of complex atrioventricular conduction rhythms in humans with use of the atrioventricular nodal recovery curve. *Circulation* 1987;76:1196–1205.
9. Guevara MR, Schier A, Glass L: Phase–locked rhythms in periodically stimulated heart cell aggregates. *Am J Physiol (Heart Circ Physiol 23)* 1988;254:H1–H10.
10. Bak P: The devil's staircase. *Physics Today* 1986;22:38–45.
11. Spear JF, Moore EN: The effects of changes in rate and rhythm on supernormal excitability in the isolated Purkinje system of the dog. *Circ Res* 1974; 50:1144–1149.
12. Chialvo DR, Michaels D, Jalife J: Different modes of recovery of excitability as determinants of periodic and non–periodic dynamics in cardiac Purkinje fibers. *Physiologist* 1988;31(IV):A178.
13. Chialvo DR, Michaels D, Jalife J: Supernormal excitability as a mechanism of chaotic dynamics of activation in cardiac Purkinje Fibers. *Circ Res* 1990;66:525–545.
14. Chialvo DR: Towards very simple generic models of excitable cells. Order and chaos in cardiac tissues. Facts and conjectures. In: *Mathematical Approaches to Cardiac Arrhythmias,* Jalife J (ed), Annals of the New York Academy of Sciences, New York, 1990, vol 591, pp 351–366.
15. Chialvo DR, Gilmour RF Jr, Jalife J: Low dimensional chaos in cardiac tissue. *Nature* 1990;343:653–657.

CHAPTER 13:

FRACTAL MECHANISMS IN CARDIAC ELECTROPHYSIOLOGY

Ary L. Goldberger

ABSTRACT

The nonlinear concept of fractals offers insights into complex branching structures (e.g. His–Purkinje network, tracheo–bronchial tree) that lack a characteristic single length scale and into certain complex processes (e.g. physiologic heart rate regulation) that lack a single time scale. Heart rate control is perturbed by alterations in neuroautonomic function in a number of important clinical syndromes, including sudden cardiac death, congestive failure, fetal distress, space sickness and physiologic aging. These conditions are associated with a loss of the normal fractal complexity of interbeat interval dynamics. Such changes, which may not be detectable using conventional statistics, can be quantitated using new methods derived from "chaos theory."

This brief overview discusses applications of the concept of fractals to selected aspects of cardiovascular physiology. Emphasis here will be on certain of the controversies which have arisen since the "fractal hypothesis" of cardiac electrical stability was first proposed.

"WHAT IS A FRACTAL?"

The term fractal, coined by the mathematician B. Mandelbrot [1], is currently used in three contexts: geometric, temporal (dynamical) and statistical. In the most general terms, fractals are defined by a property called *self–similarity* [1, 2]. Fractal objects are composed of subunits that resemble the larger scale shape. These subunits in turn are composed of yet smaller units that also look similar to the larger ones, and so on. For example, trees are fractal in their repetitive branching structure. Fractals, therefore, do not have a single length scale, but rather have structure on multiple scales of length (Fig. 1).

The term fractal also relates to the fact that these nonlinear structures have a noninteger, fractional, dimension, that is, a dimension other than 1, 2 or 3. For example, the branching tracheo–bronchial tree, a fractal–like structure, has a dimension between 2 and 3 since it converts a volume of gas in the trachea (D = 3) into something approaching a surface area (D = 2) in the alveoli.

The notion of self–similarity has also been extended into temporal and statistical domains. A temporal fractal is a process that does not have a characteristic scale of time, analogous to a fractal structure that lacks a characteristic scale of length. Instead, fractal processes have self–similar fluctuations on multiple scales of time. This is reflected by a type of broad–band frequency spectrum, i.e. one having multiple frequencies. The concept of temporal fractals is closely related to that of "chaos." Finally the fractal concept has been applied in a statistical context. An example is the structure of the mammalian lung where we have reported that there is a self–similar distribution of scale sizes across multiple generations of branchings [3, 4].

FRACTAL GEOMETRY OF THE HEART

A number of cardiac structures have a self–similar or fractal–like appearance [5–7]. Examples of this nonlinear architecture include the coronary arterial and venous trees, the chordae tendineae, certain muscle bundles and the His–Purkinje network (Fig. 1). The latter provides an efficient way of distributing the depolarization stimulus to the ventricles. Recently, there has been interest in modeling the electrogenesis of the QRS complex using a fractal–like conduction system, as well as for studying alterations in the frequency content of the normal QRS due to changes in His–Purkinje geometry or in myocardial conduction [5, 8, 9]. Berenfeld, Sadeh and Abboud [8, 9] have shown that slow conduction in myocardial cells activated by such a fractal network can lead to "late potentials" or to selective attenuation of higher frequency content of the QRS, simulating changes seen in clinical investigations.

The controversy surrounding the fractal hypothesis of QRS electrogenesis centers on two questions: 1. Is the His–Purkinje system really a fractal? 2. Even if the geometry of the His–Purkinje system is fractal–like, does its macroscopic structure actually relate in any way to the frequency content of the QRS complex?

Idealized (computer–generated) fractals have infinite scales of length and literally have no smallest scale. Physiologic fractals are obviously bounded at both the upper and lower ends. However, the definition of a fractal does not require infinite scales of length. Furthermore, it is also apparent that physiologic fractals are not identical on different scales of magnification. However, structures such has the tracheo–bronchial tree and the His–Purkinje system do maintain a similarity of dichotomous branching

Figure 1: The self-similar branchings of the His–Purkinje system constitute a fractal–like network.

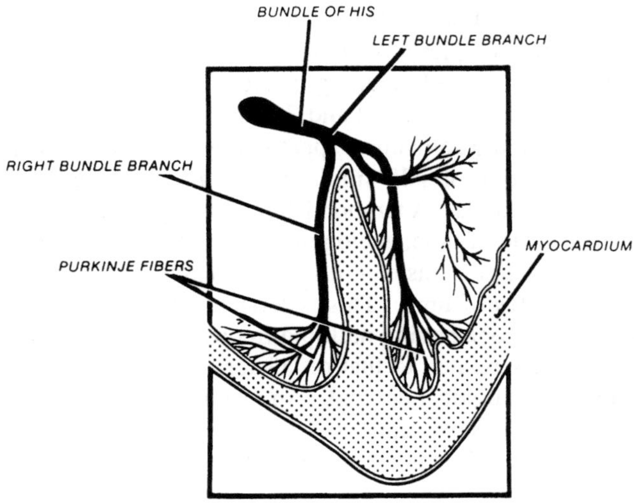

for which the term "fractal–like" is mathematically appropriate [1, 5]. Interconnections between branches of the His–Purkinje system, while making this more than a simple branching structure, also do not undermine the fractal–like nature of the geometry.

Spectral analysis of normal QRS complexes reveals a broad–band frequency spectrum (Fig. 2). While most of the frequency content of the QRS is comprised of frequencies below 20–30 Hz, there is a small but important contribution of higher frequencies that go up to several hundred Hz.

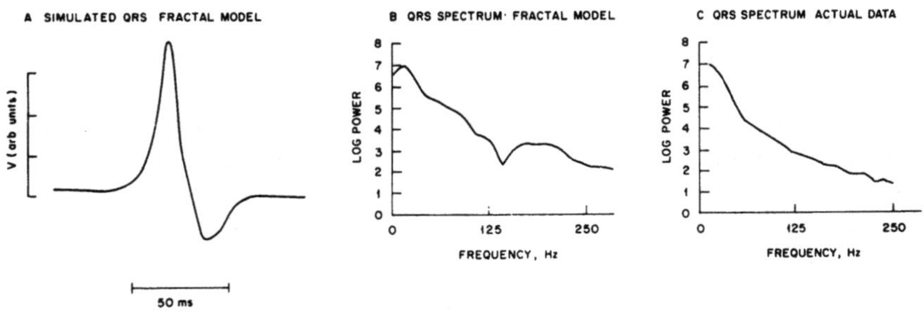

Figure 2: Activation of a 3–dimensional network of myocardial "cells" by a self–similar conduction system in a computer model generates realistic QRS complexes (left panel) with a broad–band frequency spectrum (middle panel) comparable to that obtained from actual ECG data in healthy men (right panel). Computer model QRS and spectrum from [8, 9]; clinical data from [4]. Note that the normal QRS complex is not simply a triangular waveform and that the broad–band frequency content of the QRS is richer than that of the discrete spectrum of a triangular signal.

Furthermore, spectral analysis has indicated an inverse power–law distribution to the frequency components of the normal QRS. That is to say, if one makes a plot of log QRS frequency versus log power, a good fit to a straight–line plot with a negative slope is obtained [5, 6].

A theoretical argument has been made that this broad–band spectrum with its inverse power–law distribution is consistent with depolarization of the myocardium via an irregular, self–similar branching network [5]. Therefore, according to this fractal theory, the frequency content of the QRS complex is significantly related to the macroscopic structure of the His–Purkinje system and not exclusively to the microscopic nature of the Purkinje–myocardial cell interactions and local wavefront propagation. Support for this counterintuitive notion has come from computer modeling studies in which a self–similar branching network has been used to depolarize a 3–dimensional network of cells [9, 10]. Such experiments reveal that with nine or ten generations of conduction system branchings one can generate QRS complexes that are essentially indistinguishable from those seen clinically. Furthermore, the simulated QRS complexes have a broad–band frequency spectrum comparable to that observed physiologically (Fig. 2). Such models also confirm that changes in the geometry of the branching conduction system may alter the frequency content of the QRS complexes, *independent* of any changes in myocardial conduction. This macroscopic fractal model of QRS electrogenesis, based on myocardial activation via an irregular conduction network, is not inconsistent with microscopic observations on the nature of the Purkinje–subendocardial muscle cell interface [10].

THE HEALTHY HEART BEAT IS A TEMPORAL FRACTAL

As noted, the fractal concept can be extended from geometry to dynamics. In this latter context, one can describe certain complex processes that do not have a characteristic scale of time. We have proposed that regulation of the heart rate may be one such fractal process [5–7]. This notion has also proven controversial, in part because it runs counter to the conventional dictum that the normal heart beat is highly regular ("regular sinus rhythm"). Palpation of the pulse and observation of the electrocardiogram in a healthy individual give the appearance of metronomic regularity. However, actual measurements of interbeat interval fluctuations reveal quite a different impression. Normal subjects, even those at rest, show a high degree of heart rate variability that is not subjectively perceptible. Furthermore, these fluctuations are not simply those associated with respiration. Indeed, spectral analysis of heart rate data from healthy subjects shows a broad–band spectrum with a so–called 1/f–like distribution (Fig. 3). Note that the term 1/f–like is synonymous with the *inverse power–law* defined above.

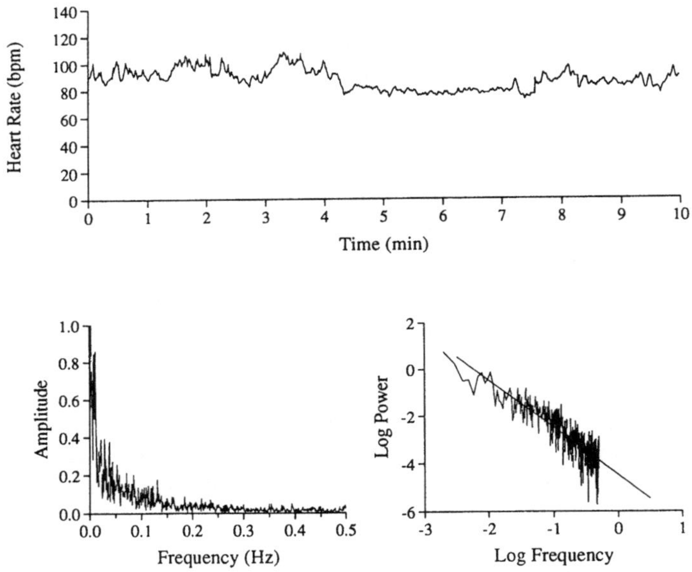

Figure 3: Normal sinus rhythm heart rate time series (top) from 61 year old woman reveals erratic fluctuations. The frequency spectrum (lower left) is broad-band with a 1/f-like or inverse power-law distribution, evident when the spectral data are replotted on log-log axes (lower right).

Another controversial aspect of the concept of the fractal heartbeat relates to its mechanism. We have proposed, in particular, that the interbeat interval fluctuations of the healthy heart are due, in part, to intrinsic variability of autonomic control ("chaos") [7]. This hypothesis stands in direct contrast to the theory of *homeostasis* enunciated by Walter B. Cannon [11] and others which states that apparently erratic fluctuations of variables such as heart rate are due primarily to *external* influences and that the normal condition of the cardiovascular system, and of other physiologic systems, is that of a *steady state*.

There are a number of lines of evidence to support the countervailing theory that deterministic chaos, *not* homeostasis, is the "wisdom of the body." The broad-band spectrum of the healthy heart beat is consistent with but not diagnostic of, deterministic chaos. Additional diagnostic tests for chaos include the measurement of a finite correlation dimension and of a positive Lyapunov exponent. Measurement of these nonlinear metrics from biologic data sets is fraught with potential problems that have been discussed in detail elsewhere [12]. However, preliminary attempts to make such measurements from heart rate data have been consistent with the hypothesis that these fluctuations do in fact represent deterministic chaos [13–15]. Finally, phase

Figure 4: Heart rate time series from a healthy subject shows complex variability. Two-dimensional phase space plot of this data set reveals a complex trajectory suggestive of a so-called strange attractor. The delay map plots the heart rate in beats per minute (bpm) at given time against the heart rate after a fixed delay (in this case 4 s) and then tracks the evolution of this heart rate vector after an arbitrary time (also 4 s in this case).

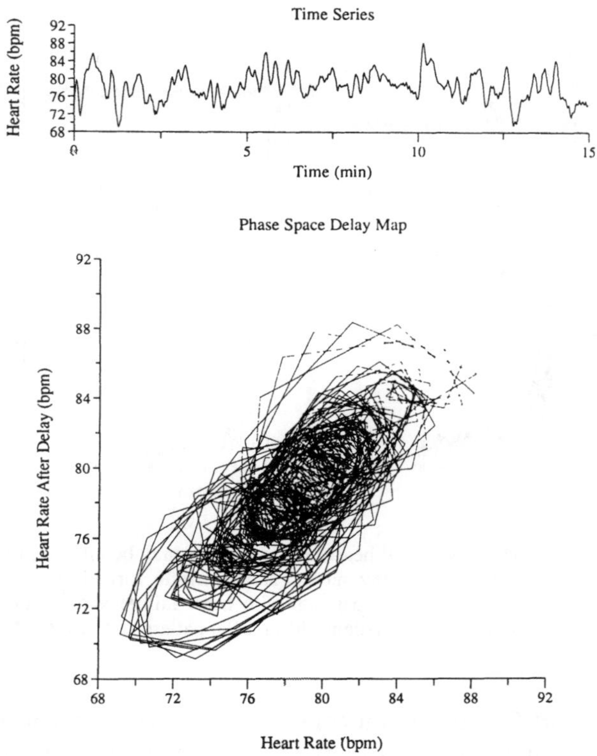

space portraits (delay maps) of interbeat interval time series are also consistent with those of so-called *strange* (chaotic) *attractors* (Fig. 4).

The mechanism for such physiologic chaos of the heart beat, if it exists, is not certain. However, it is clear that heart rate fluctuations are primarily due to autonomic nervous system control and therefore, any chaos of the heart beat must reflect chaos in nervous system dynamics. There is growing evidence for this kind of deterministic chaos even in the nervous systems of evolutionarily of more simple organisms [16].

LOSS OF CHAOS (COMPLEXITY) WITH DISEASE AND AGING

A corollary of the classical notion of homeostasis relating health to constancy is that disease and other perturbations are likely to cause a loss of regularity. The chaos hypothesis advanced above predicts just the opposite, namely, that a variety of disease states that alter autonomic function lead to a loss of physiologic complexity and therefore to greater, not less, regularity [7].

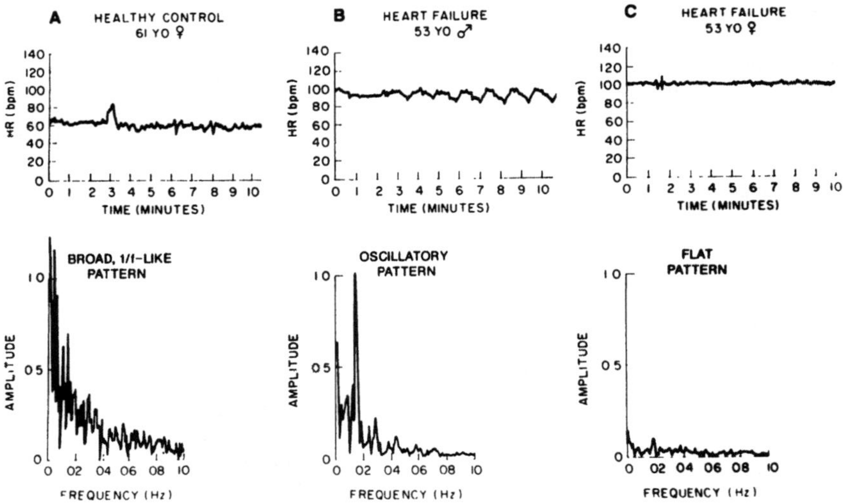

Figure 5: Healthy fractal heart rate variability may be altered in a variety of pathologic states. *(A)*, data set from a healthy subject shows erratic variability represented by a broad, 1/f–like spectrum. Data sets from patients with heart failure show a loss of fractal complexity with either low frequency oscillations *(B)* or a flat pattern *(C)* adapted from [20]).

Support for this notion comes from the comparison of heart rate time series from patients with a variety of different clinical syndromes, including those at high risk of sudden death and those with heart failure, whose sinus rhythm dynamics are typically *less* complex than those seen normally (Fig. 5) [17]. The term "complexity" is used here to include the fractal type of variability described above. It should be emphasized that quantitation of a loss of this type of nonlinear complexity cannot be accomplished by use of traditional statistics such as variance. An illustration of this principle comes from comparing two signals, one a large amplitude sine wave and the other a lower amplitude, highly erratic signal. Clearly the sine wave is less complex despite its greater variance. This observation is of more than theoretical import since there has been a flurry of interest in recent years in the analysis of heart rate variability using conventional statistics.

Aging is also associated with a loss of physiologic complexity [18, 19]. We have recently observed a reduction in both the approximate dimension and approximate entropy of heart rate and blood pressure signals in healthy old versus young subjects, despite an increased blood pressure variance in the older subjects [19]. Loss of cardiovascular complexity also appears to be an important feature of a number of other pathologic conditions including fetal distress and space sickness [5–7]. We have proposed, therefore, that the loss of physiologic complexity may provide a new prognostic and diagnostic indicator that may greatly expand the accuracy and utility of physiologic monitoring systems.

SUMMARY

This brief review has attempted to highlight key aspects of the application of fractals to cardiac physiology and to discuss some of the more controversial aspects of this theory. The fractal concept has been applied at 2 specific levels of cardiac function: 1) ventricular depolarization and 2) heart rate regulation. In the first case, that of ventricular depolarization, we have attempted to relate the self–similar and irregular branching structure of the His–Purkinje system to the dynamics of ventricular activation and to the resultant frequency spectrum of the QRS complex. Computer modeling studies have supported the counterintuitive notion that the macroscopic geometry of the conduction system may play a more important role than previously acknowledged in the genesis of the frequency content of the QRS complex under both normal and pathologic conditions. In the second instance, that of heart rate variability, we have suggested that complex fluctuations in healthy heart rate may be due in part to deterministic chaos in the neuroautonomic control system. Furthermore, perturbations of this control system, such as those seen in heart failure and in sudden cardiac death syndromes, may lead to a loss of complexity of heart rate dynamics that may be detectable with new nonlinear metrics based on calculations of dimension, entropy and Lyapunov exponents.

Acknowledgement

This work was supported in part by National Heart, Lung and Blood Institute Grant R01HL42172, by National Aeronautics and Space Administration Grant NAG2–514, and by a grant from the G. Harold and Leila Y. Mathers Charitable Foundation.

DISCUSSION

Dr. Kléber: You are actually showing us the branching of the Purkinje fibers. I think one should remember that in most of the ECG, excitation through the myocardium is continuous and that the contribution of the Purkinje system to current flow is rather small.

Dr. Goldberger: The fractal simulations show that the frequency content of the QRS complex is related to the macroscopic nature of the branchings. The complex, irregular, branchings of the His–Purkinje system *decorrelate* the arrival times of impulses at the Purkinje–myocardial interface and the statistics of this decorrelation process appear to be one important determinant of QRS frequency content *(see refs. [5, 6, 8, 9])*. Therefore, activation of the same network of myocardial cells by geometrically different "distributors" may generate QRS complexes with different frequency spectra.

Dr. Joyner: The problem is that these simulations ignore all the data from the Purkinje ventricular junctions which have been around since 1960–1970. The Purkinje strands branch so they are fractal. Everything is fractal, so that is not a valid conclusion. Actually there is a layer of interconnected Purkinje strands on the endocardial surface. The Purkinje conduction pattern is pretty continuous in all the mappings we have done; we have mapped papillary muscle with hundreds of points and can draw beautiful isotones on the layers. And yet, there are points of discrete coupling between the Purkinje layer and the ventricular muscle. Because something branches and looks like a fractal pattern, you are jumping to conclusions and obscuring the mechanism.

Dr. Goldberger: A common strategy in modeling is to start with the simplest model. One deliberately ignores certain details ("coarse–graining") to see how far you can get by taking a minimalist approach. The fractal model just described does in fact omit the detailed microscopic anatomy of the Purkinje–myocardial interface, elegantly described by Dr. Joyner and others. What is of interest is that if you start with such a simple branching self–similar network and depolarize an array of "cells," you can in fact generate QRS complexes that have the type of broadband frequency content and as well as the morphology that one sees physiologically. This supports the importance of macroscopic as well as microscopic anatomy to QRS electrogenesis. Further, if you perturb this system by slowing conduction, you can see other findings such as "late potentials" and selective loss of high frequency content, as demonstrated by Berenfeld et al. *[8, 9]*. I have not said that the His–Purkinje system is an idealized mathematical fractal, but it certainly has features of a fractal–like network, a finding that does appear relevant to the mechanism of cardiac depolarization.

Dr. Janse: Have you said that late potentials are caused by disturbances in the activation of the Purkinje network? I think that a major cause for late potentials is slow conduction in myocardial fibers surviving within the infarct, or overlying the infarct, as Dr. Wit showed us.

Dr. Goldberger: In the fractal model of Berenfeld et al. *[8, 9]* the "late potentials" appear when conduction velocity is reduced in two regions: at the distal part of the anterobasal conduction system and at the left ventricular apex. I would disagree with the statement that the *macroscopic* branching of the His–Purkinje tree is irrelevant to the genesis of the QRS spectrum and the findings I reviewed from our own work *[5]*, as well as those of Berenfeld, Sadeh and Abboud *[8, 9]* from Tel Aviv University and of Thomas Nelson from the University of California, San Diego *[Preprint]*, strongly support the importance of the fractal–like tree in QRS electrogenesis.

REFERENCES

1. Mandelbrot BB: *The Fractal Geometry of Nature*. New York, Freeman, 1982.
2. Feder J: *Fractals*. New York, Plenum, 1988.
3. West BJ, Bhargava V, Goldberger AL: Beyond similitude: renormalization in the bronchial tree. *J Appl Physiol* 1986;60:1089–1097.
4. Nelson TR, West BJ, Goldberger AL: The fractal lung: universal and species-related scaling patterns. *Experientia* 1990:46:251–254.
5. Goldberger AL, Bhargava V, West BJ, Mandell AJ: On a mechanism of cardiac electrical stability: the fractal hypothesis. *Biophys J* 1985;48:525–528.
6. Goldberger AL, West BJ: Fractals in physiology and medicine. *Yale J Biol Med* 1987;60:421–435.
7. Goldberger AL, Rigney DR, West BJ: Chaos and fractals in human physiology. *Sci Am* 1990;262:42–49.
8. Berenfeld O: Simulation of the ventricular depolarization process using a three dimensional heart model with a self-similar conduction system. *Master's Thesis*, Tel Aviv University, 1989.
9. Berenfeld O, Sadeh D, Abboud S: Simulation of late potentials using a computerized three dimensional model of the heart's ventricles with fractal conduction system. *Computers in Cardiology*, 1990. In Press.
10. Rawling DA, Joyner RW: Characteristics of junctional regions between Purkinje and ventricular muscle cells of canine ventricular subendocardium. *Circ Res* 1987;60:580–585.
11. Cannon WB: Organization for physiological homeostasis. *Physiol Rev* 1929;9:399–431.
12. Goldberger AL, Rigney DR: Nonlinear dynamics at the bedside. In: Glass L, Hunter P, eds. *Theory of Heart*. New York/Berlin, Springer, 1990, in press.
13. Destexhe A, Babloyantz A: Is the normal heart a periodic oscillator? *Biol Cybern* 1988;58:203–11.
14. Mayer-Kress G, Yates FE, Benton L, Keidel M, Tirsch W, Pöppl SJ, Geist K: Dimensional analysis of nonlinear oscillations in brain, heart and muscle. *Math Biosci* 1988;90:155–182.
15. Skinner JE, Carpeggiani C, Landisman CE, Fulton KW: The chaotic correlation dimension of the heartbeat is reduced by ischemia, stress or sleep in the pig. *Proceedings of the Sixth Annual Industry Conference and Exhibition*, Biotech USA. 1989;425–434.
16. Mpitsos GJ, Burton MR Jr, Creech HC, Soinila SO: Evidence for chaos in spike trains of neurons that generate rhythmic motor patterns. *Brain Res Bull* 1988;21:529–538.
17. Goldberger AL, Rigney DR, Mietus J, Antman EM, Greenwald S: Nonlinear dynamics in sudden cardiac death syndrome: heartrate oscillations and bifurcations. *Experientia* 1988;44:983–987.
18. Lipsitz LA, Mietus J, Moody GB, Goldberger AL: Spectral characteristics of heart rate variability before and during postural tilt: relations to aging and risk of syncope. *Circulation* 1990;81:1803–1810.
19. Kaplan D, Furman MI, Pincus S, Ryan SM, Lipsitz LA, Goldberger AL: Aging and the complexity of cardiovascular dynamics. *Preprint*, 1990.
20. Goldberger AL: Nonlinear dynamics, fractals, cardiac physiology and sudden death. In: *Temporal Disorders in Human Oscillatory Systems*, Rensing L, an der Heiden U, Mackey MC (eds), Springer-Verlag, New York/Berlin, 1987, pp 118–125.

Cell Physiology and Pathophysiology

CHAPTER 14:

CARDIAC SODIUM CHANNEL KINETICS

Harry A. Fozzard

ABSTRACT

The elemental unit of cardiac excitability is the Na channel. The channel is an integral transmembrane protein that forms a pore for entry of Na ions downhill into the cardiac cell. The channel is gated by membrane potential, and it is the site of action of most of our effective antiarrhythmic drugs. While we were long obliged to depend on studies of Na currents in nerve to model the cardiac Na channel, recently it has been possible to record Na currents as accurately as in nerve. Those studies show that the cardiac Na current is not well described by the Hodgkin–Huxley model. Kinetic behavior of the channel can be directly studied by use of single channel recordings, resulting in demonstration of stochastic properties described by a Markov chain model. Furthermore, the cardiac Na channel has been cloned and sequenced, and its primary structure has been deduced from the cDNA. Structure–function studies of the cloned channel should give us great insight into its molecular mechanism of gating, including its voltage dependence and its binding sites for drugs.

INTRODUCTION

The Na current is primary source of electrical current for excitability and conduction in atrial and ventricular muscle. This current results from the time– and voltage–dependent behavior of the Na channels. These channels are membrane proteins that provide a selective pore for Na to move down its electrochemical gradient into the cell, depolarizing it. Most pathological processes resulting in arrhythmias include alterations in the Na current, and most of our clinically effective antiarrhythmic drugs block this channel. The channel's behavior can be viewed in mechanistic terms, or as a mathematical model of configurational changes, or as a large molecule. Great progress has been made in all three approaches to the characterization of the Na channel. This article will examine these and briefly discuss the ways in which they contribute to our understanding of Na channel kinetics.

BIOPHYSICAL MODEL OF THE Na CHANNEL

In the last decade it has become possible to study the cardiac Na current under excellent voltage clamp control [1, 2]. Figure 1 shows a family of Na currents obtained in response to steps to various voltages in the action potential range. The currents activate after a delay, and then decay slowly. This behavior is similar to nerve, except that the currents are slower to activate and to decay. These data can be modeled by the Hodgkin–Huxley model fairly well, and differ only in subtle ways. The voltage–peak current relationship and the voltage–dependence of activation are also shown in Fig. 1.

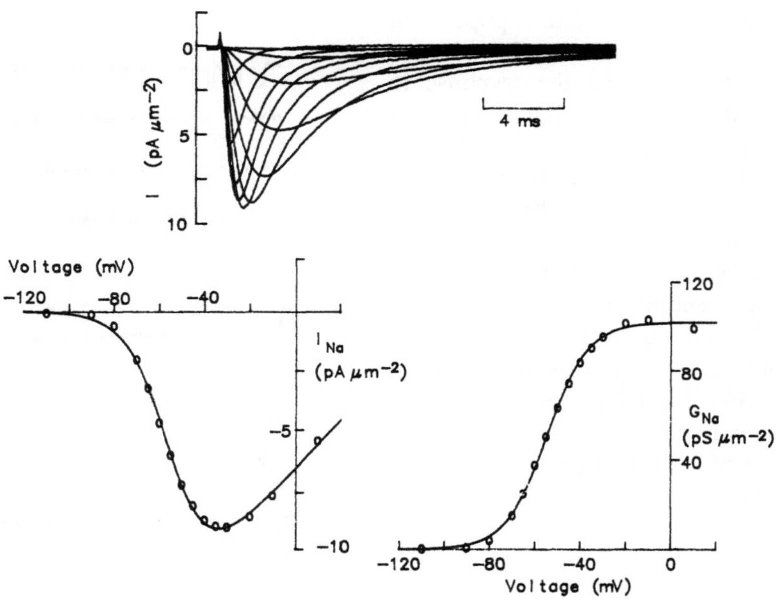

Figure 1: Na currents obtained from voltage clamp studies of a single cardiac Purkinje cell that was intracellularly perfused. A family of current recordings after leak and capacity subtraction from voltage steps to various voltages above threshold is shown on top. Voltage–peak current (left) and voltage dependence inactivation (right) are also shown. Extracellular solution was TMC Cl with 30 mM Na and 2 mM Ca. Intracellular solution was Cs F. T = 13°C. The cell had 80 pF and a seal resistance of 224 MΩ.

STOCHASTIC MODEL OF THE Na CHANNEL

Recently, single channel opening and closing events can be recorded with the patch clamp technique [3]. Such recordings show that a single channel behaves in a probabilistic manner. Upon depolarization from a negative level to one above threshold, the channel opens after a variable delay. The pattern of the delay is shown by the latency histogram plotted in the lower trace of Fig. 2. The channel remains open for a variable time, with short openings being the most common. A histogram plot of the open durations can be fitted to a declining exponential, as if the closing was a first order process. Sometimes the channel reopens once or twice before closing for the last time to an unavailable state. Frequently, the channel fails to open at all, yet enters an unavailable state, a process called *inactivation without opening*.

Single channel currents are typically about 1 pA (single channel conductance of 15–20 pS at room temperature), and mean open time is about 1 ms. Therefore, typically the opening of one channel allows about 10,000 Na ions (about 1 x 10^{-18}M) to enter the cell. The Na current of a cell is the composite of thousands of Na channels, each asked once to open in its stochastic way. This population behavior can be simulated with one channel by averaging the response to many voltage steps, producing what is called *an ensemble*. If the kinetic properties of the whole cell current are entirely derived from the stochastic properties of single channels, and the channels are identical and independent, then the ensemble should be identical with the whole cell current. The ensemble of a single channel is indeed exactly like the whole cell current, so that we may conclude that the complex behavior of the Na current is inherent in each channel [4, 5].

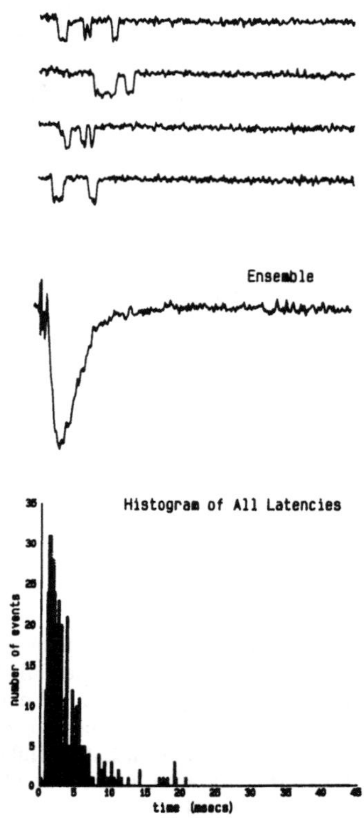

Figure 2: Single channel studies. The upper section shows examples of single channel openings in response to a depolarizing step. The middle section is an average of many single channel recordings, producing an ensemble current. The lower section shows a histogram plot of times-to-opening (latency) for the recorded channels.

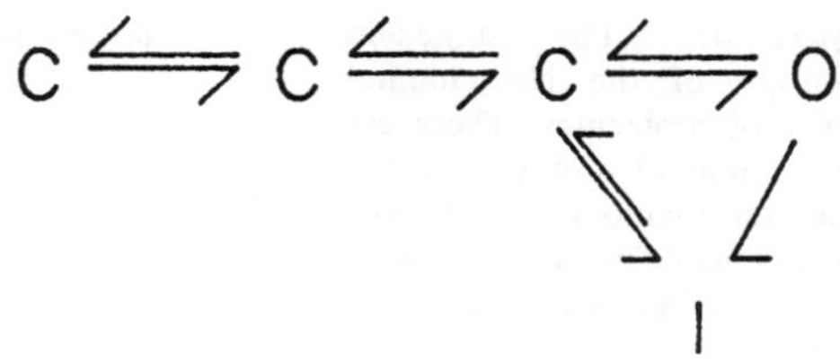

Figure 3: A Markov chain model of Na channel state transitions. The C symbols represent a sequence of closed states, through which the channel must pass on the way to opening. O is the open state, which can be recorded directly. I represents the inactivated or absorbing state.

The ensemble itself is the product of the latency distribution times the mean open time. Compare the stochastic behavior of single channels in the upper part of Fig. 2 to the ensemble in the middle trace and to the latency distribution in the lower trace. It is apparent that some channels open after a long delay, after the ensemble current has begun to decline. The decay of the whole cell current is a consequence of fewer channels opening during that time, rather than the Hodgkin–Huxley explanation of channel inactivation. This result is the clearest demonstration that the Hodgkin–Huxley model does not correspond to the actual behavior of Na channels [6].

Single channel behavior has been modeled well by assuming that the channel undergoes voltage–dependent conformational changes between three types of functional states, closed (but available to open upon depolarization), open, and inactivated (closed and not available to open until recovery at hyperpolarized potentials). These interrelated states are themselves represented by a Markov chain (Fig. 3), where transitions are allowed between a series of memoryless states [7]. Analysis of actual single cardiac channel behavior using various methods results in estimation of the set of transition rates that best fits single channel behavior. The voltage dependence of these transition rates can then be estimated [5, 8, 9]. Such models predict well both the single channel properties and the whole cell currents, and presumably they would also predict the behavior of channels engaged in generating an action potential.

Na CHANNEL STRUCTURE

In 1986 Numa and colleagues cloned three Na channels from rat brain [10]. Last year my colleagues successfully cloned the neonatal rat cardiac Na channel [11]. The derived amino acid structures of this family of Na channels shows that they all contain about 2000 amino acids and share 70–90% of their exact structure (Fig. 4). While modeling the three–dimensional structure of the channel is only beginning, some topological studies suggest a consistent pattern that can be used to infer structure [12]. There are four highly

Figure 4: The suggested topology of the Na channel family of molecules. There are four regions of highly homologous transmembrane segments. The extracellular and cytoplasmic domains are the most divergent.

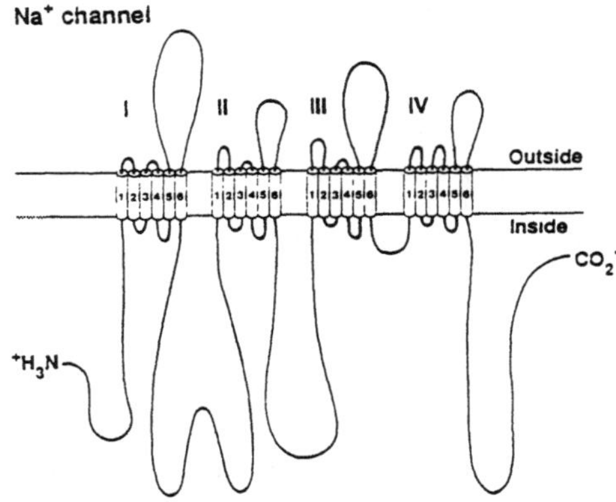

homologous regions of the Na channels, each containing at least six α–helical transmembrane segments. The fourth α–helical transmembrane segment of each region has positively charged amino acids at each third position. It is the best candidate for the channel's voltage sensor. The intracellular loop between regions III and IV is implicated by mutation and antibody studies to be involved in the inactivation process. The tetrodotoxin binding site appears to be just at the extracellular end of the fifth transmembrane segment of region I.

FUTURE DIRECTIONS

The cardiac Na channel can now be studied by high quality voltage clamp in intracellularly perfused cells, the channel gating can be directly measured, and recordings can be made of single channel opening and shutting. Further, a cardiac channel has been cloned, sequenced, and expressed. This is the only mammalian cell where all these types of studies can be made. This combination of cellular and molecular approaches makes the cardiac Na channel the best model for understanding voltage–gated ion channels, and to provide a complete picture of ion channel kinetics.

DISCUSSION

Dr. Pollack: If you think of a sodium channel as a kind of a tube that opens and allows, say 10,000 sodium ions flow through, why is it that smaller ions don't come pouring through as well?

Dr. Fozzard: The problem of selectivity of the channel, at least the way the great thinkers in the field are thinking about it, is like this: Think of a sodium ion and a potassium ion floating around in the solution. Both of them are almost the same size, and have exactly the same charge. Besides that, they have a huge cloud of water all around them. When I look at that ion from a distance, I can not tell whether it is a sodium ion or a potassium ion. How am I going to tell? I am going to have to sit there and peel off the water molecules and look inside and one is a bit of an angstrom different. The only way that the channel can presumably do that is to do what I have said: peel off the waters. The best model for this is to have carboxyl groups which can replace the oxygens of the ion water interaction at this narrow channel. Only if they are at the proper space to allow the electrons to incorporate those oxygens in their orbits will the ion fit. If the ion is either too small or too large it will not fit; it must be exact to fit energetically into that spot. That will allow the channel to be highly selective for sodium over potassium, or comparably for a potassium channel, potassium over sodium. But the differences at the selectivity region are only a part of an angstrom.

Dr. Pollack: What about very small ions like the hydrogen ions?

Dr. Fozzard: Hydrogen ions are a real problem in channels because your calculations say that they would fall through, but they do not. Probably what happens is that they are very interactive, and interact with some 50 sites on their way through and so they are slowed down by their interaction with other parts of the molecule. I am sorry I can not give a better description.

Dr. Joyner: The big difference between nerve sodium channels and heart sodium channels is the lowered cardiac sensitivity to tetrodotoxin (TTX) and that TTX has much more voltage and time dependence in heart than in nerve. Is there any indication from the differences that you have seen in the channel structure as to why there would be such a difference in the TTX effects?

Dr. Fozzard: For reasons that were plausible to me, but I can not explain them exactly, all of us who think about where in the molecule the toxins bind have focused on this large group between transmembranc segments 5 and 6 of the first repeat. For sure, this is where the Scorpion toxins bind because we have photoaffinity for that. Noma in Japan *[13]* and Stühmer in Germany *[14]* have made a point mutation that changes one charge right there at position 376. Glutamic acid was changed to glutamine, a negative charge. They did it for an entirely different reason, but when that channel was expressed in oocytes, it expressed a rat brain channel, which is normally very sensitive to TTX, to be completely resistent. They compared this sequence to the heart clone and found that the glutamic acid is also present in the heart channel. We have known for a long time that the toxin binding site has a negative charge. In the brain channel the next amino acid up is neutral. In the heart it has a positive charge and so the heart has a very different electrical field in that region, with a negative and a positive charge against each other. Dr. Stühmer *[14]* proposed that the reason why cardiac muscle is moderately toxin sensitive instead of exquisitely sensitive is that the field of that positive charge is repelling the toxin or changing the structure in that region. With some trepidation, we have made that point mutation, changing the cardiac positive charge at 377 to a neutral amino acid and sometime soon around mid–June we will have the cRNA to test and we might be able to tell you what the toxin site is in the heart.

Dr. Gettes: Could you explain what the double open state refers to and how that happens when you have it looks like a double opening or two steps in the open state?

Dr. Fozzard: Our thinking is fairly straightforward. Once the channel has made its traverse from the closed state to the open, it really does have a choice. It can inactivate or it can go back to the closed state. If it goes back to the closed state it can reopen. What we are really saying is that at voltages within 20–30 mV of threshold, this return rate is quite a reasonable one and so a substantial number of channels go back to the closed state. All of us, and we actually add up to 5 or 6 people now who have looked at this sort of thing in cardiac cells, see a great deal of this shuffling between closed and open in the threshold region. I think this is the consensus. It is a simple way of describing it.

Dr. van der Vusse: Is it possible to identify the string of amino acids in the total protein molecule that regulates closing and opening of the gate?

Dr. Fozzard: Yes, we have an impression. The impression is that the loop between repeats 3 and 4 is primarily responsible for inactivation. Amongst other things one of the first experiments Stühmar *[14]* did with Numa's clone of rat brain II channels was to change that little segment. All he had to do was to make a cut at this point and the inactivation disappeared. There are other ways to modify inactivating, but that was a very dramatic experiment. In addition Catterall *[15]* has added a monoclonal antibody to a sequence in this loop and that also blocks inactivation. It is interesting to note that in cardiac muscle this region is about an 80% homologous to the rat brain channel. It is very similar and we do have similar inactivation properties to that brain channel but that is as far as we have gotten on the identification on the part of the molecule that controls the inactivation. The activation part is much more complicated and appears to involve, among other things, several of the transmembrane loops and several of the cytoplasmic loops judged by the mutations. It is going to be a while before we sort that one out.

Dr. Mowbray: When you are discussing the length of time that the channels are open or closed, you imply they can open, go back to closed and then open again; but when you talk about the inactivation, you imply that it is of a much longer duration and so you can not go back in the same open–closed cycle time.

Dr. Fozzard: Many of you have probably heard of something called the window current. In a narrow voltage range, at least in the Hodgkin–Huxley scheme, there is steady state m and h overlap. In that region there is a steady but very small sodium current, which we think is simply continuous cycling of the channel through closed–open–inactivated and back to closed and back to open. But, essentially, at all other voltages, the channel will eventually end up in this final inactivated state. This is more or less a matter of definition. If we go to a voltage and wait awhile and further depolarizations can not open the channel, we define that as having gone to channels the inactivated state. It is a little arbitrary, but we have this state where eventually all end up until we repolarize, at which time the return transitions allow recovery.

REFERENCES

1. Brown AM, Lee KS, Powell T: Sodium currents in single rat heart cells. *J Physiol* 1981;318:479–500.
2. Makielski JC, Sheets MF, Hanck DA, January CT, Fozzard HA: Sodium current in voltage clamped internally perfused canine cardiac Purkinje cells. *Biophys J* 1987;52:1–11.
3. Hamill OP, Marty A, Naher E, Sakmann B, Sigworth FJ: Improved patch–clamp techniques for high–resolution current recordings from cells and cell–free membrane patches. *Pfluegers Arch (Eur J Physiol)* 1981;391:85–100.
4. Kunze DL, Lacerda AE, Wilson DL, Brown AM: Cardiac Na currents and the inactivating, reopening, and waiting properties of single sodium channels. *J Gen Physiol* 1985;86:691–720.
5. Scanley BE, Hanck DA, Chay T, Fozzard HA: Kinetic analysis of single sodium channels from canine cardiac Purkinje cells. *J Gen Physiol* 1990;95:411–437.
6. Aldrich RW, Corey DP, Stevens CF: A reinterpretation of mammalian sodium channel gating based on single channel recording. *Nature* 1983;306:436–441.
7. French RJ, Horn R: Sodium channel gating: models mimics and modifiers. *Ann Rev Biophys Bioeng* 1983;12:319–356.
8. Aldrich RW, Stevens CF: Voltage–dependent gating of single sodium channels from mammalian neuroblastoma cells. *J Neurosci* 1987;7:418–431.
9. Horn R, Vandenberg CA: Statistical properties of single sodium channels. *J Gen Physiol* 1984;84:505–534.
10. Noda M, Ikeda T, Kaytano T, Suzuki H, Tukeshima T, Kurasaki M, Takahashi H, Numa S: Existence of distinct sodium channel messenger RNAs in rat brain. *Nature* 1986;320:188–192.
11. Rogart RB, Cribbs LL, Muglia LK, Kephart DD, Kaiser MW: Molecular cloning of a putative tetrodotoxin–resistant rat heart Na+ channel isoform. *Proc Natl Acad Sci* 1989;86:8170–8174.
12. Catterall WA: Structure and function of voltage–sensitive ion channels. *Science* 1988;242:50–61.
13. Noda M, Suzuki H, Numa S, Stühmer W: A single point mutation confers tetrodotoxin and saxitoxin insensitivty on the sodium channel II. *FEBS Letters* 1989;259:213–216.
14. Stühmer W, Conti F, Suzuki H, Wang X, Noda M, Yahagi N, Kubo H, Numo S: Structural parts involved in activation and inactivation of the sodium channel. *Nature* 1989;339:597–603.
15. Vassilev P Scheuer T, Catterall WA: Intubation of inactivation of single sodium channels by a site–directed antibody. *Proc Natl Acad Sci USA* 1989;86:8147–8151.

CHAPTER 15:

REGULATION OF CATIONIC FLUX THROUGH CARDIAC SARCOPLASMIC RETICULUM

Qi–Yi Liu and Harold C. Strauss

ABSTRACT

Contraction of mammalian cardiac muscle is regulated by the rapid release of Ca^{2+} from the sarcoplasmic reticulum (SR). Recent data indicate that $[Ca^{2+}]$ in the SR is decreased when $[Ca^{2+}]$ is measured at peak concentration relative to control. Recent data suggest that the intracellular K^+ flux through the SR K^+ channel is an important countercurrent for Ca^{2+} release. We show that increasing Ca^{2+} decreases SR K^+ channel conductance with a $K(0)_{trans}$ = 3.5 mM and $K(0)_{cis}$ = 5.1 mM at $[K^+]$ = 50 mM.

INTRODUCTION

Activation of contractile elements in mammalian skeletal and cardiac muscle is triggered by an increase in myoplasmic Ca^{2+} resulting from its influx across the sarcolemma and release from the sarcoplasmic reticulum (SR) [1–3]. If the SR were permeable only to Ca^{2+}, the rate and magnitude of Ca^{2+} release across the SR are sufficient to rapidly and markedly polarize the SR membrane, drive the potential toward E_{Ca}, and reduce Ca^{2+} efflux [3]. Simultaneous countermovement of other ions could minimize the voltage change and maintain Ca^{2+} release from the SR.

The permeability of the skeletal and, to a lesser extent, the cardiac SR membrane has been studied using different experimental approaches. One approach has been to monitor ion flux across SR vesicles using a variety of techniques [4–10]. These studies indicate that the skeletal SR is selectively permeable, demonstrating a low permeability to sucrose and divalent cations, while monovalent ions such as K^+, Na^+, anions and H^+ readily pass through the membrane [5–9]. Data from these studies also suggest that two types of vesicles exist. Type I vesicles are very permeable to small cations such as K^+, whereas Type II vesicles are relatively impermeable to K^+ and Na^+. Although

K^+ efflux in Type I vesicles is too rapid to resolve, McKinley and Meissner [6] have estimated a very high K^+ permeability (> 10^{-6} cm/sec) in these vesicles. Measurement of resting K^+ permeability in the *in situ* membrane has also been performed using the isotope tracer flux technique and permeability was an order of magnitude lower [11].

Reconstitution studies in planar lipid bilayers [12–14], have demonstrated that K^+ flux is likely mediated via ion channels that are ideally selective for K^+ over Cl . Studies with an SR K^+ channel blocker as well as ion selectivity experiments using intact SR vesicles and ion channels fused with planar lipid bilayers strongly suggest that K^+ flux through the SR membrane is mediated by the SR K^+ channel [10]. Meissner [9] has calculated the specific conductance of the SR membrane to be about $1.5 \cdot 10^{-1}$ S/cm², assuming conductance of 120 pS, open probability of 0.25 at 0 voltage, and channel density of 50 channels/μm². Garcia and Miller [10] made similar assumptions and obtained similar values. These data suggest that the lower conductance values reported previously [4, 15] are probably too low, and additional confirmation of these more recent conductance measurements in the *in situ* membrane would be desirable [16, 17]. Because K^+ is present in such high concentration within the SR and myoplasm, one would expect it to be the principal counterion. Thus, there are data to support the view that K^+ and other countercurrent pathways can effectively short–circuit the charge build–up following Ca^{2+} release; however, controversy exists with regard to the magnitude of the different short–circuit currents and the magnitude of voltage change across the SR during Ca^{2+} release.

Another approach that has been used in investigating the role of K^+ as a counterion during Ca^{2+} release from the SR has utilized skinned skeletal muscle fibers [3]. A Ca^{2+}–sensitive fluorescent indicator was used to monitor changes in Ca^{2+} levels, while caffeine was used to trigger release of Ca^{2+} from the SR. An SR K^+ channel blocker (1,10–bis–guanidino–n–decane, bisG10) and ionic substitution experiments (choline for K^+) were used to decrease SR K^+ conductance so that its effect on the magnitude and rate of Ca^{2+} release could be assessed. Both bisG10 and choline substitution caused a concentration–dependent decrease in the Ca^{2+} release rate, supporting the view that K^+ plays a major role, physiologically, in maintaining charge balance during Ca^{2+} release.

Although much information is available on the SR K^+ channel from skeletal muscle, relatively few studies have been performed on the SR K^+ channel from cardiac muscle [18–20]. Our laboratory has been actively characterizing the SR K^+ channel from cardiac muscle for two reasons. First, we would like to discern whether the SR K^+ channel in skeletal and cardiac muscle differ. Second, we would like to know if the SR K^+ channel merely serves as a background shunt pathway or whether it actually changes its conductance during Ca^{2+} release.

METHODS

Canine cardiac membrane vesicles were isolated according to a previously described procedure [19,20]. Briefly, left ventricular tissue was isolated, and the endocardium and epicardium were dissected away. The cells were disrupted and intracellular proteins were extracted. Cellular membranes were vesiculated by vigorous homogenization and then separated by differential density centrifugation.

SR K^+ channels were reconstituted in lipid bilayers (1:1 mixtures of phosphatidylethanolamine/phosphatidylserine), and transmembrane current was monitored with a high–input impedance electrometer. This current signal was low–pass filtered (100 Hz) and stored for subsequent digital analysis. Event analysis was accomplished by defining discriminators of opening and closing midway between the means of the open– and closed–state amplitude distributions. An event was defined as "open" when there were two or more digitized points greater than the open discriminator.

RESULTS

A typical record shown in Fig. 1 indicates that the channel cycles between three gating states: a closed state and two open states, O_1 and O_2 [19, 20]. We have interpreted O_1 to represent a subconductance state of the "fully open" channel, O_2. The three gating states intercommunicate freely, and each

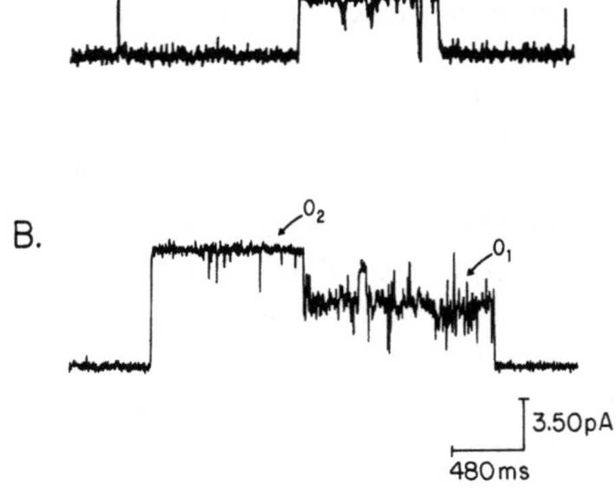

Figure 1: Unitary current recordings of SR K^+ channel with upward deflections indicating channel opening.
(A) 100 mM K^+ symmetrical, V_{hold} = +25 mV.
(B) 300 mM K^+ symmetrical, V_{hold} = +40 mV. Note that O_1 is noisier than O_2 (from [20], with permission).

has direct access to the other two. In addition, close examination of the records indicates that there are at least two closed states, one a relatively long and the other a relatively short closed state. Open channel probability was relatively voltage–insensitive (Fig. 2), and it was deduced that 1.2 units of gating charge move *trans* to *cis* on channel opening [20].

Current–voltage relationships for O_1 and O_2 were linear over a ±70 mV range [20]. The single–channel slope conductance of the substate O_1 was 105 ± 2 pS (n = 15, 0.1 M K^+), while that of O_2 was 189 ± 4 pS (n = 15, 0.1 M K^+). The relative conductance O_1/O_2 was remarkably constant at 56.1% (n = 15) even when measured over a wide variety of experimental conditions. Other subconductance states were observed much less frequently.

Selectivity as measured by zero–current reversal potentials revealed a sequence, corresponding to Eisenman sequence I (low field strength) [17]. The selectivity sequence for O_1 was identical to that for O_2, suggesting that selectivity determined from biionic potential measurements is not influenced by the mechanism whereby the channel switches between O_1 and O_2.

Although there is a resting K^+ flux across the SR, permeability might change during Ca^{2+} release. One factor that might affect the permeability of the SR K^+ channel is voltage. However, open probability is weakly voltage dependent, suggesting that other factors may also serve to modulate K^+ flux across the SR membrane during Ca^{2+} release. Recent data [21] indicate that SR [Ca^{2+}] is decreased by 42% at peak contraction relative to control (12.4 ± 1.6 mmol/Kg dry wt). Hence, to determine whether a change in SR and/or cytoplasmic [Ca^{2+}] could modulate SR K^+ permeability, we examined the effects of Ca^{2+} on SR K^+ channel in planar lipid bilayers [22]. Ca^{2+} blocked the channel from either the trans or cis side, and blockade was slightly more effective when Ca^{2+} was present on both sides (Figs. 3 and 4). Ca^{2+} was impermeable, and block was voltage–dependent (data not shown) and competitive with K^+ (Table 1).

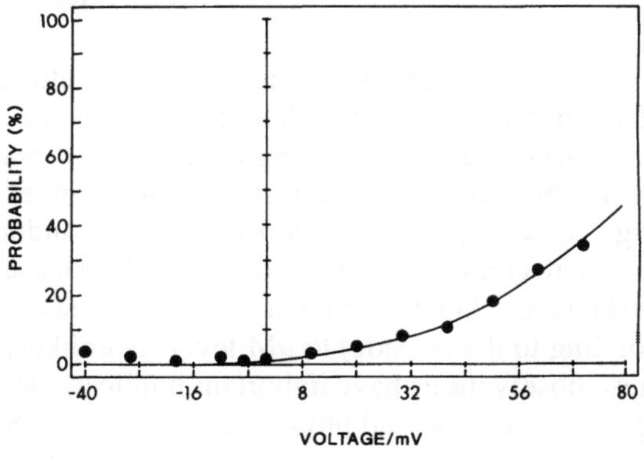

Figure 2: Voltage dependence of SR K^+ channel gating. Open probability is plotted vs. holding potential. Smooth curve is a Boltzmann relation with following parameters: z = −1.2; ΔG_i 1.9 kcal/mol, where z is gating charge and ΔG_i is the chemical energy term (from [20], with permission).

Figure 3: Single channel currents from cardiac SR K^+ channel shown as upward (V_{hold}>0) or downward (V_{hold}<0) deflections were recorded in symmetric 50 mM KAc, 10 mM histidine, pH 7.1. [Ca^{2+}] <0.1 μM (top trace), 1.6 mM in *trans* chamber (second trace), 1.6 mM in both chambers (third trace) and 40 mM in both chambers (bottom trace).

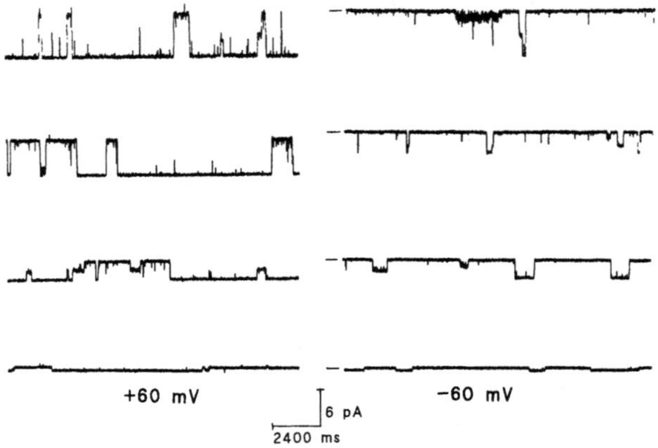

Table 1: Zero voltage dissociation constant for *cis* and *trans* block.

	$K(0)_{trans}$	$K(0)_{cis}$
[K^+] = 50 mM	3.5 mM	5.1 mM
[K^+] = 150 mM	15.0 mM	21.0 mM

Thus, the reported [Ca^{2+}] in the SR, corresponding to the *trans* side in the reconstitution experiments, is sufficiently high to modulate SR K^+ channel conductance, and the decrease in SR [Ca^{2+}] during contraction is sufficient to increase SR K^+ channel conductance.

DISCUSSION

The important new findings in this study pertain to the demonstrated effects of Ca^{2+} on SR K^+ channel conductance. Ca^{2+} reduced single–channel conductance, and the blocking action of Ca^{2+} was [K^+]– and voltage–dependent. Further, our data suggest that the channel is impermeant to Ca^{2+}. The range in [Ca^{2+}] over which a change in conductance was observed suggests that the decrease in SR Ca^{2+} during Ca^{2+} release could increase single channel conductance. On the other hand, the increase in cytoplasmic [Ca^{2+}] (corresponding to the *cis* side) to μM levels is unlikely to affect SR K^+ conductance. These observations have important implications with regard to our understanding of the function and physiologic role of this SR K^+ channel protein.

Figure 4: K^+ current blockade by different [Ca^{2+}] present in trans chamber. Experimental conditions, 150 mM KAc, 1.2 mM Mg^{2+}, 10 mM histidine, pH 7.2 solution with neutral PE bilayer.

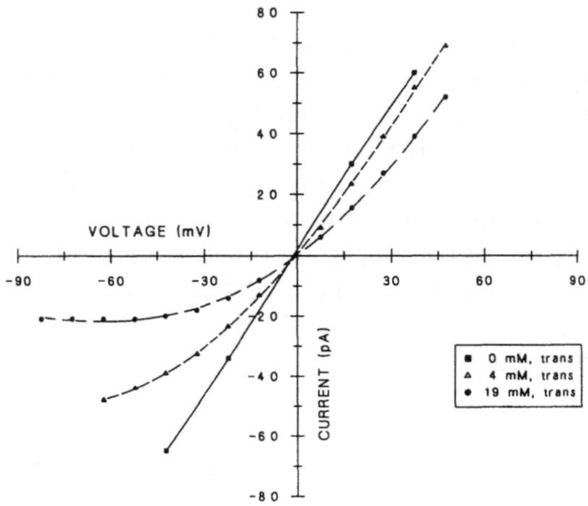

The SR K^+ channel from rabbit and frog skeletal muscle has been studied extensively [12–14, 23]. There are a number of important similarities between the cardiac and skeletal muscle SR K^+ channels: conductance value in symmetrical K^+ solutions; linear I/V relation under symmetrical K^+ over a 100 mV range; greater selectivity for monovalent cations over anions; lack of Ca^{2+} dependence or permeability; selectivity sequence measured by conductance ($K^+ > Rb^+ > Na^+ > Cs^+$); identical selectivity characteristics of the main conductance state and the substate.

Despite these similarities, important differences are apparent. For example, the skeletal muscle SR K^+ channel exhibits the characteristics of a single–ion channel [13, 24]. First, slope conductance saturates with ionic concentration and is well described By Michaelis–Menten enzyme kinetics. Second, the Na^+:K^+ permeability ratio is independent of concentration between 20 mM and 2.5 M. Third, the fractional electrical distances of Cs^+ block is less than one [25, 26]. The cardiac SR K^+ channel, on the other hand, exhibits the characteristics of a multi–ion channel [19, 20]. First, single–channel conductance plotted as a function of ionic concentration demonstrates an increase in conductance at low ionic concentrations, which then levels off but does not really saturate and continues to rise even at [K^+] = 1 M. These data are not well fit by a rectangular hyperbola, suggesting that the assumptions underlying Michaelis–Menten formalism do not hold. Second, the Cs^+:K^+ permeability ratio is concentration–dependent. Third, when the effects of different Cs^+ and K^+ mixtures were examined in greater detail in a mole fraction series of experiments, a minimum was observed in the conductance vs Cs^+ mole fraction plot, strongly suggesting that multiple occupancy of the channel may occur under these conditions. Fourth, the permeability ratio as measured by reversal potential differs from the ratio measured by conductance. Additional differences concern the degree of asymmetry of block with Cs^+ and Ca^{2+}, wherein the cardiac SR K^+ channel exhibits similar values for K_d from both the *cis* and *trans* sides of the bilayer, and the skeletal SR K^+ channel does not [12, 13, 19, 25, 26].

These observations argue that the SR K^+ channel from cardiac muscle is a multi–ion channel whose permeation properties differ appreciably from the skeletal muscle SR K^+ channel, implying that structural differences exist between these two proteins. Cloning and sequence studies should be of great interest in elucidating the basis of the difference in functional properties observed.

Acknowledgements

This research was supported in part by NIH grant HL–19216 from the National Heart, Lung and Blood Institute. We gratefully acknowledge Steffani H. Webb for her assistance in the preparation of this manuscript.

DISCUSSION

Dr. Fozzard: Rather than assume that the calcium blocks, you can actually see that the channel opens, with a lower conductance. What is happening is that the negative charges are concentrating K in the vestibule. If you alter the field by absorbing calcium, you simply lower K concentration in the vestibule.

Dr. Strauss: Preliminary data that I did not show suggest that calcium is impermeant.

Dr. Fozzard: It would not have to be impermeant. It would just shield the charge in the vestibule and reduce the K concentration in the vestibule; with K being the carrier, it would lead to a decline in conductance.

Dr. Strauss: That is one possibility. On the other hand, the lack of Ca^{2+} permeation through the channel and the voltage dependence of the block suggest that Ca^{2+} is binding to a site within the pore. At this point, we cannot exclude binding to negative charges in the vestible.

Dr. Kammermeier: If I understood right there is an intracellular countercurrent system that requires a concentration gradient within the cell.

Dr. Strauss: There clearly is a Ca^{2+} concentration gradient across the SR membrane at rest, although the actual Ca^{2+} activity in the SR is unknown, and in fact might be considerably lower than that predicted from electron probe x-ray microanalysis measurements. The total calcium in the SR is estimated to be 5 mmolar at rest; free calcium could be perhaps somewhere in the range of 2–3 mmolar. There is no evidence for a resting K^+ gradient across the SR membrane. Hence, K^+ flux across the SR would depend on the electrochemical gradient.

Dr. Marban: If you were designing this system *de novo* you would be worried about counterbalancing the calcium efflux with monovalents. If you are dealing with fluxes of calcium of several millimolars over tens of milliseconds, what about osmotic shifts that might occur in exchanging monovalents for divalents? Wouldn't it just seem inefficient that the cell would choose to do that, rather than exchange calcium for magnesium or something else? Is anything known about the potassium concentration in the SR that would make this a particularly energetically favorable way of doing that or is it perhaps good for any other reason?

Dr. Strauss: First of all, K^+ is not the only ion that is able to move and compensate for calcium release. Na^+, H^+, and Cl^- are also potential candidates. In addition, Meissner *[8]* has proposed that the function of the shunt pathways is to minimize osmotic effects as well as charge build-up during Ca^{2+} release. However, in the absence of specific blockers, it is not possible to unequivocally conclude that the SR K^+ channel is the dominant ion pathway through which charge neutralization occurs.

REFERENCES

1. Caillé J, Ildefonse M, Rougier O: Excitation– contraction coupling in skeletal muscle. *Prog Biophys Molec Biol* 1985;46:185–239.
2. Oetliker H: An appraisal of the evidence for a sarcoplasmic reticulum membrane potential and its relation to calcium release in skeletal muscle. *J Muscle Res Cell Motility* 1982;3:247–272.
3. Abramcheck CW, Best PM: Physiological role and selectivity of the *in situ* potassium channel of the sarcoplasmic reticulum in skinned frog skeletal muscle fibers. *J Gen Physiol* 1989;93:1–21.
4. Kometani T, Kasai M: Ionic permeability of sarcoplasmic reticulum vesicles measured by light scattering method. *J Memb Biol* 1978;41:295–308.
5. Meissner G, McKinley D: Permeability of sarcoplasmic reticlum membrane: The effect of changed ionic environments on Ca^{2+} release. *J Memb Biol* 1976;30:79–98.
6. McKinley D, Meissner G: Evidence for a K^+, Na^+ permeable channel in sarco–plasmic reticulum. *J Memb Biol* 1978;44:159–186.
7. Meissner G, Young RC: Proton permeability of sarcoplasmic reticulum vesicles. *J Biol Chem* 1980;255:6814–6819.
8. Meissner G, McKinley D: Permeability of canine cardiac sarcoplasmic reticulum vesicle to K^+, Na^+, H^+, and Cl^-. *J Biol Chem* 1982;257:7704–7711.
9. Meissner G: Monovalent ion and calcium ion fluxes in sarcoplasmic reticulum. *Mol Cell Biochem* 1983;55:65–82.
10. Garcia AM, Miller C: Channel–mediated monovalent cation fluxes in isolated sarcoplasmic reticulum vesicles. *J Gen Physiol* 1984;83:819–839.
11. Best PM, Abramcheck CW: Potassium efflux from single skinned skeletal muscle fibers. *Biophys J* 1985;48:907–913,
12. Miller C: Voltage–gated cation conductance channel from fragmented sarco–plasmic reticulum: steady–state electrical properties. *J Membr Biol* 1978; 40:1–23.

13. Coronado R, Rosenberg RL, Miller C: Ionic selectivity, saturation, and block in a K^+–selective channel from sarcoplasmic reticulum. *J Gen Physiol* 1980;76:425–446.
14. Labarca PP, Miller C: A K^+–selective, three–state channel from fragmented sarcoplasmic reticulum of frog leg muscle. *J Memb Biol* 1981;61:31–38.
15. Vergara J, Bezanilla F, Salzberg B: Nile blue fluorescence signals from cut single muscle fibers under voltage or current clamp conditions. *J Gen Physiol* 1978;72:775–800
16. Hals GD, Stein PG, Palade PT: Single channel characteristics of a high conductance anion channel in "sarcoballs." *J Gen Physiol* 1989;93:385–410.
17. Tang JM, Wang J, Eisenberg RS: K^+–selective channel from sarcoplasmic reticulum of split lobster muscle fibers. *J Gen Physiol* 1989;94:261–278.
18. Tomlins B, Williams AJ, Montgomery RAP: The characterization of a monovalent cation–selective channel of mammalian cardiac sarcoplasmic reticulum. *J Memb Biol* 1984;80:191–199.
19. Hill JA, Coronado R, Strauss HC: Potassium channel of cardiac sarcoplasmic reticulum is a multi–ion channel. *Biophys J* 1989;55:35–46.
20. Hill JA Jr, Coronado R, Strauss HC: Open–channel subconductance state of K^+ channel from cardiac sarcoplasmic reticulum. *Am J Physiol (Heart Circ Physiol 27)* 1990;258:H159–H164.
21. Moravec CS, Bond M: X–ray microanalysis of subcellular calcium distribution in contracted and relaxed cardiac muscle. *Biophys J* 1990;57:503a.
22. Liu Q–Y, Strauss HC: Voltage–Dependent Blockade of Cardiac SR–K Channel by Calcium (abstract). *Circulation* 1991;82: in press.
23. Coronado R, Miller C: Conduction and block by organic cations in a K^+–selective channel from sarcoplasmic reticulum incorporated into planar lipid bilayers. *J Gen Physiol* 1982;79:529–547.
24. Bell JE, Miller C: Effects of phospholipid surface charge on ion conduction in the K^+ channel of sarcoplasmic reticulum. *Biophys J* 1984;45:279–287.
25. Coronado R, Miller C: Voltage–dependence caesium blockade of a cation channel from fragmented sarcoplasmic reticulum. *Nature* 1979;280:807–810.
26. Cukierman S, Yellen G, Miller C: The K^+ channel of sarcoplasmic reticulum. A new look at Cs^+ block. *Biophys J* 1985;48:477–484.

CHAPTER 16:

VOLTAGE CLAMP TESTS OF A MODEL OF CALCIUM HANDLING BY MYOCARDIAL CELLS

Per Arlock, Bjorn Wohlfart and Mark I.M. Noble

ABSTRACT

Right ventricular papillary muscles from ferrets were studied in a single sucrose gap apparatus. The preparations were basically paced at 1.0 Hz. A train of voltage clamp pulses was then introduced until twitch force was steady. A test depolarization of variable duration was then introduced, followed by a test diastolic interval which could also be varied in duration. Prolongation of the test depolarization caused potentiation of the following contraction which decayed on subsequent beats. The behavior of contractile force during this decay period (linearly related to the force of the preceding beat), is consistent with the concept of recirculation of a contractile activator from one beat to the next. Second inward current is suppressed on the first potentiated beat after prolonged test depolarization, and increased during the period of decaying force. This confirms negative feedback of second inward current by the released activator. It is not compatible with proportionality between trigger calcium and released calcium (in calcium induced calcium release), if second inward current is assumed to be an index of trigger calcium and force is assumed to be an index of released calcium. The relationship between the force of the first beat after the test depolarization and the duration of that depolarization showed a "dead time" of approximately 120 ms. Variation of the diastolic interval following the test depolarization showed that recovery of force of the subsequent beat ("mechanical restitution", an index of recovery of internal calcium releasability), began with repolarization of the cell membrane, and not with relaxation (taken to indicate sarcoplasmic reticulum uptake). This suggests a membrane source for trigger calcium in calcium–induced calcium release.

INTRODUCTION

Ringer [1] has shown that calcium ions exchange rapidly between intracellular and extracellular space in myocardium and that they are essential for contractility. Now Rich et al. [2] have shown that the exchange and the effect on contractility occurs within one diastole. Our ideas on this subject have grown from the experience of studies of the force–interval relationship [3]. These studies are relevant to the study of calcium release because it has been shown that with changes in interval between myocardial contractions, tension and the rate of rise of tension are linearly related to the calcium release, as measured with aequorin [4]. A number of postulates previously put forward have now been tested by the voltage clamp method.

METHODS

Preparations

Right ventricular papillary muscles (diameter 0.5–0.7mm) from ferrets were studied in a single sucrose gap apparatus as previously described [5, 6]. The tendon of the papillary muscle was in the test compartment and tied to a hook fastened to an AKER electronics semiconductor strain gauge (AE 802P) by means of a short length of 6.0 surgical suture.

The normal perfusate consisted of (in mM): NaCl 145, KCl 4.0, $CaCl_2$ 1.8, $MgCl_2$ 1.05, glucose 5.5, Tris maleate 5.0. The pH of the solution was adjusted to 7.40 before use and gassed with 100% O_2. Temperature was maintained at 37 ± 0.1°C.

The muscle was stretched at the start of the experiment to produce 90% maximum force. The preparations were equilibrated for at least 30 min before being superfused with the sucrose solution (middle compartment) and isotonic KCl solution (current injection compartment). Another equilibration period of 30 min was observed before data was collected. Preparations showing spontaneous activity were discarded.

The preparations were basically paced at 1.0 Hz using square wave current pulses with a duration of 3 ms and with an amplitude 10% above threshold value. The preparations were stimulated with Tecmar (Cleveland, Ohio) Labmaster board, controlled by a Commodore AT programmed in Pascal calling assembler routines. Membrane potentials were recorded from the tip of the muscle in the test chamber by means of conventional microelectrodes (10–20 Mohm) filled with 3M KCl. During voltage clamp the holding potential was held at –70mV or –40 mV. The clamp step to +20mV was chosen according to previous experience [6] showing that this produced maximal force.

Protocols

The voltage clamp protocols are illustrated in Fig. 1. After a 20 s rest, the muscle was "primed" with 10 clamps from either –70 or –40mV holding potential to +20mV depolarization potential. The priming clamps were 200 ms in duration separated by 800 ms electrical diastolic intervals. In the first series of experiments, the "priming period" was followed by a test clamp of variable duration – test clamp 1 (Fig. 1). The subsequent clamps were identical to the priming clamps and were numbered 2,3 etc. In the next series of experiments an additional intervention was the variation (–40 or –70mV) of the holding potential between test clamps 2 and 3 (Fig. 1B). In the third series, the "priming period" was followed by a test period consisting of a test clamp (1) of 200 or 500 ms duration followed by a test electrical diastolic interval which was varied from 100 ms to 10 s. This was followed by a second test clamp (2) of 200 ms duration. The contractile force induced by test clamp 2 was measured and its relationship to test cycle duration and test electrical diastolic interval determined.

Recording

The current and voltage signals were displayed on an Elema-SchØnander ink jet recorder and also on a Tektronix oscilloscope where they could be photographed on 35mm film. Current, voltage and force were recorded.

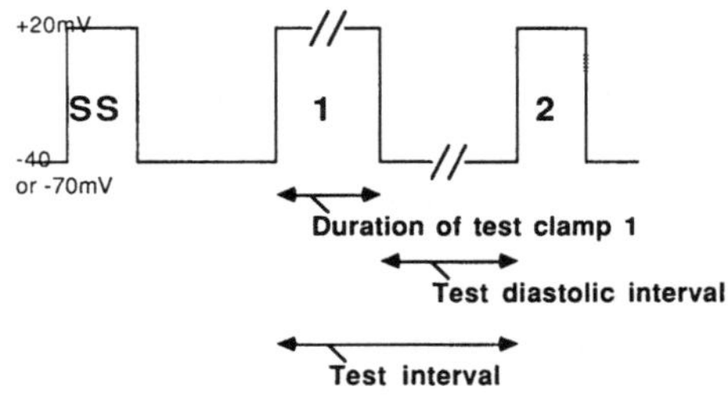

Figure 1A: Protocol for sequence of depolarization clamps. SS = last steady state depolarization. 1 = test clamp of variable duration. Note additional variation of diastolic interval between 1 and 2.

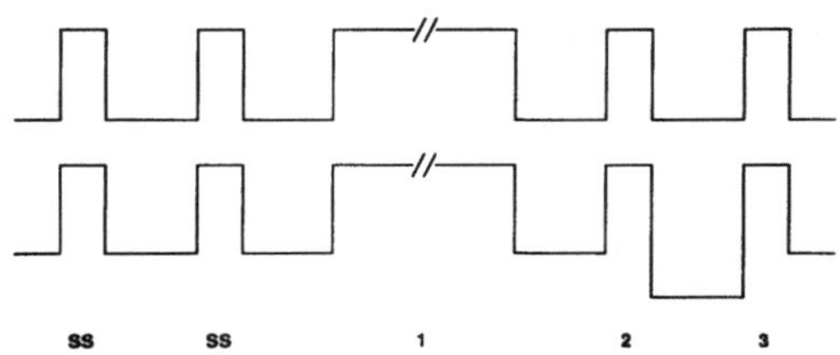

Figure 1B: Protocol for sequence of depolarization clamps. 2, 3 = first two of a series of test clamps with the same characteristics of duration and interval as the SS period. Note additional variation of diastolic potential between 2 and 3.

Analysis

Mechanical restitution curves were plotted by computer as the relationship between F2 and test cycle length, and plots of F2 against test electrical diastolic interval were also made. A similar analysis was carried out for differences between 200 and 500 ms test clamp for each muscle and the mean results calculated.

RESULTS AND DISCUSSION

Contraction and Relaxation

It is well accepted that contraction results from a reaction between the contractile proteins, adenosine triphosphate (ATP) and calcium ions released into the myocardial cell cytoplasm, and the relationship between tension produced and calcium ion concentration (a non–linear sigmoid relationship), has been accurately defined [7]. Changes in calcium release upon activation can be monitored [8], using the relationship between calcium ion concentration and light emitted by aequorin (which is also non–linear).

Tension as a calcium ion transducer is limited by the fact that the entire relationship between calcium ion concentration and tension is altered by sarcomere length [7] and intracellular cyclic adenosine monophosphate [9], resulting from the action of catecholamines [8]. If these 2 variables are kept constant and calcium release is altered only by changing the interval between excitations, tension and the rate of rise of tension turn out to be remarkably linearly related to calcium ion concentration as assessed with aequorin [4].

Relaxation occurs through removal of calcium from the contractile proteins. To a certain extent this can occur through diffusion into the cytoplasmic space and absorption on to the surfaces of organelles. Active removal of calcium occurs through the action of the sarcoplasmic reticulum (SR) [10]. What happens to the calcium after it has been taken up into the SR?

Calcium Recirculation

Calcium recirculation follows from the fact that calcium released on a particular stimulus is proportional to the calcium released on the preceding beat, as long as each of these stimuli occur after a constant interval. This has been shown to be the case in a number of studies [3]. With other variables held constant, it is found that the tension (or rate of rise of tension, or aequorin light signal) of one beat (F_n) is linearly related to the tension of the preceding beat (F_{n-1}). The slope of this linear relationship (B_f) is the proportion of

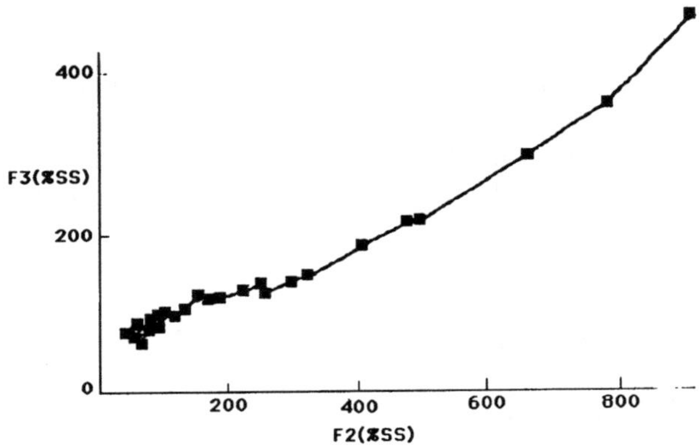

Figure 2: Plot of the force on test clamp 3 as a function of force of test clamp 2. Force is expressed as a percentage of steady state force (%S).

calcium released on one beat which is recirculated to the next beat, i.e. the "recirculation fraction". This behavior is described by the equation:

$$F_n = B_F(F_{n-1}) + C \tag{1}$$

where F_n = the force of beat n, F_{n-1} = force of the preceding beat n–1, B_F = recirculation fraction, C = constant.

The voltage clamp method allows this relationship to be explored over a much wider range of values than previous methods because much prolongation of test clamp 1 caused enormous potentiation of the force of test clamp 2. Recirculation was studied by plotting the force of test clamp 3 against that of test clamp 2, as shown in Fig. 2.

Non–linearity is now shown up at the extreme of the range, and linearity is confirmed in the physiological range previously studied.

Calcium Extrusion

The model already takes account of calcium extrusion by virtue of the fact that only a proportion (B_F) of calcium is recirculated from one beat to the next, leaving the remainder ($1 - B_F$) to be extruded. We postulated that this extrusion would occur in systole when calcium is released and the increased intracellular calcium ion concentration would be expected to drive calcium extrusion in proportion to that concentration. What would happen to the intracellular calcium release on the following beat if sodium/calcium exchange continues to extrude calcium in diastole when intracellular calcium ion concentration is extremely low?

This was tested in the voltage clamp preparation by the use of the protocol illustrated in Fig. 1B, which enables us to increase calcium extrusion by sodium/calcium exchange by lowering diastolic potential. When the potential in the diastole between clamps 2 and 3 was held at –40 mV, we obtained the result shown in Fig. 2. When this potential was reduced to –70 mV, we obtained a parallel downward shift of the relationship, indicating no change in recirculation fraction, but a fixed amount of calcium extrusion.

These results are compatible with the idea of calcium extrusion being divided into (a) an amount in diastole dependent on membrane potential but not on the amount released in the previous systole and (b) an amount in systole proportional to the amount released, i.e. the non–recirculated fraction $1 - B_F$. Recirculation of calcium is an old idea that was persuasively advocated by Morad and Goldman [11]. Does this recirculation take place within the SR, or does the recirculated calcium have to move somewhere else to be released on the subsequent beat? The former proposition implies that SR is the release as well as the uptake site [12, 13]; the latter suggestion implies a separate release site, e.g. sub–sarcolemmal (sub–tubular) cysternae [14] or sarcolemma [15]. This brings us to the fundamental issue in the subject of cardiac excitation–contraction coupling.

Potentiation

Before addressing this critical issue directly, the phenomenon whereby an extrasystole is followed by a strong beat is worthy of comment, because of its possible relevance. This phenomenon occurs whether the action potential of the extrasystole is prolonged (rabbit) or shortened (dog, man) and the potentiation can produce more than double the force of the control beats. It therefore implies a negative feedback control of released calcium on calcium inflow rate during the action potential [16]; i.e. a premature beat which is weak and associated with a small calcium release lets in more calcium ions when the second inward channel opens [3]. This is confirmed by the experiments described in the next section (Fig. 3). The negative feedback effect of released calcium upon calcium current is more easy to imagine if the calcium is released close under the sarcolemma (sub–sarcolemmal cysternae, sarcolemma) than if it released far in the cell interior at the SR.

Calcium Release

Calcium release is currently thought to be derived from the sarcoplasmic reticulum, triggered by the transarcolemmal calcium carried by the second

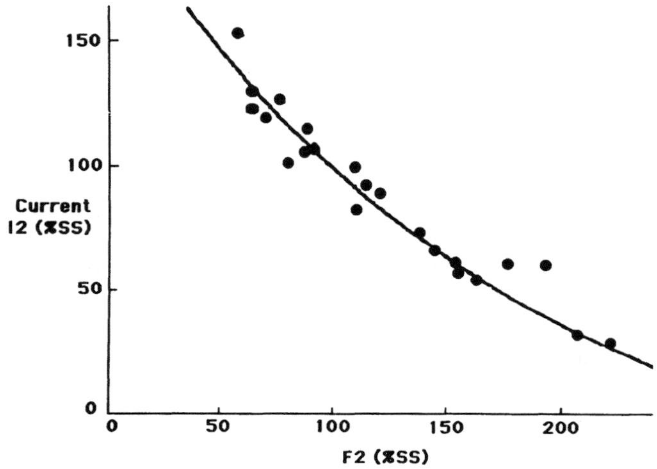

Figure 3: Plot of second inward current versus force of test clamp 2.

inward current [17–19]. This conventional theory proposes that calcium is released from the SR when a little trigger calcium entering through the calcium channel triggers the main release from the SR calcium store by the process of calcium–induced calcium–release, a phenomenon that can be demonstrated when sarcolemma is removed [13]. Does it logically follow that calcium entering by the calcium channel is the trigger for normal calcium release?

In our experiments, potentiation of beat 2 was achieved with the prolonged clamp 1. We then studied the decay of this potentiation from beats 2 onward with respect to force and second inward current. We found that current was suppressed on the first potentiated beat F2, and rose as force decayed on the subsequent beats. When F2 was varied by changes in test clamp duration 1, we recorded the relationship between second inward current amplitude and F2, as shown in Fig. 3.

The relationship was inverse. This is discordant with the idea that calcium release from the SR is proportional to the trigger calcium entering through the second inward current [17–19]. It is more consistent with calcium release being dependent on the amount in the SR, which is then released by a threshold amount of calcium in an all–or–none manner.

There are some other discordant facts: External calcium can rapidly exchange with the calcium of the internal store as shown by Langer [e.g., 2]. This group has identified the fast pool of contractile calcium in both rat and rabbit (with evidence that it is sarcolemmal) and also a slower pool in the rat, which they identify as being located in sarcoplasmic reticulum. The fast pool of calcium is not extracellular because it is rapidly affected (full effect within one diastole) by change in extracellular sodium ion concentration via Na–Ca exchange [2]; the very rapid time course for complete exchange confirms a sarcolemmal site for the fast calcium pool. Calcium exchange had previously been measured in cultured cells by Langer [20] and the measurement repeated after gas dissection of the cells to remove all the cell contents including SR – only the sarcolemma remains. Repeat measurements show almost identical calcium exchange and pH dependence, indicating that the exchangeable calcium store was still present and located in sarcolemmal phospholipids. Treatment of cardiac tissue with phosphate buffered fixative and precipitation of calcium as an antimony electron–dense complex shows that almost all of the tightly bound calcium is located in the sarcolemma. None is found in the SR [21].

In order to accommodate these findings within the theory of calcium induced calcium release from the SR, let us consider the phenomenon of mechanical restitution.

Time Course of Calcium Availability After Relaxation

Just after the end of the refractory period, a premature excitation elicits a weak contraction. The contraction gains in strength as the interval prior to excitation is prolonged. The resulting positive curvilinear relationship between contractile tension and interval between excitations [22, 23] called *mechanical restitution* has appeared within the last few years to hold the key to the controversy concerning calcium release.

The most popular idea, as has been described above, is that calcium, as well as being taken up by the SR during relaxation, is released again by the SR upon excitation [17–19]. One might well ask at this point the following questions: "If all the activator calcium is already in the SR after relaxation, why should there be a delay before it can be released again?", "What biochemical process within the SR is involved?"

As far as we are aware, no answers to all these questions are to be found in the literature. We do not include the subsarcolemmal and subtubular cysternae in "SR". Upon repolarization, we postulate that the calcium affinity of one or other of the cysternae or sarcolemma becomes maximal in the cell, so that calcium diffuses from other sites back to the cell surface and T tubules. This diffusion will take time and will occur across calcium concentration gradients within the cell in which the absolute level of concentration never exceeds threshold for the contractile proteins (i.e. 10^{-7} to 10^{-11} M), i.e. the tissue remains relaxed. An accumulation of calcium in the superficial release sites as a function of time would result in mechanical restitution.

Does Mechanical Restitution Start With Relaxation or Repolarization?

In recent meetings on excitation–contraction coupling, this question has appeared crucial because, if calcium merely requires time to become releasable within the SR and merely needs second inward current calcium to trigger release, the time course of mechanical restitution should begin with relaxation; whereas if trigger calcium has to accumulate at the polarized sarcolemma [15] to be released upon depolarization, the time course of mechanical restitution should begin with repolarization.

The latter was shown by Franz et al. [24] in pacing experiments, but the variation in repolarization time achieved (dependent on action potential duration) was sufficiently small for this result to be ignored.

In order to test this point more critically, we varied the duration of the diastolic interval between clamps 1 and 2 in Fig. 1 and recorded F2 as a function of (a) the test interval and (b) the test diastolic interval (Fig. 1A). We obtained the usual positive curvilinear relationship. We studied this for two durations of test clamp 1, namely 200 and 500 ms. If calcium merely requires time to become releasable within the SR, that time course should begin with calcium release and re–uptake, i.e. the test interval determines mechanical restitution. However, if trigger calcium has to accumulate at the polarized sarcolemma to be released upon depolarization, the time course of mechanical restitution should begin with repolarization, i.e. the test diastolic interval determines mechanical restitution. The result, in unpublished experiments of Arlock, Noble and Wohlfart [25], point clearly to the latter in that the time course for a test clamp 1 of 500 ms was delayed 300 ms with respect to a test clamp of 200 ms when test interval was the independent variable whereas the onset of mechanical activity was identical for the two test clamp durations when test diastolic interval was the independent variable.

Is the new information still compatible with the conventional hypothesis? This is indeed the case, if one assumes that the SR has calcium concentration dependent channels which do not deactivate at the level of intracellular calcium concentration present during prolonged depolarization, and that the resultant refractoriness of the SR is removed when repolarization turns on sodium/calcium exchange [26] with consequent further lowering of intracellular calcium concentration.

If such behavior of the SR could be shown, a further difficulty arises because both during mechanical restitution and during the decay of potentiation [6] second inward current amplitude is inversely related to calcium release. Therefore, second inward current calcium could only be the trigger if the trigger mechanism was an all or none phenomenon. However, the work of Fabiato [17–19] shows that this is clearly not the case; released calcium is graded in proportion to the amount of trigger calcium.

"Dead Time"

Variation in the duration of the action potential affects the following beat rather than the concomitant beat, i.e. extra calcium entering in a particular depolarization does not contribute to the concomitant reaction of calcium with the contractile elements but enters the internal calcium store and is released on the subsequent depolarization. In previous intact dog experiments we found no correlation between action potential duration (recorded monophasically) [27], but did find a correlation with contractility of the subsequent beat [28]. However there was a residual negative constant after correlation, so that the data fitted the equation:

$$F_n = B_F(F_{n-1}) + B_{AP}(AP_{n-1} - D) \qquad (2)$$

where F = the contractility of a given beat; AP_{n-1} = action potential duration of the preceding beat; B_{AP} = coefficient of proportionality; D = "dead time". The dead time averaged 120 ms and implied that the first 120 ms of depolarization was "dead" for calcium entry. Clearly, the second inward current begins during this time. How could it not be carrying calcium inward?

We have therefore tested this point more formally by varying the duration of test clamp 1. F2 was recorded as a function of the duration of preceding depolarization, i.e. test clamp 1.

The well known potentiating effect of duration of depolarization upon subsequent force is illustrated in Fig. 4. However there is a clear time delay before this potentiation begins. F2 does not exceed the control force (100%) for about 120 ms. This is remarkably similar to the results obtained in dogs.

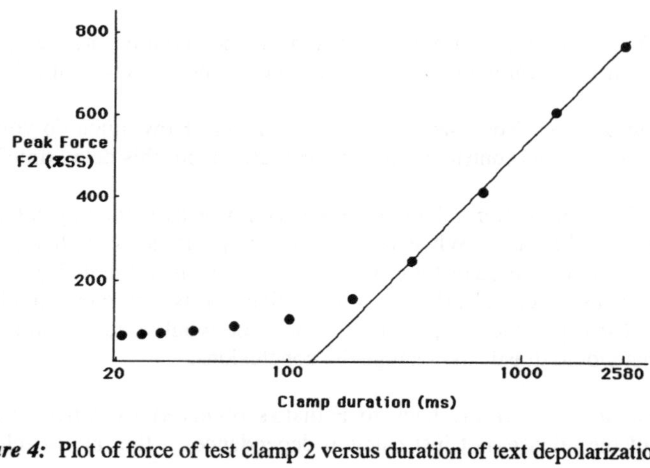

Figure 4: Plot of force of test clamp 2 versus duration of text depolarization 1.

CONCLUSION

The most plausible way in which calcium–induced calcium–release can be reconciled with the new findings is to postulate that the trigger calcium is derived from sarcolemma, and that this trigger calcium is dependent on electric field in the manner originally postulated by Lullmann and Peters [29], namely by calcium binding to proton dependent anionic phospholipids placed exclusively on the inner leaflet. The prediction by these authors of asymmetry of phospholipid composition between inner and outer sarcolemmal leaflets, has now been proved [30]. Such complete fulfillment of a hypothetical prediction constitutes in our opinion the most remarkable example of the proper working of the scientific method in myocardial physiology of recent years.

Acknowledgements

MIM Noble is supported by the Garfield–Weston Trust as Weston Professor of Cardiovascular Medicine.

DISCUSSION

Dr. Fozzard: It seems to me as though a plausible alternative is that the latter part of the longer depolarization, step 1, could be the time when sodium–calcium exchange is operating in a reverse mode, since the intracellular calcium is going down and the membrane potential remains high. That would say that it would be a function of the voltage in that region. You can test this or indeed you can take away extracellular sodium which would obviously block the sodium–calcium exchange, and it would remove that aspect of the phenomena.

Dr. Noble: Yes. Indeed our current idea is that the potentiation picked up on F2 is a result of calcium going through sodium–calcium exchange during clamp 1.

Dr. Bassingthwaighte: You used a linear base model. How much do you think that the linearity assumptions contribute to your estimates about this delay time?

Dr. Noble: The recirculation relationship is very linear over the physiological range that we studied in the past. When we extend it way out as we do here, it begins to curve. In the case of the dead time which we previously estimated by a linear assumption, we can see quite clearly that it is not linear; it is very close to a logarithmic relationship. Clearly those are problems which one would need to sort out. But I don't think that would drastically change the conclusions.

Dr. Bassingthwaighte: Going back to Fabiato's observations, where there is time dependence of the release, not just calcium dependence – the rate of release would depend on the derivative of the rate of change of calcium concentration of the SR and not just the free calcium concentration. How would you figure that into your analyses?

Dr. Noble: It would seem reasonable to suppose that if the trigger calcium is coming from the surface of the cell either through the calcium channel or from sarcolemma, the rate of rise would be proportional to the amplitude, but I guess that does not necessarily follow.

Dr. Adam: Could there be any contribution of the contraction itself to the changes that you see?

Dr. Noble: The difficulty here is that you are probably thinking in terms of the feed-back loop part of it. We have previously found that that does not influence the action potential duration in other preparations. In this preparation you do not allow that to happen. It is quite unlikely that the force *per se* would cause the inhibition of the second inward current.

Dr. Ben Haim: Is this an isometric preparation?

Dr. Noble: Yes. This is an isometric preparation. It is very likely to be a true phenomenon but I am not sure that is likely to apply here since we are keeping the length of the preparation constant.

REFERENCES

1. Ringer S: A further contribution regarding the influence of the different constituents of the blood on the contraction of the heart. *Pflügers Archiv ges Physiol* 1883;308:91–110.
2. Rich T, Langer GA, Klassen MG: Two components of coupling calcium in single ventricular cell of rabbits and rats. *Am J Physiol* 1988;254:937–946.
3. Wohlfart B, Noble MIM: The cardiac excitation–contraction cycle. *Pharmac Ther* 1982;16:1–43.
4. Wier WG, Yue DT: Intracellular calcium transients underlying the short- term force–interval relationship in ferret ventricular myocardium. *J Physiol* 1986;376:507–530.
5. Arlock P, Katzung BG: Effects of sodium substitutes on transient inward current and tension in guinea–pig and ferret papillary muscle. *J Physiol* 1985; 360:105–120.
6. Arlock P Wohlfart B: Force production following transient transmembrane potential changes in ferret and guinea pig papillary muscles. *Acta Physiol Scand* 1990; in press.
7. Kentish JC, ter Keurs HEDL, Riccardi L, Bucx JJ, Noble MIM: Comparison between the sarcomere length–force relations of intact and skinned trabeculae from rat right ventricle. *Circ Res* 1986;58:755–768.
8. Allen DG, Kurihara S: Calcium transients in mammalian ventricular muscle. *European Heart J* 1980;1:Suppl A, 5–15.
9. Herzig JW, Kohler G, Pfizer G, Ruegg JC, Woffle G: Cyclic AMP inhibits contractility of detergent–treated glycerol extracted cardiac muscle. *Pflügers Arch Europ J Physiol* 1981;391:208–212.

10. England PJ: Phosphorylation of cardiac muscle contractile proteins. In: *Cardiac Metabolism* Drake–Holland AJ, Noble MIM (eds), J Wiley & Sons, Chichester, 1983;365–390.
11. Morad M, Goldman Y: Excitation–contraction coupling in heart muscle: membrane control of development of tension. *Prog Biophys Molec Biol* 1972;27:257–313.
12. Fabiato A, Fabiato C: Contractions induced by a calcium–triggered release of calcium from the sarcoplasmic reticulum of single skinned cardiac cells. *J Physiol* 1975;249:469–495.
13. Fabiato A: Calcium–induced release of calcium from the cardiac sarcoplasmic reticulum. *Am J Physiol* 1983;245:C1–14.
14. Kaufman R, Bayer R, Furniss T, Krause H, Tritthart H: Calcium– movement controlling cardiac contractility – II. Analog computation of cardiac excitation–contraction coupling on the basis of calcium kinetics in a multi–compartment model. *J mol cell Cardiol* 1974;6:543–559.
15. Lullman H, Peters T, Preuner J: Role of plasmalemma for calcium homeostasis and for excitation–contraction coupling in cardiac muscle. In: *Cardiac Metabolism*, Drake–Holland AJ, Noble MIM (eds), J Wiley & Sons, Chichester, 1983, pp 1–18.
16. Bassingwaighte JB, Fry CH, McGuigan JAS: Relationship between internal calcium and outward current in mammalian ventricular muscle; a mechanism for control of the action potential duration? *J Physiol* 1976;262:15–37.
17. Fabiato A: Rapid ionic modifications during the aequorin–detected calcium transient in skinned canine cardiac Purkinje cell. *J Gen Physiol* 1985;85:189–246.
18. Fabiato A: Time and calcium dependence of activation and inactivation of calcium–induced release of calcium from sarcoplasmic reticulum of a skinned canine cardiac Purkinje fibre. *J Gen Physiol* 1985;85:247–289.
19. Fabiato A: Simulated calcium current can both cause calcium loading in and trigger calcium release from the sarcoplasmic reticulum of a skinned canine cardiac Purkinje cell. *J Gen Physiol* 1985;85:291–320.
20. Langer GA: The effect of pH on cellular and membrane calcium binding and contraction of myocardium. A possible role for sarcolemmal phospholipid in EC coupling. *Circ Res* 1985;57:374–382.
21. Borgers M, Thone F, Verheyen A, ter Keurs HEDJ: Localisation of calcium in skeletal and cardiac muscle. *Histochem J* 1984;16:295–309.
22. Allen DG, Jewell BR, Wood EH: Studies of the contractility of mammalian myocardium at low rates of stimulation. *J Physiol* 1976;254:1–17.
23. Edman KAP, Johannsson M: The contractile state of rabbit papillary muscle in relation to stimulation frequency. *J Physiol* 1976;254:565–581.
24. Franz MR, Schaefer J, Schottler M, Seed WA, Noble MIM: Electrical and mechanical restitution of the human heart at different rates of stimulation. *Circ Res* 1983;53:815–822.
25. Arlock P, Noble MIM, Wohlfart B: Cardiac cell membrane repolarization is required for the onset of mechanical restitution in isolated ferret ventricular muscle. *J Physiol* 1989;420:87P.
26. Chapman RA, Coray A, JAS McGuigan: Sodium–calcium exchange in mammalian heart: the maintenance of low intracellular calcium concentration. In: *Cardiac Metabolism* Drake–Holland AJ, Noble MIM (eds), J Wiley & Sons, Chichester, 1983;117–150.
27. Drake AJ, Noble MIM, Schouten V, Seed A, ter Keurs HEDJ, Wohlfart B: Is action potential duration of the intact dog heart related to contractility or stimulus rate. *J Physiol,* 1982;331:499–510.

28. Drake–Holland AJ, Noble MIM, Pieterse M, Schouten VJA, Seed WA, ter Keurs HEDJ, Wohlfart B: Cardiac action potential and contractility in the intact dog heart. *J Physiol* 1983;345:75–85.
29. Lullman H, Peters T: Plasmalemmal calcium in cardiac excitation–contraction coupling. *Clin Exp Pharmcol Physiol* 1977;4:49–57.
30. Post JA, Langer GA, Op den Kamp JAF, Verkleij AJ: Phospholipid asymmetry in cardiac sarcolemma: Analysis of intact cells and "gas–dissected" membranes. *Biochim Biophys Acta* 1988;943:256–266.

CHAPTER 17:

MODULATION OF VENTRICULAR IMPULSE INITIATION AND REPOLARIZATION BY SYMPATHETIC INNERVATION

Michael R. Rosen, Susan F. Steinberg, Gabriella Malfatto and Tove S. Rosen

ABSTRACT

A neonatal rat model was prepared for studying the development of α-adrenergic actions on cardiac rhythm. Newborn rats were administered either nerve growth factor (NGF), its antibody (Ab), or placebo for the first 10 days of life. As compared to the placebo group, those treated with NGF had accelerated sympathetic nerve growth, an increase in the presence of a 41 kDa GTP regulatory protein that is a pertussis toxin substrate and that transduces α–1 adrenergic actions on cardiac automaticity, and a response to α–1 agonist effects on automaticity consistent with that in the mature rat. In contrast, the Ab treated animals had minimal innervation, a lower level of the GTP regulatory protein and an immature response to α–1 agonists. Moreover, the Ab-treated animals had an abnormally prolonged Q–T interval on ECG. α–1 and β–adrenergic receptor number and affinity were not significantly different among the 3 groups. These data indicate that the failure of normal sympathetic innervation to develop is associated with abnormalities in repolarization and in the response to α–1 adrenergic stimulation. Whether these abnormalities are associated with arrhythmias, as in the congenital long Q–T syndrome, remains to be tested.

INTRODUCTION

α–1 adrenergic stimulation is increasingly appreciated as a mechanism whereby cardiac pacemaker function and repolarization can be modulated [1–6]. We previously have demonstrated that α–1 stimulation tends to decrease automaticity in the canine [4–6] and the rat ventricle [7] and in the human atrium [8]. This effect on automaticity, however, is seen primarily in the mature heart. In this setting, innervation is complete, as are receptor–effector coupling mechanisms. We have shown, as well, that the prolongation of action potential duration induced by α agonists occurs regardless of the developmental state [9].

The action of α agonists to decrease automaticity is the result of stimulation of the sodium–potassium pump through a receptor–effector pathway that involves a 41 kDa pertussis toxin sensitive GTP regulatory protein (G_i) [10, 11]. The functional expression of this protein seems to depend on the sympathetic nervous system [12]; that is, the development of innervation is associated with an increased level of protein as measured by pertussis toxin catalyzed ADP–ribosylation. At present the precise identity of this G_i protein coupled to the inhibitory α–1 adrenergic response is uncertain since pertussis toxin ADP–ribosylates a family of homologous 41 kDa proteins.

In the functional absence of the relevant pertussis toxin sensitive G protein, the effect of α agonists is to increase automaticity [12, 13], possibly via stimulation of phosphoinositide hydrolysis [14, 15]. This signaling pathway could accelerate the rate of impulse initiation by increasing intracellular calcium which in turn would increase nonspecific monovalent cation conductances across the membrane.

To determine the extent to which sympathetic modulation might be implicated in the control of cardiac rhythm in the developing animal, we devised a model in the rat, in which we could accelerate nerve growth using nerve growth factor (NGF) or slow nerve growth using the antiserum to nerve growth factor (Ab) [16]. Using this model we have been able to demonstrate that by altering sympathetic nerve growth we can accelerate or retard the development of the pertussis toxin sensitive GTP regulatory protein and induce changes in the repolarization of the heart recorded electrocardiographically and in the effects of α–adrenergic stimulation on impulse initiation. The data reviewed herein have been reported previously [16, 17].

METHODS

Neonatal Wistar rats were divided randomly into three groups on the day of birth and injected subcutaneously for the first 10 days of life with nerve growth factor (1.0 μg/day until day 5; then 1.5 μg/day), placebo (saline, 10 μl/day until day 5; then 15 μl/day), or the antiserum to NGF (10 μl/day until day 5; then 15 μl/day) [15]. Electrocardiograms were recorded on days 1 and 10 of life and the animals were then narcotized with CO_2 and decapitated. The hearts were removed via a sternotomy, placed in cold Tyrode's solution bubbled with 95% O_2 and 5% CO_2, and containing (mM) NaCl (131), $NaHCO_3$ (18), NaH_2PO_4 (1.8), $MgCl_2$ (0.5), KCl (2.7), $CaCl_2$ (2.7, and dextrose (5.5). The pH was maintained at about 7.3. The hearts were dissected and a preparation of the ventricular septum, alone, was placed in a Lucite tissue bath perfused with Tyrode's solution at a rate of 12 ml/min, such that bath volume was changed 3 times a minute.

Standard techniques were used to record the spontaneous rates of these preparations via bipolar silver electrodes [18]. Records were made on a strip chart recorder. Only preparations whose spontaneous rates varied by less than ± 5% during a 20 minute control period were used for further study. These were superfused for consecutive 10 minute intervals with Tyrode's containing phenylephrine in the following concentrations, 10^{-9} and 10^{-8} M [16], or isoproterenol in 10^{-10} through 10^{-5} M concentrations [16]. These superfusion periods permitted the preparations to attain a steady state rate, which was recorded as above.

The septal preparations described and the right and left ventricular free walls were studied for their GTP regulatory proteins and α and β–adrenergic receptors. ADP–ribose is covalently incorporated into the stimulatory G protein, G_s, in the presence of cholera toxin, and into a family of the 41 kDa G_i proteins in the presence of pertussis toxin. To study the G proteins, we measured the incorporation of (^{32}P) ADP–ribose from (^{32}P) NAD into appropriate molecular weight proteins according to methods previously described by us [13, 17]. Tissue protein was determined using the method of Lowry, et al. [19].

α– and β–adrenergic receptors were characterized using radioiodinated BE2254 and cyanopindolol (CYP) respectively. Previously described methods were used to make membrane preparations representative of each experimental group [20, 21]. Specific binding, defined as that component of (^{125}I)IBE2254 binding inhibited by prazosin, 10^{-6} M and (^{125}I) ICYP binding inhibited by propranolol, 10^{-6} M, constituted at least 85% of total binding at ligand concentrations in the range of the equilibrium dissociation constant.

All data are expressed as Means ± SE. The statistical methods used were ANOVA and Bonferroni's test [22]. A p value <0.05 was considered significant. Data from the ligand binding experiments were analyzed by the method of Scatchard [23].

RESULTS

The electrocardiographic data recorded on days 1 and 10 are summarized in Table 1. Note that for all three groups on day 1 the heart rate, P–R interval and Q–T interval were identical. By day 10 the following changes had occurred: heart rate was significantly faster in the NGF group than in the placebo and antiserum groups; and the rate corrected Q–T interval was significantly longer in the antiserum group than in the other two groups.

The effects of phenylephrine on automatic rate of isolated septal preparations are reported in Fig. 1. Note that on day 0 all septal preparations showed increased automaticity. On day 10, approximately 50% of the placebo treated animals showed increased automaticity and 50%, a decrease. In

Table 1: ECG Characteristics

	0–1 DAY			10 DAY		
	Ab	Placebo	NGF	Ab	Placebo	NGF
Heart Rate (bpm)	299 ± 10	302 ± 4	292 ± 15	353 ± 11*↕	364 ± 11*↕	401 ± 10*╫↑
P–R (ms)	55 ± 1	52 ± 2	53 ± 2	44 ± 2*	46 ± 1*	44 ± 1*
QRS (ms)	17 ± 1	18 ± 1	18 ± 1	15 ± 1	15 ± 1	15 ± 2
QT (ms)	113 ± 4	110 ± 5	114 ± 5	91 ± 2*↑↕	84 ± 2*╫	78 ± 2*╫
QT_c (ms)	243 ± 6	242 ± 7	249 ± 7	222 ± 3*↑↕	200 ± 5*╫	199 ± 4╫

Legend: *, p<0.05 vs its control at 0–1 day; ╫, p<0.05 vs Ab; ↑, p<0.05 vs Pl; ↕, p<0.05 vs NGF. n = 25–26 at 0–1 day; 18–26 at 10 days (from [15] with permission).

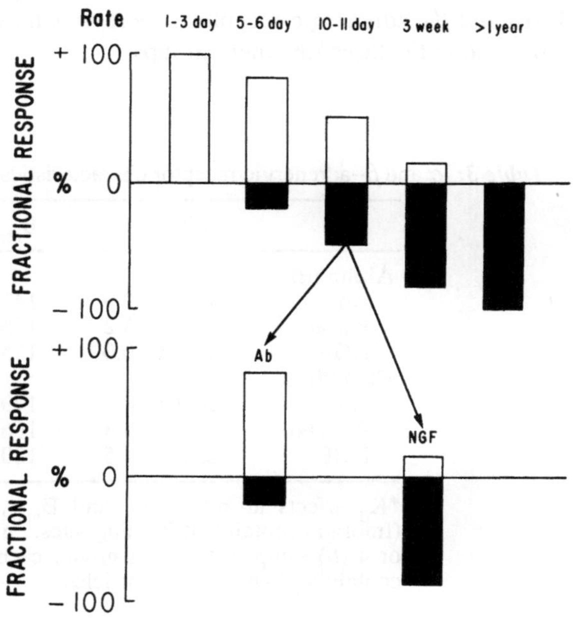

Figure 1: Relationship of fractional response to phenylephrine to age and treatment in rats. The normal developmental change in the chronotropic response to α–1 adrenergic stimulation (upper panel) is compared with the changes in this response induced by treatment of 10–11 day rats (lower panel). The chronotropic response is expressed as "fractional response," i.e. the percent of preparations at each age or in each treatment group showing either an increase or a decrease in rate. Treatment with NGF antiserum rendered the chronotropic response to phenylephrine of 10–11 day old septa more like that of septa from less mature hearts; conversely, septa from NGF treated rats responded like preparations from more mature animals. (reprinted from [15], by permission).

Table 2: Tissue levels of G_i and G_s in 10 day old rat septa.

	NGF	Placebo	Antibody
G_i (fmol/mg)	157 ± 27*	128 ± 24	112 ± 19*
n	5	5	5
G_s (fmol/mg)	191 ± 18	198 ± 40	160 ± 21
n	6	5	7

* p<0.05 cf placebo

contrast, 85% of NGF treated animals manifested decreased automaticity, a result not usually seen in the normal rat until 3 weeks of age. Furthermore, only 23% of antiserum treated rats showed decreased automaticity at 10 days, a result usually seen in the 5–6 day control rat. In contrast to our results with phenylephrine, all three treatment groups showed increased automaticity on superfusion with isoproterenol, and there was no difference in the magnitude of the response among them.

The G protein data for these studies are summarized in Table 2. Note that there was no significant difference in the stimulatory G_s protein. In contrast, the pertussis toxin sensitive G_i protein was significantly increased above control in the NGF group and lower than control in the antiserum group.

The α– and β–adrenergic receptor data are presented in Table 3. For both α– and β– adrenergic receptors, there was no difference in number and affinity among the three treatment groups.

Table 3: α and β–adrenergic receptor characteristics of neonatal rat hearts.

	Kd*	B_{max}*
Alpha (α)		
Ab	45.8 ± 2.9	142 ± 5
Placebo	46.6 ± 5.2	155 ± 14
NGF	45.1 ± 10.0	175 ± 33
Beta (β)		
Ab	26.3 ± 4.9	179 ± 21
Placebo	30.2 ± 6.3	166 ± 20
NGF	25.0 ± 4.5	120 ± 18

*K_d affects affinity (pM) and B_{max}, density (fmol/gm protein) of binding sites. n = 3 (α) or 4 (β) samples for each group, each sample containing 6–8 pooled ventricles.

DISCUSSION

The coupling of α–adrenergic receptors to their effectors has been a major focus of our research. We and our colleagues have previously noted that α agonists stimulate the Na/K exchange pump [10, 11]. The result is a decrease in intracellular sodium activity, as well as the generation of an outward current that suppresses pacemaker function in normally polarized fibers of the ventricular conducting system. This pathway is transduced by a GTP regulatory protein that is a substrate for pertussis toxin [10]. Of interest is the fact that the pertussis toxin sensitive G protein is demonstrable in limited quantities prior to sympathetic innervation, and is more prevalent after innervation is established [12]. This was demonstrated in experiments in which cardiac myocytes in monolayer tissue culture showed little substrate for pertussis toxin–dependent ADP–ribosylation and a positive chronotropic response to α–adrenergic stimulation. In contrast, myocyte cultures of the same age in which coculture techniques had been used to establish sympathetic innervation manifested far more substrate for pertussis toxin and a decrease in automaticity in response to phenylephrine.

It has also been demonstrated that α agonists can increase automaticity via a different signal transduction pathway. Recent studies establish that the neonatal cardiac α–1 adrenergic receptor stimulates the hydrolysis of membrane phosphoinositides [15]. This response is dependent on a GTP regulatory protein, but one that is not a substrate for pertussis toxin. This pathway provides a plausible mechanism for the action of α–1 agonists to increase automaticity. Thus, second messengers generated via the breakdown of phosphoinositides may increase the cytosolic calcium ion concentration which, in turn, may enhance the monovalent cation conductance of the membrane. Recent studies from our laboratory have suggested that activation of each of these pathways is the result of occupancy of a different α–1 adrenergic receptor subtype [24, 25]. That which is associated with Na/K pump stimulation and a decrease in automaticity is blocked by the alkylating agent, chloroethylclonidine (CEC), but not by the competitive α–1 antagonist, WB 4101. That which is associated with the hydrolysis of membrane phosphoinositides and an increase in automaticity is blocked by WB 4101 but not by CEC.

We also have learned that the action of α–1 agonists to prolong repolarization requires agonist concentrations two to three orders of magnitude greater than those needed to increase or decrease automaticity [9]. The effect on repolarization is not dependent on signal transduction by a pertussis toxin substrate, and is blocked by WB 4101 but not chloroethylclonidine. Whether this receptor represents a low affinity state of the same α–1 receptor that is responsible for the increase in automaticity or is a completely different receptor is as yet uncertain.

In evaluating the results in the present experiments, we needed assurance that we had, indeed, effected the anticipated changes in sympathetic innervation. This was tested in three different ways. First, in studies using tyrosine hydroxylase staining [16], our colleague, Dr. Ursell, demonstrated that 10 day old rats treated with NGF had, in fact, a marked increase in tyrosine hydroxylase positive neural processes as compared to 10 day old placebo–treated animals and untreated adults. Antibody–treated animals showed an abnormal pattern of staining and no clearly defined nerve processes. Second, tissue norepinephrine levels in NGF treated animals were significantly higher than placebo (32 ± 7.6 and 23 ± 3.6 ng/mg protein, respectively, $p<0.05$), which in turn were higher than in antibody treated animals (15 ± 2.5 ng/mg; $p<0.05$) [16]. This, too, suggests an increase in innervation in the NGF group and a decrease in the antibody group. Finally, superfusion of ventricular septa with tyramine induced a marked increase in automaticity in the NGF group and no change in the antibody group [15]. The placebo group was intermediate. Hence, three different types of experiments all supported the conclusion that NGF increased and antibody delayed sympathetic innervation. Moreover, it appears that in the intact animal, as in our previous isolated tissue and cell culture studies sympathetic innervation is associated with the developmental expression of a pertussis toxin sensitive G protein and with a change in the response to α–adrenergic stimulation from excitation to inhibition of automaticity. It is of interest that all these changes are at the post–receptor level: no differences in α receptor number or affinity were demonstrable across the three groups [16].

Our β–adrenergic receptor–effector studies offer a contrast to the above. The response to isoproterenol across the groups was equivalent, as were the levels of the stimulatory G protein, G_s, and β receptor number and affinity [17]. Hence, changes in innervation of the magnitude effected by NGF and by antibody induced no change in β–adrenergic receptor–effector coupling.

Considering the α– and β–adrenergic components of the autonomic response together, in the mature animal or in younger animals where sympathetic innervation is accelerated, β–adrenergic stimulation can be viewed as enhancing and α–adrenergic stimulation as suppressing automaticity. In contrast, in the neonatal animal or under pathological conditions where sympathetic innervation is delayed, both adrenergic pathways excite automaticity. Hence, the spectrum of rhythmic activity in response to catecholamines can differ importantly depending on the developmental age and factors that delay or accelerate innervation.

The clinical implications of these studies are as follows: imbalances of sympathetic innervation have been proposed as a basis for the congenital long Q–T syndrome [26] and for some instances of sudden infant death [27]. The identification of a long Q–T interval in rats treated with NGF antiserum is of interest in light of the above proposal and may implicate a mechanistic link between abnormal innervation and the clinical syndromes. In addition, the heightened propensity to catecholamine–induced increases in automaticity in animals with delayed innervation may provide a trigger mechanism for arrhythmias in the face of delayed repolarization. Needless to say, these latter comments are highly speculative. Nonetheless, they provide a mechanism for the further consideration of the role of sympathetic modulation in the suppression and the development of lethal arrhythmias.

Acknowledgements

These studies were supported by USPHS–NHLBI grant HL–28958 and a Glaxo Cardiovascular Discovery Award

DISCUSSION

Dr. Joyner: Why is the G protein distributed in a spatially inhomogeneous way?

Dr. Rosen: Sympathetic nerves are distributed in an inhomogeneous fashion and it is the sympathetic nerves that determine the development of G proteins.

Dr. Corr: Is the impulse initiation increased by the inhibition of the IP_1, or IP_2 phosphatase with lithium?

Dr. Rosen: We have turned on PI metabolism using phospholipase C, which also turns on automaticity. A subthreshold concentration of Phospholipase C, together with lithium, 10 mm, which in its own right has no effect, increase the IP3 present and turn on automaticity.

Dr. Corr: Is it truly an IP_3 as opposed to a diacylglycerol response?

Dr. Rosen: Yes.

REFERENCES

1. Giotti A, Ledda F, Mannaioni PF: Effects of noradrenaline and isoprenaline, in combination with α– and β–receptor blocking substances, on the action potential of cardiac Purkinje fibers. *J Physiol (Lond)* 1973:299:99–113.
2. Ledda F, Marchetti P, Manni A: Influence of phenylephrine on transmembrane potentials and effective refractory period of single Purkinje fibers of sheep heart. *Pharmacol Res Commun* 1971;3:195–206.
3. Rosen MR, Hordof AJ, Ilvento J, Danilo P Jr: Effects of adrenergic amines on electrophysiologic properties and automaticity of neonatal and adult canine cardiac Purkinje fibers. *Circ Res* 1977;40:390–400.
4. Posner P, Farrar E, Lambert C: Inhibitory effects of catecholamines in canine cardiac Purkinje fibers. *Am J Physiol* 1976;231:1415–1420.

5. Reder R, Danilo P Jr, Rosen MR: Developmental changes in α–adrenergic effects on canine Purkinje fiber automaticity. *Devel Pharmacol Ther* 1984;7:94–108.
6. Rosen MR, Weiss RM, Danilo P Jr: Effects of α–adrenergic agonists and blockers on Purkinje fiber transmembrane potentials and automaticity in the dog. *J Pharmacol Exp Ther* 1984;231:566–571.
7. Drugge ED, Rosen MR and Robinson RB: Neuronal regulation of the development of the α–adrenergic chronotropic response in the rat heart. *Circ Res* 1985;57:415–423.
8. Mary–Rabine L, Hordof A, Bowman FO, Malm Jr, Rosen MR: Alpha and beta-adrenergic effects on human atrial specialized conducting fibers. *Circulation* 1978;57: 84–90.
9. Lee JH, Steinberg SF, Del Balzo U, Rosen MR: Modulation of Purkinje fiber repolarization by an α–1 adrenoceptor subtype. *Circulation* 1989;80:II-1206.
10. Shah A, Cohen IS, Rosen MR: Stimulation of cardiac alpha receptors increases Na/K pump current and decreases g_K via a pertussis toxin–sensitive pathway. *Biophys J* 1988;54:219–225.
11. Zaza A, Kline R, Rosen MR: Effects of alpha–adrenergic stimulation on intracellular sodium activity and automaticity in canine Purkinje fibers. *Circ Res* 1990;66:416–426.
12. Steinberg SF, Drugge ED, Bilezikian JP, Robinson RB: Acquisition by innervated cardiac myocytes of a pertussis toxin–specific regulatory protein linked to the α–1 receptor. *Science* 1985;230:186–188.
13. Rosen MR, Steinberg SF, Chow YK, Bilezikian JP, Danilo P, Jr: The role of a pertussis toxin sensitive protein in the modulation of canine Purkinje fiber automaticity. *Circ Res* 1988;62:315–323.
14. Han HM, Robinson RB, Bilezikian JP, Steinberg SF: Developmental changes in guanine nucleotide regulatory proteins in the rat myocardial alpha_1 adrenergic receptor complex. *Circ Res* 1989;65:1763–1773.
15. Steinberg SF, Kaplan LM, Inouye T, Zhang JF, Robinson RB: Alpha_1–adrenergic stimulation of 1,4,5–inositol trisphosphate formation in ventricular myocytes. *J Pharm Expt Therapeutics* 1989;250:1141–1148.
16. Malfatto G, Rosen T, Sun L, Steinberg S, Danilo P Jr, Rosen MR: Sympathetic innervation in neonatal rats induces a GTP regulatory protein that modulates α–adrenergic effects on ventricular automaticity. *Circ Res* 1990;66:427–437.
17. Steinberg SF, Rosen TS, Malfatto G, Rosen MR: Beta adrenergic modulation of cardiac rhythm in a rat model of altered sympathetic neural development. *J Mol Cell Cardiol*, in press, 1990.
18. Sun LS, Roberts LA, Rosen MR, Robinson RB: The positive chronotropic effect of acetylcholine has muscarinic and nicotinic components in the neonatal rat heart. *J Pharmacol Exp Ther* 1988;247:585–589.
19. Lowry OH, Rosebrough NJ, Farr Al, Randall RJ: Protein measurement with folin phenol reagent. *J Biol Chem* 1951;193:265–275.
20. Steinberg SF, Bilezikian JP: Identification and characterization of alpha_1–adrenergic receptors in rat myocardium with a new iodinated radioligand $[^{125}I]$IBE2254. *J Mol Cell Cardiol* 1982;14:601–610.
21. Engel G, Hoyer D, Berthold R, Wagner H: (±)125 iodocyanopindolol, a new ligand for beta–adrenoceptors: identification and quantification of subclasses of beta–adrenoceptors in guinea pig. *Naunyn Schmiedebergs Arch Pharmacol* 1981;317:277–285.
22. Zar, JH: *Biostatistical Analysis.* Englewood Cliffs, NJ: Prentice Hall Publishing Company, 1983.
23. Scatchard G: The attractions of proteins for small molecules and ions. Ann NY *Acad Sci* 1949;51:660–672.

24. Del Balzo U, Rosen MR: Physiologic effects of two α–1 adrenoceptor subtypes in adult canine Purkinje fibers. *Circulation* 1989;80:II–68.
25. Steinberg SF, Kaplan L, Inouye T: Chloroethylclonidine reveals distinct α–1 adrenergic receptor subtypes in the heart. *Circulation* 1988;78:II–560.
26. Schwartz PJ, Teriti M, Malliani A: The long QT syndrome. *Am Heart J* 1975; 89:378–390.
27. Schwartz PJ: The quest for the mechanism of the sudden infant death syndrome: doubts and progress. *Circulation* 1987;75:677–683.

CHAPTER 18:

RELATING MEMBRANE RECEPTORS TO DRUGS

Arnold M. Katz

ABSTRACT

Comparison of the interactions of propranolol, nimodipine, and amiodarone with membrane lipids, and the duration of their pharmacological effects suggests that the pharmacokinetics of a drug may be partly determined by its location within the phospholipid bilayer.

INTRODUCTION

It has been known for almost a century that anesthetic potency of a variety of organic chemicals varies in proportion to their partitioning between fats and water; the more lipid soluble the substance, the greater its narcotizing effect [1]. These and other findings indicate that absorption of a drug within the cell membrane plays an important role in its clinical effects [2, 3], and that interactions of drugs within the membrane bilayer may influence their pharmacological activity.

The clinical effects of most cardiac drugs are more specific than those of the anesthetics; however, like the latter, most of these drugs are lipid–soluble molecules that readily partition into the membrane bilayer. While specificity of action implies that most cardiac drugs react with specific classes of membrane protein, several lines of evidence indicate that some drug interactions with their protein receptors may require that the drug first dissolve in the membrane bilayer.

The approach of a drug molecule to a membrane receptor by diffusion within the phospholipid bilayer has been calculated to be several orders of magnitude faster than diffusion to the receptor through the aqueous environment outside the membrane [4]. This greater rapidity of approach of a drug to its membrane receptor by way of the bilayer pathway would be especially advantageous in membranes where the receptor site density is low

[5]. More direct evidence that some cardiac drugs approach their receptors through the bilayer has been obtained in studies of the effects of dihydropyridines on single calcium channels. Rapid interactions of these highly lipophilic drugs with calcium channels have been seen in studies where access of the drug to the channel via the aqueous medium would not have been possible because the microelectrode used to record channel opening was sealed tightly to the membrane [6, 7].

A substantial body of evidence is now developing which suggests that interactions of lipophilic drugs with cardiac sarcolemmal membrane lipids can play an important role in determining both the pharmacokinetic and pharmacodynamic properties of the ligands. This chapter compares the interactions of members of 3 classes of cardiac drugs: a beta-adrenergic receptor blocker (propranolol), a calcium channel blocker (nimodipine) and an antiarrhythmic drug (amiodarone) with membrane lipids. These data indicate that the durations of the clinical actions of these drugs are influenced by hydrophobic interactions between the drug and the membrane lipids. They also support the view that the interactions of these lipophilic drugs with their membrane receptors may occur after the drugs first dissolve in the membrane bilayer.

INTERACTIONS OF DRUGS WITH MEMBRANE LIPIDS

Washout From Membrane Lipids

The rates at which nimodipine, propranolol and amiodarone could be washed out of sarcoplasmic reticulum membranes, which lack specific receptor sites for any of these drugs, differed considerably. Amiodarone could not be removed from these membranes even after 20 washes, whereas both propranolol and nimodipine were rapidly eliminated, with most of the drug being eliminated after 2–3 washes [8]. The rate at which propranolol washed out of these membranes was about 3 times faster than that of nimodipine [8].

Partition Coefficients

While propranolol, nimodipine and amiodarone all partitioned into the membrane lipids, their partition coefficients differed markedly. The partition coefficient of amiodarone was more than 100 times greater than that of nimodipine and propranolol; nimodipine's partition coefficient was approximately 4 times that of propranolol [8]. Overall, therefore, these partition coefficients correlated inversely to the rates of washout of these drugs; amiodarone, which had the slowest washout having the highest partition coefficient.

Drug Localization in Biological Membranes

X–ray and neutron diffraction studies have shown that propranolol and nimodipine, when incorporated into membranes, are located approximately 6 Å from the membrane surface, within the phosphate headgroup region of the bilayer near the hydrocarbon core/water interface [9, 10]. This location is similar to that proposed for tricyclic antidepressants and phenothiazines [11].

Amlodipine, a positively charged dihydropyridine drug that has a longer washout time than nimodipine, was recently found to have a membrane location similar to that of the other dihydropyridines [12]. Amlodipine also has a higher partition coefficient that nimodipine, which may be explained in part by the unique shape of this molecule that allows its charged amino group, as well as the hydrophobic dihydropyridine rings, to interact with the membrane bilayer [12].

Amiodarone, which has by far the longest clinical half life of these drugs, partitions more deeply in the membrane bilayer. Thus, the iodine moieties of amiodarone are located approximately 17 Å from the phosphate headgroup region in membrane bilayers made of dimyristoyl–phosphatidyl choline [13]. This location of amiodarone, near the terminal methyl groups of the fatty acyl chain region, is consistent evidence obtained in an earlier study of diphenylhexatriene depolarization in synaptic membranes [14].

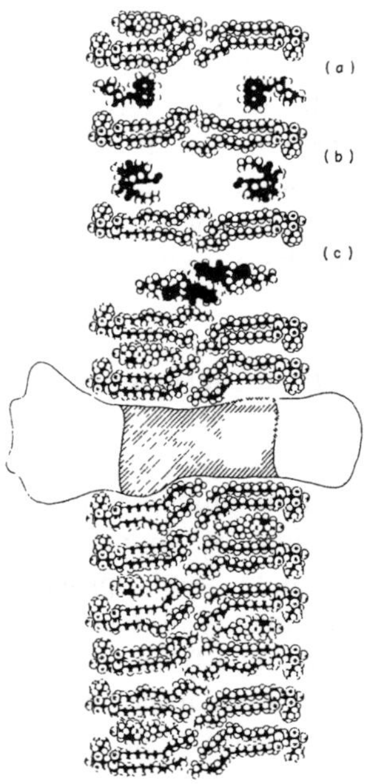

Figure 1: Tentative locations of propranolol, nimodipine and amiodarone in the plane of the membrane bilayer as determined by x–ray and neutron diffraction. From Herbette et al. [8] with permission.

CONCLUSIONS

These data are consistent with the model shown in Fig. 1, which tentatively localizes propranolol, nimodipine and amiodarone in the membrane bilayer. These membrane locations appear likely to contribute to the different pharmacokinetics of these drugs. Amiodarone, which has the longer clinical half–life, highest partition coefficient and greatest resistance to washout, is located most deeply within the membrane bilayer, nearest to its hydrophobic core where this drug would be relatively inaccessible to the surrounding extracellular space. Propranolol and nimodipine, which have substantially shorter clinical half–lives, are located closer to the membrane surface, at the hydrocarbon core/water interface, where they would have more ready access to the surrounding aqueous medium and so would be expected to have shorter clinical half lives.

These findings are consistent with the view that the physical characteristics of amphipathic cardiac drugs, which contain both polar and nonpolar regions, influence their location within the membrane bilayer. As these properties would be expected also to influence the approach and interaction of a drug with its membrane receptor, the physical location and orientation of the drug in the membrane might also influence the interaction of the drug molecule with its binding site on a receptor protein. Thus, drug interactions with the membrane lipid bilayer may play a role in the pharmacokinetics and pharmacodynamics of clinical drug actions.

Acknowledgements

I thank Drs. L.G. Herbette, D. Rhodes and R.P. Mason for their many contributions to the concepts discussed in this Chapter, and Ms. C. Cronin for her expert help in composing the manuscript. Preparation of this Chapter was supported in part by NHLBI Program Project HL–33026.

REFERENCES

1. Straub W: Die Bedeutung der Zellmembran fuer die Wirkung chemischer Stoffe auf den Organismus. *Verh Ges Deutsch Naturf Arzte.* 1912;1–25.
2. Seeman P: The membrane actions of anesthetics and tranquilizers. *Pharm Rev* 1972;24:583–655.
3. Hille B: Local anesthetics: Hydrophilic and hydrophobic pathways for the drug–receptor reaction. *J Gen Physiol* 1977;69:497–515.
4. Rhodes DG, Sarmiento JG, Herbette LG: Kinetics of binding of membrane–active drugs to receptor sites. *Mol Pharm* 1985;27:612–623.
5. McClosky M, Poo MM: Rates of membrane associated reaction: reduction of dimensionality revisited. *J Cell Biol* 1986;102:88–96.

6. Brown AM, Kunze DL, Yatani A: The agonist effect of dihydropyridines on Ca channels. *Nature* 1984;311:570–572.
7. Kokubun S, Reuter H: Dihydropyridine derivatives prolong the open state of Ca channels in cultured cardiac cells. *Proc Nat Acad Sci USA* 1984;81:4824–4827.
8. Herbette LG, Trumbore M, Chester DW, Katz AM: Possible molecular basis for the pharmacodynamics of three membrane active drugs: propranolol, nimodipine, amiodarone. *J Mol Cell Cardiol* 1988;20:373–378.
9. Herbette LG, Katz AM, Sturtevant JM: Comparisons of the interaction of propranolol and timolol with model and biological membrane systems. *Mol Pharm.* 1983;24:259–269.
10. Chester DW, Herbette LG, Mason RP, Joslyn AF, Triggle DJ, Koppel DE: Diffusion of dihydropyridine calcium channel antagonists in cardiac sarcolemmal lipid multibilayers. *Biophys J* 1987;52:1021–1030.
11. Frenzel J, Arnold K, Nuhn P: Calorimetric, 13C NMR, and 31P NMR studies on the interaction of some phenothiazine derivatives with dipalmitoyl phosphatidylcholine model membranes. *Biochim Biophys Acta* 1978;507:185–197.
12. Mason RP, Campbell SF, Wang S–D, Herbette, LG: Comparison of location and binding for the positively charged 1,4–dihydro–pyridine calcium channel antagonist amlodipine with uncharged drugs of this class in cardiac mambranes. *Mol Pharm* 1989;36:634–640.
13. Trumbore MD, Chester DW, Moring J, Rhodes D, Herbette LG: Structure and location of amiodarone in a membrane bilayer as determined by molecular mechanisms and quantitative X–ray diffraction. *Biophys J* 1988:54:535–543.
14. Chatelain P, Laruel R, Gillard M: Effect of amiodarone on membrane fluidity and Na+/K+ ATPase activity in rat–brain synaptic membranes. *Biochem Biophys Res Comm* 1985;129:148–154.

CHAPTER 19:

EXCITATION–CONTRACTION COUPLING IN STUNNED MYOCARDIUM

Hideo Kusuoka, William C. Rose and Eduardo Marban

ABSTRACT

Calcium transients were measured before and after ischemia in perfused hearts using gated ^{19}F NMR and the NMR detectable Ca^{2+} indicator 5F–BAPTA. In stunned myocardium, peak $[Ca^{2+}]_i$ was paradoxically higher than that in control. These results indicate that contractile failure in stunned myocardium is due to a decrease in the myofilament sensitivity to Ca^{2+} as well as to the decrease in maximal Ca^{2+}–activated force; failure of activator Ca^{2+} delivery cannot be implicated.

INTRODUCTION

Stunned myocardium is defined as contractile dysfunction in heart muscle reperfused after a period of ischemia brief enough to avoid necrosis [1, 2]. Although the phenomenon has important implications for clinical reperfusion therapy, the cellular mechanism of the decrease in force after ischemia is not yet clear. This presentation summarizes our work on stunned myocardium; the reader is referred to the primary citations [3–9] for a complete treatment.

The nature of the impairment in force generation in stunned myocardium can best be characterized by breaking down the final common pathway of the excitation–contraction process into three elementary parameters. The first of these is the pulse of intracellular free Ca^{2+} concentration, $[Ca^{2+}]_i$, that occurs during each cardiac cycle (the "Ca^{2+} transient"). The Ca^{2+} transient triggers the sequence of biochemical events that lead to force generation by the contractile proteins. The two remaining parameters characterize the responsiveness of the contractile machinery to Ca^{2+}: myofilament Ca^{2+} sensitivity describes the range of $[Ca^{2+}]_i$ which activates contraction, while maximal Ca^{2+}–activated force represents the amplitude of the contractile response.

Most theories regarding the pathogenesis of stunned myocardium imply that force falls as a consequence of a decrease in Ca^{2+} transients [10–13]. Nevertheless, each of these specific theories is controversial, and the issue of whether or not Ca^{2+} transients are reduced has not been addressed directly. Indeed, recent work from our laboratory hints otherwise. Maximal Ca^{2+}–activated pressure is decreased in postischemic ferret hearts, indicating that at least one parameter of the myofilament responsiveness to Ca^{2+} is decreased in stunned myocardium [3]. We also found an abnormally low sensitivity of contractile pressure to extracellular Ca^{2+} ("Ca_o–sensitivity"), which could reflect either a decrease in Ca^{2+} transients or a shift in myofilament Ca^{2+} sensitivity to higher $[Ca^{2+}]_i$. Direct quantitation of Ca^{2+} transients is required to distinguish between these two possibilities.

It has previously been impossible to determine Ca^{2+} transients in stunned myocardium, because the methods available to measure $[Ca^{2+}]_i$ were not suitable for perfused hearts. Recently, a new approach has been developed that enables the determination of $[Ca^{2+}]_i$ at various times during the cardiac cycle in isolated perfused hearts using gated NMR spectroscopy [14] and the NMR–detectable Ca^{2+} indicator 5F–BAPTA [15]. We have used this approach to determine whether Ca^{2+} transients are decreased in stunned myocardium induced by reperfusion after 15 min of global ischemia.

METHODS

Hearts of male ferrets were isolated and perfused at a constant coronary flow rate at 30°C. Isovolumic pressure was measured with a latex balloon in the left ventricle. As previously described [14, 16], 5F–BAPTA was loaded into cells by perfusion with its tetraacetoxymethyl ester derivative (5F–BAPTA–AM).

The methods to measure Ca^{2+} transients in whole hearts using gated ^{19}F NMR are described in detail elsewhere [14]. Intracellular free calcium concentration, $[Ca^{2+}]_i$, is calculated according to the equation $[Ca^{2+}]_i = K_d*([B]/[F])$, where [B] is the area under the Ca–5F–BAPTA peak, and [F] is the area under the Ca–free 5F–BAPTA peak [15, 16]. We have used a dissociation constant (K_d) of 285 nM for Ca–5F–BAPTA to calibrate our signals [8].

After loading with 5F–BAPTA, $[Ca]_o$ was increased from 2 mM to 8 mM to antagonize the Ca buffering effect of 5F–BAPTA. The temperature was initially set at 30°C to minimize the slow extrusion of 5F–BAPTA that can occur at higher temperatures [16]. After developed pressure reached steady state, gated ^{19}F spectra were collected at sufficient temporal resolution to define clearly end–diastolic and peak systolic $[Ca^{2+}]_i$. The temperature was then increased to 37°C and the heart subjected to 15 minutes of total global ischemia. This temperature was maintained during ischemia to mimic as closely as possible the conditions during myocardial stunning *in situ* [8, 9]. After 5 min of reperfusion at 37°C, the temperature was returned to 30°C. To define the Ca^{2+} transient in stunned myocardium, another set of gated ^{19}F spectra was obtained in the reperfused heart when developed pressure had recovered to a new steady state.

RESULTS

Calcium Transients in Stunned Myocardium

Figure 1 shows $[Ca^{2+}]_i$ before and after ischemia, plotted as a function of the time from the pacing stimulus. Before ischemia *(left)*, $[Ca^{2+}]_i$ in this particular heart increased from 0.08 μM before the stimulus to a peak of 0.65 μM in early systole. The plot of the Ca^{2+} transient during reflow *(right)* indicates the increase in $[Ca^{2+}]_i$, particularly during systole. The overall amplitude of the Ca^{2+} transient (i.e., systolic $[Ca^{2+}]_i$ minus diastolic $[Ca^{2+}]_i$) was paradoxically larger in the stunned heart than in control, although developed pressure fell in a typical manner in this experiment (from 45 mm Hg before ischemia to 31 mm Hg afterwards).

The paradoxical increases in systolic $[Ca^{2+}]_i$ and in the amplitude of the Ca^{2+} transient after stunning were observed quite consistently. Figure 2 summarizes the data from 7 hearts: the slight increase in diastolic $[Ca^{2+}]_i$ was not significant (P>0.10), but $[Ca^{2+}]_i$ at peak systole did increase (P<0.02). Concomitantly, the amplitude of Ca^{2+} transients in stunned myocardium was significantly higher than that before ischemia. We also ascertained whether the time course of the Ca^{2+} transients was altered. Neither the average time-to-peak-$[Ca^{2+}]_i$ nor the time constant of the declining phase of the Ca^{2+}

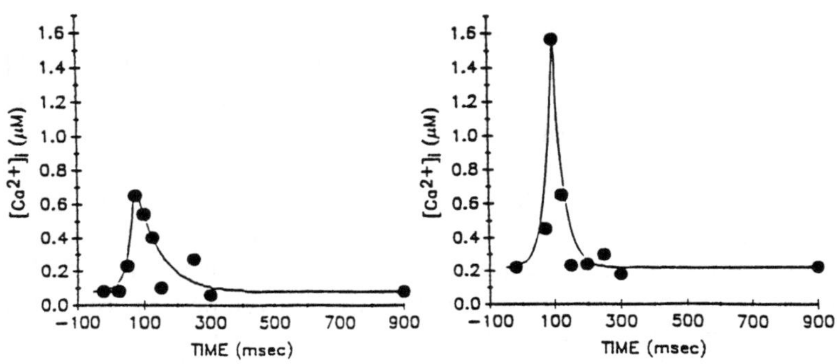

Figure 1: Changes in $[Ca^{2+}]_i$ during a cardiac cycle observed before (left) and after (right) ischemia.

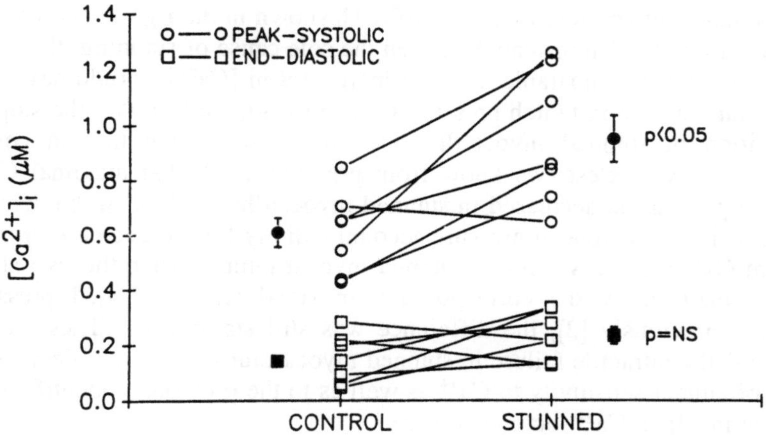

Figure 2: Paired data showing $[Ca^{2+}]_i$ in each heart at end–diastole (□) and at peak systole (o), before (left, CONTROL) and after (right, STUNNED) ischemia. The data obtained in each individual heart in the two phases of each experiment are connected; means ± SE are also shown.

transient was changed, thus simplifying the interpretation of the results [17]. These findings demonstrate that contractile failure during reflow is not due to a critical failure of any of the mechanisms that regulate cytoplasmic activator Ca^{2+}.

Implications of the Disparity Between Ca^{2+} Transient Amplitude and Developed Pressure

The increase in the amplitude of Ca^{2+} transients is in the direction opposite to that which would help explain the decrease of developed pressure in stunned myocardium. Despite a larger chemical trigger, the myofilaments generate less force after reflow. This dissociation between Ca^{2+} and contractile force indicates there must be a decrease in the responsiveness of the myofilaments to activator Ca^{2+}. The new information regarding the amplitude of Ca^{2+} transients now allows us to dissect out the individual components of this decrease in myofilament Ca^{2+} responsiveness. As summarized in the Introduction, this could either represent a shift in the range of activation (myofilament Ca^{2+} sensitivity) or a decrease in the amplitude of the maximal response (maximal Ca^{2+}–activated pressure), or both.

Table 1 summarizes the values for developed pressure (DP) and Ca^{2+} transient amplitude ($\Delta[Ca^{2+}]_i$) in control and after stunning. Assuming there is no active force generation in diastole, we calculated the slopes for the ap–

parent relationship between DP and $\Delta[Ca^{2+}]_i$ shown in the right–hand column. If a decrease in Ca^{2+} transients had been the sole cause of stunning, the slopes should have been comparable: a given increment in $[Ca^{2+}]_i$ should have produced a similar rise in twitch pressure [17]. Instead, we find that the slope of the relation in stunned myocardium is much lower than that in control ($P<0.05$). Nevertheless, we know from previous work that maximal Ca^{2+}–activated pressure is decreased in stunned myocardium [3]. Can the decrease in maximal Ca^{2+}–activated pressure account entirely for the decrease in myofilament Ca^{2+} responsiveness in stunned myocardium? When the two slopes were normalized by the corresponding maximal Ca^{2+}–activated pressures measured previously [3], the difference was still significant. These results indicate that contractile failure in stunned myocardium is due to a decrease in the myofilament sensitivity to Ca^{2+} as well as to the previously identified decrease in maximal Ca^{2+}–activated force.

Table 1: Relationship between developed pressure (DP) and Ca^{2+} transient amplitude ($\Delta[Ca^{2+}]_i$) before and after ischemia.

	DP (mmHg)	$\Delta[Ca^{2+}]_i$ (μM)	Slope (mmHg/μM)
Control	21.8 ± 5.0	0.43 ± 0.06	57 ± 17
Stunned	15.5 ± 4.2	0.78 ± 0.12	20 ± 7

Values are mean ± SEM.

DISCUSSION

Most hypotheses regarding the pathogenesis of contractile dysfunction focus on a specific initial mechanism while ignoring the convergent downstream links in contractile activation. Having dissected out three fundamental endpoints of excitation–contraction coupling, we now cast doubt on the entire category of hypotheses that imply stunning is due to a decrease in cytoplasmic activator Ca^{2+}. Instead, our findings suggest that the focus in stunned myocardium should be shifted to determining the proximate cause of the abnormally low Ca^{2+} responsiveness of the contractile proteins. Both components of myofilament Ca^{2+} responsiveness, maximal Ca^{2+}–activated force [3] and myofilament Ca^{2+} sensitivity, are abnormally low in the stunned heart. The combined effect of these two factors is severe enough to more than neutralize the observed increase in Ca^{2+} transients.

The present finding that systolic $[Ca^{2+}]_i$ is increased in stunned myocardium has escaped detection in previous estimates of time–averaged $[Ca^{2+}]_i$ in reperfused hearts [7, 8, 17]. This insensitivity of the time–averaged measurements is not surprising; the time during which $[Ca^{2+}]_i$ is much higher than the diastolic level comprises only a fraction of the duration of an entire cardiac cycle. Thus, a difference could easily be missed in the estimates of time–averaged $[Ca^{2+}]_i$. The insensitivity of the time–averaged measurements is compounded by the properties of the indicator: 5F–BAPTA can only give a lower–limit estimate for $[Ca^{2+}]_i$ when there is underlying spatiotemporal Ca^{2+} inhomogeneity [6]. Thus, the observed changes in Ca^{2+} transients in this study are not discrepant with previous observations.

The precise mechanism of the decrease in myofilament Ca^{2+} responsiveness remains to be determined. A number of lines of evidence [3–5, 9, 18] suggest that the decrease in myofilament Ca^{2+} responsiveness is a lingering after–effect of the transient increase in $[Ca^{2+}]_i$ during ischemia and early reflow. Calcium overload can activate protein kinases in myocardium [3, 19, 20] and possibly induce changes in Ca^{2+}–sensitivity and/or maximal Ca^{2+}–activated force through phosphorylation of one or more of the contractile proteins. Although this is a plausible mechanism for stunning, no evidence for or against such changes in the myofilaments has yet been reported in response to Ca^{2+}–induced phosphorylation. Covalent modification of contractile proteins by Ca^{2+}–activated proteases also merits future consideration. Such modification would help explain why stunning reverses over several days, a time course consistent with the expected rate of *de novo* protein synthesis [21].

It is worth considering whether the observed changes in Ca^{2+} transients are likely to be adaptive or maladaptive. Conventional wisdom might dictate that stunning represents a state of decreased energy demand which "rests" the myocardium after a bout of ischemia. While the observed increase in Ca^{2+} transients tends to preserve function by offsetting partially the decreased Ca^{2+} responsiveness of the myofilaments, the cycling of Ca^{2+} by the cell exacts its own cost. Excitation–contraction coupling consumes a large fraction of the energy used by the heart [22]; this fraction is even greater during the maneuvers that are known to increase Ca^{2+} transients [22, 23]. Thus, the increase in Ca^{2+} transients is adaptive in that it minimizes the degree of contractile impairment, but is possibly maladaptive in that additional energy must be spent in the cycling of intracellular Ca^{2+}.

The observation that Ca^{2+} transients are increased certainly helps explain the reported inefficiency of energy utilization in stunned myocardium [24]. Oxygen consumption in stunned myocardium is much higher than in control when normalized for the degree of force generation [25]. Preliminary calculations reveal that the free energy change available from ATP hydrolysis is actually increased in stunned myocardium despite the decrease in force generation [6]. These results suggest either that energy transfer to contraction or energy consumption in non–contractile work is inappropriately high in stunned myocardium. The energy demands imposed by the cycling of in–

creased intracellular Ca^{2+} imply that at least the latter mechanism contributes to the decreased efficiency of energy utilization in stunned myocardium.

Acknowledgements

This work was supported by the International Scientific Research Program of the Ministry of Education, Science and Culture of Japan (University–to–University Cooperative Research Program 62045020) and by the National Institutes of Health (SCOR grant HL 17655). Dr. Marban was the recipient of a Research Career Development Award of the National Institutes of Health (KO4 HL 01872).

DISCUSSION

Dr. Strauss: Can you say something as to whether the submaximal or maximal responses to calcium are disturbed in this ischemic period? Have you done a history time course of calcium to contractile force to determine where the lesion is with the tractile protein, i.e. troponin?

Dr. Marban: No is the answer to the latter question. I suppose if you could measure the disassociation of calcium from troponin or the association of calcium to troponin and you saw a change, it might imply that the calcium affinity to troponin had changed. But we do not have the tools to do that kind of experiment. Another possibility would be that calcium binding to troponin is perfectly normal and it is something about actin or myosin that is different. We do not know yet. Those are interesting points to be resolved by other techniques.

Dr. Kléber: You showed that during ischemia there was no change of the calcium transient as measured with aequorin luminescence. Morgan *[26]* has shown an immediate increase of calcium in ischemia, and Allen *[27]* described delayed increase of the systolic calcium. Clusin *[28]* observed a very early increase during the first 60 seconds. Could you comment on these discrepancies?

Dr. Marban: Tom Guarnieri's data *[29]* with aequorin does in fact show an increase at 10–20 min of ischemia. He finds that, on the average, Ca^{2+} transient amplitude roughly doubles and diastolic calcium goes up only very modestly. This is much more like the results of Allen et al. *[27]* in isolated microinjected papillary muscles, than either the Clusin *[28]* or the Morgan *[26]* results. The interpretation of Clusin's results has been put into some doubt because of the realization that endothelial cells are heavily loaded with indo1AM, and a lot of the signal that is seen in the first 30 seconds has been attributed to the endothelial cell reaction to ischemia. Clusin now admits that this is a problem, and he is trying to work out ways of getting around it. I do not know how to explain the discrepancy with Morgan's results, except to say that the Guarnieri data is from directly microinjected aequorin. Before any aequorin was injected, the cell was seen to be polarized and then it was directly injected into the endocardium. The Morgan loading is by a clysis method in EGTA–permeabilized hearts, and the metabolic condition of such hearts has been questioned by Lee and Allen *[30]*. Direct microinjection is the gold standard for aequorin and, if you fantasize what the results might look like with time–averaged calcium, it matches pretty well with our NMR data that show that for the first 10–15 minutes of ischemia there are only modest changes. After about 10 minutes you start to get statistically significant increases in time–averaged calcium.

Dr. Gettes: That was stunned myocardium. There are ways that you can bring back the stunned myocardium to normal contraction force, for example, β–adrenergic stimulation. Does that raise intracellular calcium even more?

Dr. Marban: Yes, presumably. No matter how much catecholamine you gave, you could never get the maximum response that you could have in control. But since the heart usually operates in the dynamic range of being less than 50% activated, you have got a lot of room for motion even if you have an overall downward scaling of some 20%. I would predict that although the calcium is increased in the stunned myocardium, there is still room to increase it even more and so you can still get more "bang for your buck" by recruiting more calcium delivery. That may be why you can recover function in stunned myocardium to some extent by catecholamines.

REFERENCES

1. Braunwald E, Kloner RA: The stunned myocardium: prolonged, postischemic ventricular dysfunction. *Circulation* 1982;60:1146–1149.
2. Braunwald E, Kloner RA: Myocardial reperfusion: a double–edged sword? *J Clin Invest* 1985;76:1713–1719.
3. Kusuoka H, Porterfield JK, Weisman HF, Weisfeldt ML, Marban E: Pathophysiology and pathogenesis of stunned myocardium: depressed Ca^{2+} activation of contraction as a consequence of reperfusion–induced cellular calcium overload in ferret hearts. *J Clin Invest* 1987;79:950–961.
4. Kitakaze M, Weisfeldt ML, Marban E: Acidosis during early reperfusion prevents myocardial stunning in perfused hearts. *J Clin Invest* 1988;82:920–927
5. Kitakaze M, Weisman HF, Marban E: Contractile dysfunction and ATP depletion after transient calcium overload in perfused ferret hearts. *Circulation* 1988;77:685–695.
6. Kusuoka H, Inoue M, Marban E: Decreased efficiency of energy utilization in stunned myocardium. *Circulation* 1988;78:II–261.
7. Marban E, Koretsune Y, Corretti M, Chacko VP, Kusuoka H: Calcium and its role in myocardial cell injury during ischemia and reperfusion. *Circulation* 1989;80:IV–17–IV–22.
8. Marban E, Kitakaze M, Koretsune Y, Yue DT, Chacko VP, Pike MM: Quantification of $[Ca^{2+}]_i$ in perfused hearts: critical evaluation of the 5F–BAPTA/NMR method as applied to the study of ischemia and reperfusion. *Circ Res* 1990;66:1255–1267.
9. Kusuoka H, Koretsune Y, Chacko VP, Weisfeldt ML, Marban E: Excitation–contraction coupling in postischemic myocardium. Does failure of activator Ca^{2+} transients underlie stunning? *Circ Res* 1990;66:1268–1276.
10. Levine JH, Moore EN, Weisman HF, Kadish AH, Becker LC, Spear JF: Depression of action potential characteristics and a decreased space constant are present in postischemic, reperfused myocardium. *J Clin Invest* 1987;79:107–116.

11. Krause S, Hess ML: Characterization of cardiac sarcoplasmic reticulum dysfunction during short–term, normothermic, global ischemia. *Circ Res* 1984;55:176–184.
12. Engler R, Covell B: Granulocytes cause reperfusion ventricular dysfunction after 15–minute ischemia in the dog. *Circ Res* 1987;61:20–28.
13. Kloner RA, Ganote CE, Jennings RB: The "no–reflow" phenomenon after temporary coronary occlusion in the dog. *J Clin Invest* 1974;54:1496–1508.
14. Marban E, Kitakaze M, Chacko VP, Pike MM: Ca^{2+} transients in perfused hearts revealed by gated ^{19}F NMR spectroscopy. *Circ Res* 1988;63:673–678.
15. Smith GA, Hesketh RT, Metcalfe JC, Feeney J, Morris PG: Intracellular calcium measurements by ^{19}F NMR of fluorine–labelled chelators. *Proc Natl Acad Sci USA* 1983;80:7178–7182.
16. Marban E, Kitakaze M, Kusuoka H, Porterfield JK, Yue DT, Chacko VP: Intracellular free calcium concentration measured with ^{19}F NMR spectroscopy in intact ferret hearts. *Proc Natl Acad Sci USA* 1987;84:6005–6009.
17. Yue, DT, Marban E, Wier WG: Relationship between force and intracellular $[Ca^{2+}]$ in tetanized mammalian heart muscle. *J Gen Physiol* 1986;87:223–242.
18. Kuroda H, Ishiguro S, Mori T: Optimal calcium con– centration in the initial reperfusate for post–ischemic myocardial performance (calcium concentration during reperfusion). *J Mol Cell Cardiol* 1986;18:625–633.
19. Hochachka PW: Defense strategies against hypoxia and hypothermia. *Science* 1986;231:234–241.
20. Allen DG, Eisner DA, Pirola JS, Smith GL: The relationship between intracellular calcium and contraction in calcium–overloaded ferret papillary muscles. *J Physiol (London)* 1985;364:169–182.
21. McKee EE, Cheung JY, Rannels DE, Morgan HE: Measurement of the rate of protein synthesis and compartmentation of heart phenylalanine. *J Biol Chem* 1978;253:1030–1040.
22. Suga H, Hisano R, Goto Y, Yanada O, Igarashi Y: Effect of positive inotropic agents on the relation between oxygen consumption and systolic pressure–volume area in canine left ventricle. *Circ Res* 1983;53:306–318.
23. Allen DG, Blinks JR: Calcium transients in aequorin–injected frog cardiac muscle. *Nature* 1978;273:509–513.
24. Liedtke AJ, DeMaison L, Eggleston AM, Cohen LM, Nellis SH: Changes in substrate metabolism and effects of excess fatty acids in reperfused myocardium. *Circ Res* 1988;62:535–542.
25. Stahl LD, Weiss HR, Becker LC: Myocardial oxygen consumption, oxygen supply/demand heterogeneity, and microvascular potency in regionally stunned myocardium. *Circulation* 1988;77:865–72.
26. Kihara Y, Grossman W, Morgan JP: Direct measurement of changes in intracellular calcium transients during hypoxia, ischemia, and reperfusion of the intact mammalian heart. *Circ Res* 1989;65:1029–1044.
27. Allen DG, Smith GL: The consequences of simulated ischemia on intracellular Ca^{2+} and tension in isolated ferret ventricular muscle. *J Physiol (London)* 1989;410:297–323.
28. Lee HC, Mohabir R, Smith N, Franz MR, Clusin WT: Effect of ischemia on calcium–dependent fluorescents in rabbit hearts containing Indo 1. *Circulation* 1988;78:1047–1059.
29. Guarnieri T: Direct measurement of $[Ca^{2+}]_i$ in early and late reperfused myocardium (abstract). *Circulation* 1989;80:II–241.
30. Lee JA, Allen DG: The effects of repeated exposure to anoxia on intracellular calcium, glycogen and lactate in isolated ferret heart muscle. *Pflügers Arch* 1988;413:83–89.

CHAPTER 20:

BIOCHEMICAL MECHANISMS CONTRIBUTING TO ARRHYTHMOGENESIS DURING ISCHEMIA

Peter B. Corr

ABSTRACT

Myocardial ischemia elicits marked electrophysiologic derangements which occur within minutes and are rapidly reversible with sustained reperfusion. The fact that irreversible cell damage is not present during this early ischemic interval suggests that subtle and reversible biochemical alterations, within or near the sarcolemma, occur. Here, we report our work on 2 structurally similar amphipathic metabolites: long–chain acylcarnitine (LCA) and lysophosphatidylcholine (LPC). In isolated tissue, incorporation of either metabolite into the sarcolemma at concentrations of only 1–2 mole %, as verified using electron microscopic (EM) autoradiography, elicits profound electrophysiologic derangements analogous to those seen in the ischemic heart *in vivo*. The study of these metabolites, both *in vitro* and *in vivo*, and the development of inhibitors to prevent their accumulation in the ischemic myocardium, may be helpful in modifying the incidence of sudden cardiac death in patients.

INTRODUCTION

Several clinical findings suggest that transient myocardial ischemia associated with a coronary thrombus is the likely event leading to malignant ventricular arrhythmias and sudden cardiac death [1, 2]. These lethal arrhythmias usually occur prior to the development of a distinct infarct indicating that the major mechanism responsible for sudden cardiac death in man is an acute ischemic episode.

Several findings indicate that toxic metabolites accumulating within the ischemic myocardium contribute to the electrophysiologic alterations which occur and thereby also to the development of lethal ventricular arrhythmias. Venous blood obtained from an ischemic region *in vivo* has been reported to elicit electrophysiologic derangements in isolated tissue *in vitro* [3], analogous to those changes seen within minutes of ischemia *in vivo* [4]. Likewise, perfusion of an ischemic region *in situ* with a buffer solution containing low pO_2 results in a rapid yet transient normalization of the epicardial action potentials within the ischemic region [4]. These findings suggest that brief intervals of ischemia lead to the production of metabolites or, indirectly, to the accumulation of extracellular ions which modulate ion flux through the sarcolemma of the myocyte and thereby contribute to the associated electrophysiologic derangements.

Although our efforts have been directed at 2 specific metabolites, long–chain acylcarnitine (LCA) and lysophosphatidylcholine (LPC), this does not exclude other potential factors which we have reviewed in extensive detail elsewhere [5, 6], including free fatty acids, extracellular K^+, acidosis, catecholamines, cyclic AMP, calcium and others. It is almost certain that these factors interact and in some cases share a common mechanism underlying their accumulation as well as their interaction with the sarcolemma. However, due to the brief nature of this review, we will focus on LCA and LPC. Both LCA and LPC are amphiphiles and have been shown to produce profound direct electrophysiologic alterations *in vitro* and to accumulate within minutes in ischemic myocardium *in vivo*. For review see [5] and [6]. This brief summary will focus on these two amphipathic metabolites, the biochemistry and kinetics underlying their accumulation during ischemia, the electrophysiologic effects elicited by each, and possible approaches to prevent their accumulation and thereby to modify the lethal arrhythmias associated with myocardial ischemia in man.

INTERACTIONS OF AMPHIPHILES WITH THE SARCOLEMMA

Amphiphiles, including LCA and LPC which are similar both structurally and functionally, accumulate in ischemic tissue *in vivo* and exert deleterious effects on membrane function. Both LPC and LCA possess a dipolar headgroup which contains regions of positive and negative charge. However, in contrast to membrane phospholipids which contain two aliphatic hydrocarbon chains, LPC and LCA contain a single aliphatic hydrocarbon chain. As a consequence, the incorporation of LPC or LCA into the membrane phospholipid bilayer results in a significant perturbation of the orderly packing of adjacent phospholipid molecules that could potentially modify the conformation of integral membrane proteins, including ion channels. Electron spin resonance studies of isolated myocardial sarcolemma using spin–labeled fatty acids as probes indicate that LPC or LCA incorporation into the membrane elicits significant alterations in the order parameter, a measure of the amplitude of probe motion within the membrane

[7]. At analogous membrane concentrations (2–3 mole %), quantified by electron microscopic autoradiography, LPC will elicit profound electrophysiologic derangements *in vitro* similar to those seen in ischemic tissue *in vivo* [8, 9]. Consistent with this hypothesis is the finding that the effects of LPC and long–chain acylcarnitine on membrane molecular dynamics are additive [7] as are the electrophysiologic effects of both amphiphiles [10].

Mechanisms Underlying Electrophysiologic Influences and Accumulation of LPC During Ischemia

LPC, an amphipathic phospholipid, is formed from phosphatidylcholine (PC), the predominant phospholipid of the cardiac sarcolemma, by hydrolysis of the sn–2 fatty acid by phospholipase A_2 (PLA_2). LPC is usually present in very small concentrations because of the relatively high activity of catabolic enzymes: LPC:Acyl–CoA acyltransferase, which re–acylates LPC to PC; lysophospholipase, an acyl hydrolase which catalyzes the hydrolysis of the sn–1 fatty acid to produce glycerophosphorylcholine (GPC) and free fatty acid; and lysophospholipase–transacylase, a mixed–function enzyme which catalyzes the transfer of a fatty acid from one LPC molecule to another at high substrate concentrations to produce PC and GPC.

LPC accumulates in ischemic myocardium [11–14] and in venular [15] and lymphatic effluents [16] from ischemic regions. Other metabolic influences occurring during early ischemia may likely contribute to the accumulation of LPC. Catecholamines have been shown to increase the activity of PLA_2, acting through either β–adrenergic receptors [17] or α–adrenergic receptors [18]. Acidosis reduces markedly the activity of one of the major catabolic enzymes for LPC, membrane–bound lysophospholipase; and accumulation of LCA markedly inhibits two of the other major catabolic enzymes for LPC, cytosolic lysophospholipase as well as lysophospholipase–transacylase. For review see [5].

Exogenous LPC, when incorporated into the sarcolemma at concentrations of only 1–2 mole %, induces reversible electrophysiologic alterations *in vitro*, including decreases in the resting membrane potential, maximal rate of upstroke of phase 0 ($\dot{V}_{max}$) and action potential duration [8, 10, 11, 15], alterations analogous to those which occur during early myocardial ischemia *in vivo* [4]. The electrophysiologic effects of LPC are enhanced markedly in the presence of acidosis [10,15], comparable to that observed during early ischemia *in vivo*. LPC also leads to additional electrophysiologic effects including: 1) action potentials dependent exclusively on the slow inward calcium current [11]; 2) an increased membrane resistance and internal longitudinal resistance, and increased length and time constants [19]; 3) biphasic effects on excitabili–

ty [19]; and 4) nonspecific depression of membrane channels [20]. These effects suggest that LPC may contribute to slow conduction and block and the development of reentrant arrhythmias during early ischemia *in vivo*. LPC also can directly stimulate adenylate cyclase resulting in an increase in cAMP [21] and intracellular calcium, suggesting that it may contribute to the development of triggered activity due to delayed after–depolarizations (DADs). Sawicki and Arnsdorf [22] have demonstrated that low concentrations of LPC can initiate early after–depolarizations associated with sustained rhythms. This influence of LPC was also associated with the development of two stable steady–state resting potentials, one near the plateau level, the other at more polarized potentials, an effect which was normalized by lidocaine [22].

We have recently demonstrated that LPC does induce DADs and triggered activity, an effect which is enhanced by catecholamines and which persists even in the presence of hyperkalemia and acidosis [23]. *In vivo*, significant increases in LPC were noted in ischemic myocardium within 3 min after the onset of ischemia and the magnitude of the increase was related directly to the severity of the spontaneous ventricular arrhythmias [24]. The magnitude of the increase in LPC *in vivo* within 3 min of ischemia (0.5 nmol/mg protein) [24] is nearly identical to the increase *in vitro* required to elicit the above electrophysiologic alterations (0.4 nmol/mg protein) [8]. The fact that a significant extent of the increase in LPC in ischemic myocardium occurs extracellularly, as measured in venous effluents [15] or cardiac lymph [16], suggests that nonmyocytic elements including endothelial and smooth muscle cells may contribute [25].

Mechanisms Underlying the Electrophysiologic Influences and Accumulation of LCA During Ischemia

Fatty acids are the primary metabolic fuel of the heart. The metabolism of fatty acids in the myocardial cell begins with the uptake of free fatty acids by the cardiac myocyte, followed by binding to an intracellular fatty acid binding protein and transport to the mitochondria [5]. In the outer mitochondrial membrane, the fatty acids are thioesterified to coenzyme A to produce acyl–CoA. However, acyl–CoA cannot readily cross the inner mitochondrial membrane. For transport to occur requires that the fatty acid be transesterified to carnitine, producing LCA, catalyzed by the enzyme carnitine acyltransferase I (CAT–I). After crossing the inner mitochondrial membrane in exchange for free carnitine via carnitine–acylcarnitine translocase, the LCA is then transesterified to CoA to form acyl–CoA in the mitochondrial matrix by the enzyme carnitine acyltransferase II (CAT–II). The acyl–CoA then undergoes β–oxidation, with catabolism of the long–chain fatty acyl–CoA to two–carbon acetyl–CoA groups producing NADH and $FADH_2$. The acetyl–CoA then enters the citric acid cycle with production of additional NADH, which enters the electron transport chain for oxidation and subsequent production of ATP.

Myocardial ischemia and severe hypoxia inhibit electron transport, resulting in a large increase in NADH and $FADH_2$ which, through a negative feedback mechanism, inhibits β–oxidation [5]. However, LCA continues to be produced and thereby accumulates in ischemic myocardium *in vivo*. Like LPC, exogenous LCA induces reversible electrophysiologic alterations in normoxic myocardium *in vivo* analogous to that seen during ischemia *in vivo* including a marked reduction in the resting membrane potential, V_{max} of phase 0, and action potential amplitude and duration [10].

Although studies performed *in vitro* with exogenous extracellular LCA provide evidence suggesting that this metabolite may contribute to the electrophysiologic alterations during ischemia, LCA accumulates intracellularly during ischemia or hypoxia and is not released into the extracellular space until irreversible cell damage ensues. To evaluate the effects of endogenous intracellular accumulation of LCA, isolated myocytes were exposed to hypoxia *in vitro* and the subcellular localization of LCA was delineated using electron microscopic autoradiography procedures in prelabeled cells [26]. In isolated myocytes, hypoxia induced a marked increase in total cellular LCA [26] analogous to that found during ischemia *in vivo*. Although the total cellular LCA increased over 3–fold, the content in sarcolemma increased 70–fold with marked increases also in the mitochondrial and cytoplasmic membrane structures [26]. Pretreatment of cells with sodium 2–[5–(4–chlorophenyl)–pentyl]–oxirane–2–carboxylate (POCA), an inhibitor of carnitine acyltransferase I (CAT I), prevented the accumulation of LCA and markedly attenuated the electrophysiologic alterations induced by hypoxia [26]. Thus, the sarcolemmal accumulation of long–chain acylcarnitine contributes significantly to the electrophysiologic alterations induced by hypoxia. We have recently demonstrated that inhibition of CAT–I with POCA prevents not only the accumulation of LCA but also LPC in ischemic tissue *in vivo* [14]. Inhibition of CAT–I was associated with a significant antiarrhythmic effect since prior treatment with POCA prevented the development of ventricular tachycardia and ventricular fibrillation, independent of changes in regional myocardial blood flow or hemodynamics [14]. Thus, the development of inhibitors of CAT–I to prevent the accumulation of both LCA and LPC in ischemic myocardium may be a potential and novel approach to modify the incidence of sudden cardiac death in patients with ischemic heart disease.

CONCLUSION

In isolated myocytes *in vitro*, the electrophysiologic derangements elicited by hypoxia are associated with a marked increase in the endogenous sarcolemmal accumulation of LCA. Inhibition of carnitine acyltransferase I

(CAT–I) not only prevents the accumulation of LCA in isolated myocytes exposed to severe hypoxia, but also markedly attenuates the electrophysiologic alterations. Analogous findings have recently been shown in ischemic myocardium *in vivo*. LPC accumulates to a large extent in the extracellular space during ischemia which may be secondary to release from vascular endothelium, smooth muscle or blood cell elements. In crude homogenates of myocardial tissue, the total enzymic activity for catabolism of LPC far exceeds the total activity for synthesis of LPC. Therefore, for net accumulation of LPC to occur would require inhibition of catabolism. Three enzymes responsible for the catabolism of LPC are inhibited by either long–chain acylcarnitine or acidic pH. Thus, accumulation of long–chain acylcarnitine and acidosis contribute to the increase in LPC in ischemic tissue. In concert, these and other results indicate that inhibition of CAT–I may be a promising therapeutic strategy to modify the incidence of lethal arrhythmias associated with ischemic heart disease in man.

Acknowledgement

The author's research was supported in part by National Institutes of Health grant HL–17646, SCOR in Ischemic Heart Disease, and NIH grants HL–28995 and HL–36773.

DISCUSSION

Dr. Janse: You showed that some of the electrophysiological effects of LPC *in vitro* were enhanced by a low pH. Do you know what happens if you make the extracellular milieu "more ischemic", for example when you add to the acidosis an elevated K^+ concentration and a low pO_2?

Dr. Corr: The electrophysiological effects of LPC are enhanced by a reduction in pH but it was unassociated with an increase in the net accumulation of LPC in the membrane. This surprised us because we thought that inhibiting the membrane by lysophospholipase due to the reduction in pH would result in a net increase in LPC in the sarcolemma. That was not what we found. I do not know any data indicating additive effects of other factors in the ischemic milieu to the effects of LPC.

Dr. Janse: You emphasized that accumulation of LPCs occurs *in vivo*, not in isolated perfused hearts. Yet, regional ischemia in isolated hearts does result in arrhythmias. What is the difference between *in vivo* hearts and perfused hearts in this respect?

Dr. Corr: There are several answers to that question. One is that it appears that the endothelial cells and the smooth muscle cells of the vasculature is a key component in terms of the extracellular accumulation of LPC. In other words, the extracellular accumulation is, in part, from an extra–myocytic source. We know that the endothe–

lial cells themselves and the smooth muscle cells can be stimulated to release, at least *in vitro*, an increase in LPC. In addition, Clark and coworkers *[25]* have shown that Leukotriene D4 stimulates PLA2 activity in smooth muscle cells and endothelial cells. There is a relatively large increase and it is actually blocked by cyclohexamide. The increase occurs about 3 min after Leukotriene D4. What it suggests is that other components occurring extracellulary, for example LTD4 or LTA4 from the neutrophil, could then be converted to D4 in the endothelial cell and stimulate the activity of PLA2 in the endothelial cell and smooth muscle. This sequence of events would occur *in vivo* but not in isolated buffer perfused hearts.

Dr. Janse: Are you talking about hearts perfused with Tyrode solution or blood–perfused hearts?

Dr. Corr: I am talking about a Tyrode perfused heart. I do not know whether the blood perfused heart would or would not behave similarly.

Dr. Rosen: What would the changes in fluidity affect. You have shown that for something like an α–receptor there is a change in its availability as membrane fluidity changes as a result of these events. Now what about membrane spanning protein such as channels, especially when you are talking about differential changes in fluidity and in various portions of the membrane. One can view them as no longer being able to distribute themselves in a normal way.

Dr. Corr: There is really no direct way to measure that. What we can measure is the actual fluidity in the membrane and using electron spin resonance is a direct measure of membrane fluidity. You can make certain assumptions that, as the order parameter decreases, membrane fluidity increases. As Dr. Fozzard showed us this morning, we are just beginning to learn about the structure of these ion channels within the membranes. All we could do is simply hypothesize that the change in membrane fluidity induced by LPC or long–chain acylcarnitines is responsible for the change in ion channel function.

You can delay the angle of clustering at this point. I am sure you know the data showing that long chain acylcarnitine accumulation in the membrane can alter the exposure and, most importantly, the coupling of the α–1 adrenergic receptor to phospholipase C. So it is not simply the electrophysiological effects that it can alter, but it can alter a whole variety of membrane bound proteins, G proteins presumably, and coupling to phospholipase C.

Dr. Mowbray: It seems that the accumulation of Palmitoyl Coonitive in the sarcolemma is produced in ischemia from the mitochondria and exported from there. Because, when you put your inhibitory substance (and I presume you have to do that before you stop perfusion) you do not get any accumulation. This ties in with Mcafry and Fosters observation that the transporter in the heart is more sensitive to malonyl CoA than it is in liver.

Dr. Corr: Your question is whether there is migration from the mitochondria. I showed the sum result during normoxia, in which long–chain acylcarnitines were primarily confined to the mitochondria and the sarcoplasmic reticulum. With hypox–

ia, there was a migration which was progressive into the sarcolemma of the cell, and a marked increase which occurred there. POCA would only be effective in preventing the increase prior to ischemia or hypoxia.

Dr. Gettes: I wonder how you put together this information that you have been accumulating so beautifully.

Dr. Corr: I certainly do not want to give the impression that if you inhibit the accumulation of these metabolites, you would somehow get an ischemic heart to become normal electrophysiologically. I believe that the key to arrhythmogenesis early after ischemia is one of heterogeneity and how rapidly those effects occur. If we could modify those, the progressive changes would be more uniform, less heterogeneous and less arrhythmogenic.

Dr. van der Vusse: I wonder whether lysophospholipids play an essential role in reperfusion-induced arrhythmias. Our own experimental findings do not favor this possibility. We fail to find any significant relationship between increased tissue levels of lysophospholipids in postischemic tissue of isolated rat hearts and the occurrence of fibrillation *[Van Bilsen et al., Circ Res 1989; 64:304–314]*.

Dr. Corr: In your studies, the increases were almost nonexistent. We found the same thing when we used an isolated buffer perfused heart. In other words, a lot of LPC which accumulates is extracellular. The extracellular level of LPC in blood is about 197 μM and is bound to albumin. If you remove that pool with a buffer, then even though you produce a great deal extracellularly, you have to saturate that binding pool first before you could see incorporation into the myocyte. In other words, you have a model in which you try to reproduce what happens during ischemia *in vivo*. If most of the LPC is accumulating, or is already there extracellularly, and you are washing that out initially in a buffer perfused heart, it is not analogous to ischemia. Another explanation is that these buffer perfused hearts do not show an increase in long-chain acylcarnitines analogous to what occurs during ischemia *in vivo*. Since long-chain acylcarnitines are critical in inhibiting two of the catabolic enzymes for LPC, this could be another important distinction between your isolated heart studies and ischemia *in vivo*.

Dr. van der Vusse: Maybe there are essential differences between ischemia and reperfusion related changes in cardiac lysophospholipid production and arrhythmias. We were able to confirm our rat experiments in blood perfused dog hearts *[unpublished results]*.

Dr. Corr: If you measure the increase in amount of LPC in blood, you can account for a large proportion of the total tissue level of LPC. If you are using a buffer perfused preparation, the level you are beginning at is very much lower than it would be in *in vivo*.

REFERENCES

1. Lie JT, Titus JL: Pathology of the myocardium and the conduction system in sudden coronary death. *Circulation* 1975;52 (Suppl III):41–52.
2. Nikolic G, Bishop RL, Singh JB: Sudden death during Holter monitoring. *Circulation* 1982;66:218–225.
3. Downar E, Janse MJ, Durrer D: The effect of "ischemic" blood on transmembrane potentials of normal porcine ventricular myocardium. *Circulation* 1977;55: 455–462.
4. Downar E, Janse MJ, Durrer D: The effect of acute coronary artery occlusion on subepicardial transmembrane potentials in the intact porcine heart. *Circulation* 1977;56:217–224.
5. Corr PB, Gross RW, Sobel BE: Amphipathic metabolites and membrane dysfunction in ischemic myocardium. *Circ Res* 1984;55:135–154.
6. Corr PB, Saffitz JE, Sobel BE: What is the contribution of altered lipid metabolism to the ischemic heart? Hearse DJ, Manning AS, Janse MJ (eds): *Life-Threatening Arrhythmias During Ischemia and Infarction*, Raven Press, New York, 1987, pp 91–114.
7. Fink KL, Gross RW: Modulation of canine myocardial sarcolemmal membrane fluidity by amphiphilic compounds. *Circ Res* 1984;55:585–594.
8. Gross RW, Corr PB, Lee BI, Saffitz JE, Crafford WA Jr, Sobel BE: Incorporation of radiolabeled lysophosphatidyl choline into canine Purkinje fibers and ventricular muscle: Electrophysiological, biochemical and autoradiographic correlations. *Circ Res* 1982;51:27–36.
9. Saffitz JE, Corr PB, Lee BI, Gross RW, Williamson EK, Sobel BE: Pathophysiologic concentrations of lysophosphoglycerides quantified by electron microscopic autoradiography. *Lab Invest* 1984;50:278–286.
10. Corr PB, Snyder DW, Cain ME, Craffort WA Jr, Gross RW, Sobel BE: Electrophysiological effects of amphiphiles on canine Purikinje fibers: Implications for dysrhythmia secondary to ischemia. *Circ Res* 1981;49:354–363.
11. Corr PB, Snyder DW, Lee BI, Gross RW, Keim CR, Sobel BE: Pathophysiological concentrations of lysophosphatides and the slow response. *Am J Phys* 1982;243:H187–H195.
12. Man RYK, Slater TL, Pelletier MPJ, Choy PC: Alterations of phospholipids in ischemic canine myocardium during acute arrhythmia. *Lipids* 1983;18:677–681.
13. Shaikh NA, Downar E: Time course of changes in porcine myocardial phospholipid levels during ischemia. *Circ Res* 1981;49:316–325.
14. Corr PB, Creer MH, Yamada KA, Saffitz JE, Sobel BE: Prophylaxis of early ventricular fibrillation by inhibition of acylcarnitine accumulation. *J Clin Invest* 1989;83:927–936.
15. Snyder DW, Crafford WA Jr, Glashow JL, Rankin D, Sobel BE, Corr PB: Lysophosphoglycerides in ischemic myocardium effluents and potentiation of their arrhythmogenic effects. *Am J Phys* 1981;241:H700–H707.
16. Akita H, Creer MH, Yamada KA, Sobel BE, Corr PB: Electrophysiologic effects of intracellular lysophosphoglycerides and their accumulation in cardiac lymph with myocardial ischemia in dogs. *J of Clin Invest* 1986;78:271–280.
17. Franson RC, Pang DC, Weglicki WB: Modulation of lipolytic activity in isolated canine cardiac sarcolemma by isoproterenol and propranolol. *Biochem Biophys Res Comm* 1979;90:956–962.
18. Ho AK, Klein DC: Activation of α_1–adrenoceptors, protein kinase C or treatment with intracellular free Ca^{++} elevating agents increases pineal phospholipase A_2 activity. *J Bio Chem* 1987;262:11764–11770.

19. Arnsdorf MF, Sawicki GJ: The effects of lysophosphatidylcholine, a toxic metabolite of ischemia, on the components of cardiac excitability in sheep Purkinje fibers. *Circ Res* 1981;49:16–30.
20. Clarkson CW, Ten Eick RE: On the mechanism of lysophosphatidylcholine-induced depolarization of cat ventricular myocardium. *Circ Res* 1983;52:543–556.
21. Ahumada GG, Bergmann SR, Carlson E, Corr PB, Sobel BE: Augmentation of cyclic AMP content induced by lysophosphatidyl choline in rabbit hearts. *Cardiovasc Res* 1970;13:377–382.
22. Sawicki GJ, Arnsdorf MF: Electrophysiologic actions and interactions between lysophosphatidylcholine and lidocaine in the nonsteady state: The match between multiphasic arrhythmogenic mechanisms and multiple drug effects in cardiac Purkinje fibers. *J Pharm Exp Thera* 1985;235:829–838.
23. Pogwizd SM, Onufer JR, Kramer JB, Sobel BE, Corr PB: Induction of delayed after-depolarizations and triggered activity in canine Purkinje fibers by lysophosphoglycerides. *Circ Res* 1986;59:416–426.
24. Corr PB, Yamada KA, Creer M, Sharma AD, Sobel BE: Lysophosphoglycerides and ventricular fibrillation early after onset of ischemia. *J Mol Cell Cardiol* 1987;19:45–53.
25. Clark MA, Littlejohn D, Conway TM, Mong S, Steiner S, Crooke ST: Leukotriene D_4 treatment of bovine aortic endothelial cells and murine smooth muscle cells in culture results in an increase in phospholipase A_2 activity. *J Biol Chem* 1986;261:10713–10718.
26. Knabb MT, Saffitz JE, Corr PB, Sobel BE: The dependence of electrophysiological derangements on accumulation of endogenous long-chain acylcarnitine in hypoxic neonatal rat myocytes. *Circ Res* 1986;58:230–240.

CHAPTER 21:

THE CONTROL OF CARDIAC ADENINE NUCLEOTIDES: THE DISCOVERY AND POTENTIAL FUNCTION OF HIGH-ENERGY OLIGOMERIC DERIVATIVES OF ATP

John Mowbray

ABSTRACT

Post–ischemic synchronized nucleotide oscillations in perfused hearts led us to predict, locate and identify a hitherto unrecognized major purine derivative. It is a oligomer of alternating units of phosphoglycerate and ATP. We have now confirmed the basic unit structure by unambiguous chemical synthesis and have purified from mitochondria a specific lyase which cleaves it to 3 phosphoglyceroyl–ATP thus providing the basis for a convenient enzymatic assay.

INTRODUCTION

Langendorff perfused rat hearts, judged by a number of criteria to be in a steady–state, were found to show large fluctuations in the contents of their cyclic nucleotides. To investigate this further we used HPLC to examine the contents of all purine derivatives in hearts sampled at various times over a 2–hour perfusion period, taking care to avoid introducing possible systematic errors. The results of this study were that all the purine nucleotides were observed to show large synchronous systematic variations over the first 40–60 min of perfusion [1–3]. A disturbing feature of these results was that the sum of the adenine nucleotides was not constant but showed large regular and rapid changes in size so that within a 10 minute period it could change by up to 1.3μmol/g. Furthermore there were no corresponding changes in any other purine nucleotides or nucleosides although these would have been fully ex–tracted by the procedures used. Analysis of the perfusate showed that it accu–mulated irreversibly hypoxanthine, xanthine and uric acid, and reversibly small amounts of inosine and adenosine, but could not be the source of the increased adenine nucleotide even if the known salvage pathways were 100 times faster than reported. Similarly *de novo* synthesis could only supply about 1% of new nucleotide in the time while around 120% of tissue RNA and 275% of heart pyridine nucleotide would have been required to undergo breakdown and resynthesis within 20 min to accommodate the variations. Accordingly we were driven to conclude that some previously unidentified major source of adenine nucleotide must exist in hearts.

METHODS AND PROCEDURE

Discovery of a Rapidly Exchanging TCA–Insoluble Nucleotide Derivative

Adenosine radiolabelled only in the purine ring, added to the perfusate at 0.25μM, was rapidly incorporated into heart nucleotides. Following washout perfusion to remove labeled nucleotides, all of the TCA/methanol extractable radioactivity was shown to be confined to ATP, ADP, AMP, GTP and IMP for up to a further hour of perfusion. Approximately 15% of the tissue ^{14}C was rapidly incorporated into the TCA/methanol insoluble fraction. The exchange between the TCA– insoluble and TCA–soluble nucleotide proved to be very rapid, essentially reaching specific radioactivity equilibrium within 10 min. This insoluble radioactivity could not be solubilized by repeated acid extraction though it was readily dissolved by alkali to give species which chromatography showed to be slightly more polar than ATP, ADP and AMP as well as adenosine and a non–migrating form [4].

Purification and Characterization of Phosphoglyceroyl–ATP

The TCA/methanol insoluble fractions of hearts previously labeled with 8[^{14}C]–adenosine were solubilized in detergent containing buffer and subjected to phenol extraction by a technique originally devised to separate nucleic acid from protein. The radioactive species in the nucleic acid rich fraction was purified by DEAE–cellulose chromatography under sterile conditions into one major and 3 minor species. Two of the minor forms are closely related to the major compound while the third appears to be a rapidly turning–over form of RNA [5]. The major radioactive species was somewhat more polar than ATP but proved to have the same specific radioactivity as ATP (80,000 dpm/μmol) and a phosphate: purine ratio of 4:1 [4]. ^{31}P–Nuclear magnetic resonance (NMR) analysis showed that there were 4 distinct phosphate environments present in equal proportions: two correspond to the β and α phosphate of ATP while the third was near the γ and the fourth appeared to be a typical phosphodiester [6]. Selective digestion and chromatography of this compound showed that it contained equal proportions of adenosine, ribose and glyceric acid [4, 5]. Based on the NMR data, and that mild alkaline

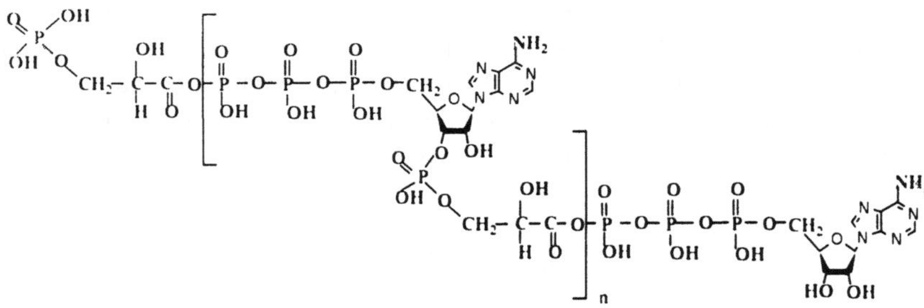

Figure 1: Oligo–phosphoglyceroyl–adenosine–triphosphate (oligo–PG–ATP). Suggested derivative structure. Slightly larger but apparently identical oligomers have also been purified from kidney [8] and rat liver [17].

hydrolysis yields enzymatically assayable ATP, we tentatively proposed a unit structure of ppp5'A3'pGri. Fast atom bombardment mass spectrometry produced a number of species in the high–field positive–ion spectrum. The most abundant corresponded to a dimer (M_r = 1329) of the proposed unit while the largest corresponded exactly to a pentamer. Digestion with snake venom diesterase which prefers 3',5' over 2',5' diesters supports the proposed 3' diester link and we propose the structure for the derivative shown in Fig. 1. An invited mini–preview of these findings has been published [7].

We have now synthesized chemically, by an unambiguous route, the compound representing the basic structural unit which is shown in Fig. 1 within the square brackets. This compound, the result of cleavage of oligo–PG–ATP at its labile mixed anhydride link, co–chromatographs with the same compound (one of the minor labeled species) purified from rat heart [9].

RESULTS AND DISCUSSION

The Metabolism of PG–ATP

In order to be able to investigate the function of oligo– PG–ATP in cells we clearly need both a simple convenient assay for it, as well as to characterize the enzymes involved in its metabolism. We chose to purify first activities which breakdown oligo–PG–ATP. Our hope, now realized, was that this might provide the basis for an enzymic assay which could be used with tissue extracts.

Two initial obstacles to doing this were first the very labile nature of the oligo–PG–ATP substrate and second the expectation that the range of nucleotidases and phosphatases present in most tissue extracts would make analysis of any specific products of the cleavage difficult and complex. The first of these, which had always been a cause for concern if oligo PG–ATP were to have any significant role in nucleotide metabolism, was overcome when we found that the oligomer was remarkably stabilized at 37°C in neutral solutions at Mn^{2+} or Mg^{2+} concentrations at or above 2 mM. Modeling studies suggest that oligo–PG– ATP can fold into a repeating dimer structure in which the mixed anhydride carbonyl shares a H-bond with the C_6 amino group of the purine [10].

The second obstacle was tackled by designing a crude assay which could be used initially to partially purify any cleavage activity away from contaminating hydrolases; the aim was to characterize the initial breakdown products and design a more specific assay based on detecting these. This initial assay relied on the ability of the oligomer to precipitate in 50% aqueous ethanol. Thus [^{14}C] oligo- PG–ATP purified from hearts prelabelled in perfusions with 8–[^{14}C]–adenosine or –adenine is precipitated with ethanol, trapped on a millipore (0.45μm) filter and the filter assayed for radioactivity. The soluble cleavage products pass through the filter.

Figure 2 shows the partial purification by FPLC of a cleavage activity from an ammonium sulphate cut of the soluble fractions from rat heart homogenate. The analogous activity from rat liver has been partially purified by identical means starting from a sonicated, washed mitochondrial fraction. Percoll density gradient analysis of this mitochondrial fraction has shown that

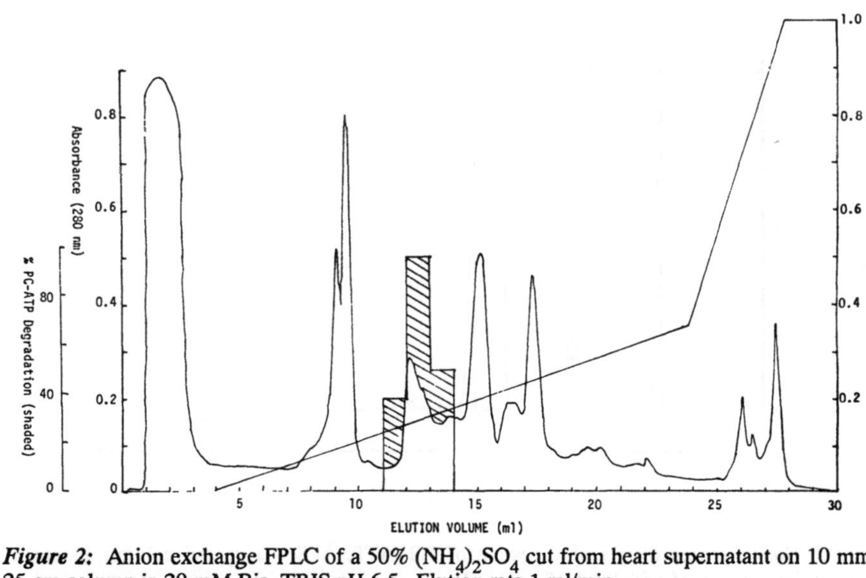

Figure 2: Anion exchange FPLC of a 50% $(NH_4)_2SO_4$ cut from heart supernatant on 10 mm x 25 cm column in 20 mM Bis–TRIS pH 6.5. Elution rate 1 ml/min.

the cleavage activity migrates preferentially with the light mitochondrial fraction which is separated from marker enzymes for lysosomes and endoplasmic reticulum (data not shown).

The ethanol/H_20 soluble radioactive product of oligo PG–ATP cleavage comigated as a single spot with authentic ATP in our 2–D thin layer chromatography (TLC) system (see [6]). However both the luciferase [11] and the less specific enzymatic assay [12] for ATP failed to detect any ATP in the product. Ion–exchange HPLC analysis showed that the compound migrated between AMP and ADP (see below). The observation that the product migrates on tlc in borate buffer strongly suggests that the 2' and 3' 0H of ribose are free [13]. Hence we propose the following reaction sequence (A = adenosine; Gril = glyceryl):

$$\text{p3Gril–[ppp5'A 3'p3Gril–]}_n\text{–ppp5'A} \rightarrow \text{m(p3Grilppp5'A)} + \text{p3Gril–[ppp5'A 3'p3Gril–]}_{n-m}\text{–ppp5'A} \quad (1)$$

The product in Eq. (1) is suggested to be PG–ATP (monomer). If so, then this would be amenable to hydrolysis by snake venom diesterase (a 5'diesterase which requires free 3'OH) yielding ATP, ADP or AMP depending on its selectivity among the three diester bonds on offer.

Figure 3 shows that partial digestion of the product yields ADP which would support the suggested identity of the reaction product. Moreover, this

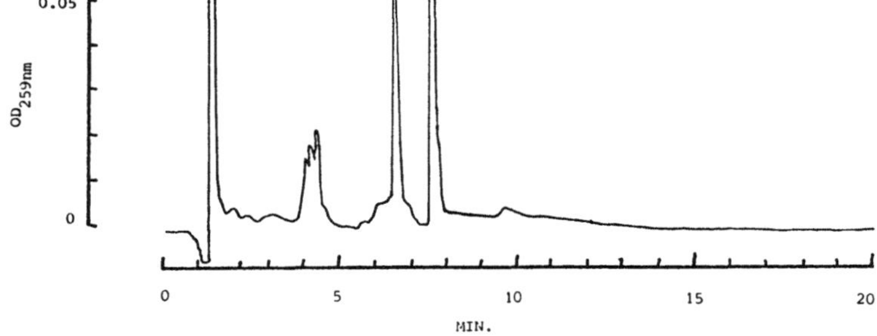

Figure 3: Snake venom diesterase digestion of PG–ATP. APS – Hypersil column 15 cm x 4 mm: elution rate 1 mM with 0.4 M Pi, pH = 3.

has now provided a means for assaying this cleavage activity (which we have provisionally called oligo–PG–ATP Lyase) by adding venom diesterase digestion and an ATP generator to the luciferase assay thus:

$$\text{p3Grilppp5'A} \xrightarrow[\text{diesterase}]{\text{venom}} \text{ADP} + \text{3Phosphoglycerate} + \text{Pi} \qquad (2)$$

$$\text{ADP} + \text{Phosphoenol pyruvate} \xrightarrow[\text{kinase}]{\text{pyruvate}} \text{ATP} + \text{pyruvate} \qquad (3)$$

$$\text{ATP} + \text{Lu Luciferase} \rightarrow \text{Enz–Lu–AMP} + \text{PPi} \qquad (4)$$

$$\text{Enz–Lu–AMP} + 0_2 \rightarrow \text{AMP} + \text{OxyLu} + \text{C}0_2 + \text{Enz} + \underline{\text{LIGHT}} \qquad (5)$$

The Function of PG–ATP in Hearts

The function of PG–ATP in the heart remains conjectural. However, with access to a specific oligo–PG–ATP lyase with which to assay the compound it should be possible to begin to tackle this question directly.

One possibility is that it may serve as a store compound either for purine nucleotides or pyrophosphate bonds or both. Similar types of nucleotide e.g diadenosine and diguanosine tetraphosphates have been observed to provide "starter" NTP and bases during embryogenesis in a range of aquatic species (see [14]) and have been reported to be produced in bacteria subjected to heat shock or oxidative stress [15]. The ability to reprovide rapidly within the cardiomyocyte the 10–20% loss (e.g. to endothelium) of ATP as nucleosides and bases which follows short term ischemia could be an important property of the healthy heart in an active animal.

At variance with a storage role is the very rapid attainment of specific radioactivity equilibration between oligo PG–ATP and soluble ATP. This suggests a more dynamic function. One possibility might be the existence of a kind of buffer for the ATP/ADP ratio which could explain the observed apparent inability of free ADP, even in stress conditions, to rise above a low threshold value (c.40 μM): this condition is clearly essential for the maintenance of unidirectional flux in pathways and vital to the cell's survival. Perhaps the post–ischemic oscillations we observed, result from the attempt of the heart's regulatory mechanisms, bereft of sympathetic innervation and blood–borne hormones, to regain optimum nucleotide ratios for stimulating energy production from oxidation without permitting the ADP to rise appreciably. Consistent with some form of rapid buffering is an observation using ^{31}P nmr of the contraction cycle in perfused rat hearts in which 75% of the matched inverse variations in ATP and Pi concentrations observed could not be accounted for by oxidative metabolism but instead appear to be replenished from an nmr invisible high energy pool [16]. The oligomeric nature of oligo PG–ATP or its possible location in mitochondria could contribute to this nmr invisibility. All soluble heart extracts do appear to contain monomeric PG–ATP [17], but at a level below the sensitivity of nmr detection. By contrast, ^{31}P nmr work with frog heart has suggested interconversion of ATP with an unidentified adenosine tetraphosphate during the early period of ischemic stress [18].

Acknowledgements

We gratefully acknowledge the support of the Medical Research Council of the UK and of the British Heart Foundation.

DISCUSSION

Dr. Noble: Absolutely fascinating. You relate to the mystery of how the increased demand and the rate of contraction in the heart produces an exactly matched metabolic rate without any systematic change in any biochemical regulator that I know of. Do you think that this new compound could have some sort of chemical role in that control mechanism?

Dr. Mowbray: I think that is possible but it would be highly conjectural at this stage, and it is not true to say that under all circumstances you do not see some respiratory control. If you start with a very low free ADP concentration, then, as demand increases in the perfused heart, free ADP rises and so too does the respiration rate. What is yet unexplained is why ADP never rises above a rather low threshold concentration, around 40 micromolar, while oxygen uptake goes on increasing as work demand increases. That is, there is now no longer any correlation with ADP concentration.

Dr. Bassingthwaighte: A comment on the differences in different tissues. You indicated that the kidney is different from the heart in having a different compound. Does this give you any insight into how the cells utilize this material?

Dr. Mowbray: Not really. I meant that the main oligomeric form that we extract from kidney is larger. We also get the same form as we get from heart, but the relative abundance of the two things is changed. As far as our analysis goes, there seems to be no real difference between those forms except in the length of the oligomer. We may be wrong about that: there may be something else attached to it which we have not seen. For a good yield you have to use slightly different techniques to extract hearts and kidneys; whether it is this slight change in our extraction procedure that gives us more of one kind than the other is not clear.

Dr. Bassingthwaighte: Long chain oligomers would be expected to take longer to come to specific activity equilibrium with soluble nucleotides than would short chain compounds. Time dependent observations could show this up. Have you ever looked at that?

Dr. Mowbray: No. We have only tried, so far, to look at this incorporation in post–ischemic tissues. This rapid incorporation and turnover of the oligomer may be an artifact of the post–ischemic oscillations in the nucleotides, possible caused by the control systems, now deprived of neural or humoral inputs, yawing a bit as they attempt to regain stable equilibrium between free and oligomeric nucleotides. Under normal metabolic circumstances, the turnover may not nearly be so high. Now that we may soon be able to measure oligo PG–ATP without prior incorporation of ^{14}C, we should be able to answer that question, but not as yet.

Dr. Kammermeier: Would you comment on the NMR aspects of this? When you do saturation transfer studies it appears that there is phosphorous which disappears and a substantial fraction of it does not return in recovery. This would imply that there is a very quick incorporation into an immobile phosphate form.

Dr. Mowbray: Yes. There are two ways of looking at this NMR invisibility. If we take this enzyme out from mitochondria with a crude preparation, it turns out to have a high reagent blank. If you now treat this preparation with a pH 8.5 buffer the blank disappears; in effect this treatment breaks down in PGATP. So the chances are that both enzyme and the PGATP are inside mitochondria (the oligomeric form). Thus this may be NMR invisible, (a) because it is in mitochondria, and there are some results recently which suggest that is not true for all of mitochondrial adenine nucleotides *[19]* and (b) the fact that as an oligomer it is not tumbling in the field and therefore you do not see it.

Now the fact that it is made and broken down very quickly is more consistent with a buffering function. There is an experiment done by Drs. Joanne Ingwall and Eric Fossel *[16]* a number of years ago in which they used NMR gated to the pulse of the heart. They measured the ^{31}P contents of the various compounds at various times during the cardiac cycle; at systole and diastole and in between. What they found was that as systole proceeds you get an increase in inorganic phosphate and then in dias–tole it goes down again. The ATP did exactly the opposite; it went down and went back up, as did creatine phosphate. There were some other changes as well. This looks very nice; this is just what we might expect to find in the cardiac cycle. In other words, ATP is consumed during the contraction and then during recovery it makes it up again. The only problem with this is the actual size of the inorganic phosphate and ATP changes, which were matched more or less. If you assume ADP (invisible) rephosphorylates to ATP, using this inorganic phosphate via oxidative phosphoryla–tion, then it would have to take up to 5 times more oxygen than the preparation took up. If you look at the difference between the ATP breakdown and the creatine phos–phate breakdown, then the very small difference between them would have been enough exactly to take up the observed oxygen. In answer to your question, it would seem that all that ATP which disappeared in those hearts, and the inorganic phosphate which appeared, were being interconverted but not by a process using oxidative phos–phorylation, not taking up oxygen, and therefore presumably via some other energy pool which is NMR invisible. So 80% of the ATP change seen under those circum–stances will actually be rephosphorylated and dephosphorylated during the cycle (200 ms) to some other compound and back again. That is a pretty rapid buffering.

REFERENCES

1. Bates DJ, Perrett D, Mowbray J: Systematic variations in the content of the purine nucleotides in the steady-state perfused rat heart. *Biochem J* 1978;176:485-493.
2. Mowbray J, Bates DJ, Perrett D: Inversely related oscillations in the contents of cyclic GMP and the total adenine nucleotides in steady-state perfused rat hearts. *Febs Lett* 1981;131:55-59.
3. Mowbray J, Perrett D, Bates DJ: Post-ischemic synchronous purine nucleotide oscillations in perfused rat heart. *Int J Biochem* 1984;16;889-894.
4. Mowbray J, Hutchinson WL, Tibbs GR, Morris PG: The discovery of a rapidly metabolized polymeric tetraphosphate derivative of adenosine in perfused rat heart. *Biochem J* 1984;223:627-632.
5. Hayworth PG, Hutchinson WL, Mowbray J: A comparison of rapid adenine nucleotide incorporation into phosphoglyceroyl-ATP and into RNA-like species in perfused rat heart. *Biochem Biophys Acta* 1987;927: 158-162.
6. Hutchinson WL, Morris PG, Mowbray J: The molecular structure of a rapidly formed oligomeric adenosine tetraphosphate derivative from rat heart. *Biochem J* 1986;234:623-627.
7. Lawson R, Mowbray J: Purine nucleotide metabolism: the discovery of major new oligomeric adenosine tetraphosphate derivative in rat heart. *Int J Biochem* 1986;18:407-413.
8. Hutchinson WL, Ratcliffe PJ, Mowbray J: Evidence for the presence of oligophosphoglyceroyl-ATP in rat kidney. *Biochem J* 1986;240:597-599.
9. Wozniak L: Personal Communication, unpublished data, 1990.
10. Mowbray J: Unpublished data, 1990.
11. Wulff K, Doppen W: ATP: Luminometric method. In: *Methods of Enzymatic Analysis*, H-U Bergmeyer (ed), 3rd Edn, Academic Press, New York, 1985;Vol VII:357-360.
12. Lamprecht W, Trautschold I: ATP: Determination with hexokinase and glucose-6-phosphatase. In: *Methods of Enzymatic Analysis*, H-U Bergmeyer (ed), 2nd Edn, Academic Press, New York, 1974;Vol IV:2101-2110.
13. Neuhard J, Randerath E, Randerath K: Ion-exchange thin layer chromatography-resolution of complex nucleoside triphosphate mixtures. *Anal Biochem* 1965;13:211-222.
14. Warner AH, McClean DK: Studies on the biosynthesis and role of diguanosine tetraphosphate during growth and development of artemia salina. *Dev Biol* 1968;18:278-293.
15. Lee PC, Bochner BR, Ames BN: AppppA heat-shock stress and cell oxidation. *1983 Proc Natn Acad Sci US* 1983;80: 7496-7500.
16. Fossel ET, Morgan HE, Ingwall JS: Measurements of changes in high-energy phosphates in the cardiac cycle by using gated 31P nmr. *Proc Natn Acad Sci US* 1980;77:3654-3658.
17. Patel B: unpublished data, 1990.
18. Keidel WD, Keidel J, Reiman V, Csatary N: Study for clarifying two unknown peaks in PAMT spectra appearing during the breakdown of high energy phosphates in ischaemic heart and skeletal muscle. *Res Exp Med* 1984;184:73-84.
19. Hutson SM, Berkich D, Williams GD, LaNoue KF, Briggs RW: ^{31}P NMR visibility and characterisation of rat liver mitochondrial matrix adenine nucleotides. *Biochemistry* 1989;28:4325-4332.

CHAPTER 22:

ON THE CONTRACTILE MECHANISM IN CARDIAC MUSCLE

Gerald H. Pollack

ABSTRACT

The sliding filament cross–bridge theory is generally invoked as the mechanism underlying contraction in skeletal and heart muscle. Yet, an increasingly large body of evidence conflicts with this theory. A review of some of this evidence, points the way toward approaches that may be potentially more productive.

INTRODUCTION

What is the molecular mechanism of contraction in the heart? After years of research, this simple question ought to be answerable, and indeed, most texts cite the sliding filament cross–bridge theory as the underlying mechanism. According to that theory, sarcomere shortening is brought about by the sliding of thick filaments past thin ones, a process driven by the action of rotating cross–bridges. This theory has now been in existence for some 35 years, and has enjoyed wide popularity. The question addressed here is simple: is the theory adequate?

The cross–bridge theory was introduced when back to back papers by A. F. Huxley and H.E. Huxley [1, 2] reported that A–band widths remained constant during contraction. Up to that time, contraction was generally thought to be mediated by a shortening of the A–band (now recognized to correspond to the thick filament). But the papers by the Huxleys challenged that view, and implied that A–bands did not shorten. These papers ushered in the modern era of thinking about muscle contraction.

Since that time, reports of thick filament shortening or A–band shortening have been abundant (for review, see [3]). The issue has become relevant enough to inspire H.E. Huxley, one of the two inventors of the theory, to reinvestigate the subject with new experimental methods [4].

The question of whether thick filaments shorten is but one of a number of issues relevant to the theory that are now being challenged. These challenges notwithstanding, cross–bridge mechanisms continue to be invoked almost by reflex, with astonishing lack of concern for the issues under debate. This situation is well articulated in a recent book on the muscle by the late Graham Hoyle [5]. Especially insightful to the reader is the section titled "Why Do Muscle Scientists 'Lose' Knowledge?"

Several of the issues under debate are briefly introduced here.

STEPWISE SHORTENING

Those who happened upon a recent review of this area [6] and the response by Sir Andrew Huxley published simultaneously [7] will be familiar with the issue to be considered: sarcomere shortening in discrete, quantal steps.

Molecular phenomena are ultimately quantal: each event stems from a single molecular action. Best known is the quantal release of acetylcholine at the myoneural junction. That quantal phenomena should occur as well in contractile systems is hardly surprising, for the interaction (irrespective of mechanism) between the single myosin molecule and the single actin subunit can be intuitively envisioned.

Quantal behavior has indeed been confirmed in several motile systems. The bending of cilia, for example, takes place in discrete steps, each swing comprising a series of micro–bends, interrupted by short pauses [8]. Another example is in the microtubule–mediated cellular transport system: constituent proteins, kinesin and tubulin, have been studied in isolated models [9]. Kinesin–studded microspheres glide past tubulin filaments in consistent size steps, punctuated by pauses. And in striated muscle, sarcomeres along a single myofibril shorten stepwise [10]. The shortening waveform resembles a staircase, much like the waveforms of ciliary bending and microtubular translation.

Why are stepwise phenomena discordant with the cross–bridge theory? Couldn't rotating bridges imply stepwise motion: one stroke, one step? Such correlation can be envisioned at microlevels. At the level of a single myofibrillar sarcomere, for example, it is easy to imagine each shortening step as driven by coordinated rotation of the million or so bridges. More difficult to understand is why stepwise behavior should persist on a larger scale. Stepwise phenomena are detectable over both millimeter– [11] and centimeter–length regions [12], containing enormous numbers of bridges. If the steps arise out of synchronized bridge rotations, why should synchrony extend over what, by molecular standards, are practically cosmic reaches? Where is the coxswain needed to synchronize the strokes?

In his response [7], Sir Andrew details the potential for artifact. Shortening might, after all, be smooth, the steps arising out of instrument-based artifact. In particular, Sir Andrew considers two potential sources of artifact arising from features of the striation pattern. However, the phenomenon has since been replicated under conditions where such potential artifacts do not apply [10]. It has been replicated by methods that are independent of the striations [11, 12]. And it has been replicated in independent laboratories using still other methods [13–15]. Depending on how one counts them, the number of confirmatory methods totals five to seven. If some common artifact is generated in all of these methods, the responsible gremlin must be lurking insidiously beneath the surface.

On the other hand, if the phenomenon is genuine, it is not obvious how the implied synchrony could square with the concept of independently acting cross-bridges [16]. The steps could arise only if bridge rotations were synchronized over vast spatial domains. But what agent is available to accomplish so formidable a task?

LENGTH–TENSION RELATION

Until the late seventies, the final word on the length–tension relation could be found in the classical paper by Gordon, Huxley and Julian [17]. This result was in good agreement with the prediction of the cross-bridge theory. Subsequently, a paper by ter Keurs and colleagues [18] challenged this conclusion: the linear relation between tension and overlap could be found only when tensions obtained with length–clamped sarcomeres were plotted (as in the classical work). When fiber–isometric tensions were plotted instead, tension scarcely varied over a considerable range of sarcomere–length variation. Fiber–isometric tension failed to reflect the number of cross-bridges available for action.

The "contradictory" result of ter Keurs and colleagues stirred a controversy. Numerous follow–up studies have been conducted. The results have fallen into two categories (for review, cf. [19]). Those carried out in a manner similar to the classical studies, with benefit of sarcomere–length control, obtained the classical result. Those carried out in a manner similar to the experiment of ter Keurs and colleagues, without benefit of length control, obtained a result similar to that of ter Keurs: a higher, flatter length–tension curve in which tension was almost invariant from full overlap to half-maximal overlap.

Why the difference? The only apparent difference between the two types of experiment is in the control of sarcomere length. When the sarcomere-length clamp is used, sarcomeres in question remain (nominally) perfectly isometric; they do not shorten or lengthen. When the fiber is held isomet–

rically, constituent sarcomeres can follow their natural tendency to shorten, and stretch others in series. Is it possible, then, that perfectly isometric sarcomeres yield the classical length–tension curve, while sarcomeres that can shorten slightly yield the higher, flatter, length–tension curve?

The answer appears to be yes. To check this, we carried out experiments in which we compared the length–tension curve for sarcomeres that were constrained to remain isometric with the length–tension curve for sarcomeres allowed to shorten slightly [20, 21]. Isometric sarcomeres gave the classical length–tension curve. Sarcomeres that were allowed to develop tetanic tension isometrically, then allowed to contract to a slightly shorter length during the tetanic plateau, produced considerably more tension. At full overlap, the increase was negligible; with decreasing overlap, the extra tension induced by shortening was appreciable; at 25% maximal overlap, the post–release tension was 167% of the tension before release.

Which length–tension curve is correct? When the length–tension curve for isometric sarcomeres is plotted, the result fits the expectation of the cross–bridge theory. However, this tension is sub–maximal. Maximal tension is obtained only after small (1–2%) release. When this, maximal, tension is plotted the result no longer fits: tension varies only little from full overlap to half–maximal overlap.

The cross–bridge theory predicts that maximal tension at each sarcomere length should vary directly with the degree of overlap. This is a central feature of the theory. The new results imply that maximal tension does not vary directly with the degree of overlap. Whether some modified cross–bridge theory can fit these results remains to be seen. Very puzzling, according to the theory, is why 1–2% pre–shortening should yield so much more isometric tension than the case without shortening. Could the bridges be endowed with memory?

CONTRACTION BEYOND OVERLAP

The discussion above leads to another aspect of the length–tension curve that I wish to focus on: whether tension can still be generated when the fiber is stretched beyond overlap. Several oft–cited works have shown that the cutoff point is very near to the sarcomere length at which overlap ceases [22–24]. I see no obvious flaws in these conclusions, save the limited volume of data, and perhaps the degree of variability found in the first of these papers.

On the other hand, an entirely different conclusion has been drawn in other studies. In similar work carried out in a laboratory in Denmark during the same period, vigorous contraction persisted to substantially longer sarcomere lengths [25–27]. Tension development was not abolished until the sarcomeres had been stretched to 4.2 μm, and active shortening remained in evidence at initial lengths as long as 4.5 μm. Overlap cuts off at 3.6 μm.

Careful checks for residual overlap of thick and thin filaments were made in these experiments, in spite of the extremity of the stretch. First, the fiber was scanned from one end to the other by optical microscopy. Sarcomere–length variation remained within 6% of the mean, irrespective of the degree of stretch, and including the end regions where sarcomeres were naturally shorter. Then, the stretched fibers were fixed, embedded, and sectioned with systematic checks for shrinkage carried out at each stage. Of several hundred fields examined in the electron microscope, fewer than 1% had sarcomere lengths as low as 3.7 to 3.8 μm. The absence of residual overlap left the authors with little alternative but to conclude that thick/thin interaction could not be essential for contraction.

More recently, similar results were found in experiments on the tail–levator muscle of the horseshoe crab, *Limulus*. Tension measurements revealed a marked ability to contract beyond overlap, though the data here are less precise due to skew of the striation pattern [27]. Some 80% of the fiber's tension–generating ability remained just beyond overlap, while modest active tension could still be elicited with as much as 55% additional stretch.

In a third study with strips of skinned cardiac and skeletal muscle fibers, de Beer et al. [28] found tension development persisting to sarcomere lengths beyond 4.0 μm. Fibers were scanned optically at rest and during contraction. Sarcomere lengths remained homogeneous upon activation, local values staying constant or decreasing by less than 0.1 μm. Checks for residual overlap were carried out with the electron microscope, though less exhaustively than in the Danish studies. Conclusions were similar: contraction beyond overlap.

For some theories, tension generation beyond overlap is tolerable. It is now clear, for example, that a third filament interconnects the tip of the thick filament with the Z–line [29], so that contractile mechanisms driven by thick filament shortening could live with tension beyond overlap. For the cross–bridge theory, tension beyond overlap is forbidden.

THICK FILAMENT SHORTENING

To some, the idea that thick filaments might shorten seems incompatible with reason. Thick filament shortening seems to violate some fundamental principle of nature. Yet, the evidence is not quite so clear cut.

In a review article written in 1983 [3], I considered the evidence. In that article I listed every paper, more than 40, that could be found on the subject, published since 1954. Though several authors doubted the authenticity of their observations, all but a few reported substantial shortening of the A–band or thick filament under at least some circumstances. Further, I note that prior to 1954, before the appearance of the two *Nature* papers that upset the established paradigm, A–band shortening had been the prevailing view. In light of the mass of contradictory evidence published before and after 1954, I find it astonishing that for most, the question of thick filament shortening has seemed closed.

But not for all. A fair number of new papers have appeared on the subject. These fall into two areas: studies on *Limulus* muscle, and studies on vertebrate striated muscle.

In the early 1980s it had been broadly accepted that during contraction from intermediate lengths, thick filaments of *Limulus* shortened (for review see [30]). Thick filament shortening was considered an adjunct to the "normal" mode of contraction, installed to accommodate the unusually large changes of sarcomere length characteristic of this beast's tail muscles.

However, this interpretation was brought into question by work from Levine's laboratory [31]. Comparison of isolated thick filaments that had shortened with those that had not shortened revealed little or no structural difference: molecular packing was identical. Failure to detect any structural change led to the hypothesis that the shortening of the thick filament was due merely to the release of myosin molecules from the ends of the filament, not to the kind of repacking that would be attendant with thick filament shortening of more consequential nature.

More recent work from Dewey's laboratory [32], on the other hand, implied that molecules were not lost from the ends of the filaments. Using the STEM at Brookhaven, Dewey and colleagues found no statistically significant difference of mass between filaments that had shortened and those that had not. The issue of how *Limulus* thick filament shorten awaits resolution.

As for the vertebrate muscle, advances have been made in several laboratories. Huxley et al. [4] froze rabbit psoas muscles at various stages of stretch and contraction. Frozen samples were then freeze–substituted, embedded, and examined in the electron microscope. Thick filaments were found to be of variable length. However, diminished thick filament length was not considered to be genuine; it was assumed to arise out of preparation–induced shrinkage. To correct for this, Huxley applied a correction factor based on the presumed constancy of myosin's axial repeat, and concluded that filament length remained constant.

In another quick–freeze study using an essentially shrinkage–free method (by substituting and embedding at low temperature), Edelmann [33] examined intact frog skeletal muscle fibers. Fibers were frozen at extended length or during the course of active shortening to intermediate lengths. A clear difference between relaxed and contracted fibers was seen: in contracting fibers, thick filaments were approximately 15% shorter.

Finally, studies similar in concept to those of Edelmann were carried out using other methods [34]. These involved single frog fibers shortening actively from extended to intermediate sarcomere lengths. A–band widths were followed by polarizing microscopy, thick filament lengths by electron microscopy (glutaraldehyde fixation). We found that thick filaments / A-bands shortened by about 15 ± 20%, with both methods. These results are in turn similar to the 15% shortening seen by Edelmann using a third method.

CONCLUSION

I have attempted to set down some of the most obvious conflicts between the cross-bridge theory's predictions and experimental observations: questions of independence of force generators (stepwise shortening); of the shape of the length-tension relation; of tension beyond thick / thin filament overlap; and of constancy of filament length. There are more [19]. In spite of such difficulties, the theory has remained popular, perhaps because of the absence of an attractive alternative.

A theory that takes account of these phenomena has been formulated [19]. In this theory, the driving force is thick filament shortening. Shortening occurs molecule by molecule along the filament, giving rise to a stepwise shortening of the filament, and hence, stepwise shortening of the sarcomere. The mechanism can account for tension beyond overlap: the thick filament merely pulls on the connecting filament (the element that interconnects either end of the thick filament with the nearest Z-line), and thus confers tension on the sarcomere's extremities. The model also explains the length-tension relation, or more precisely the two classes of length-tension relation that have been published.

The above mentioned theory is a radical departure from tradition. Many "new" phenomena are included, as are some long-forgotten "old" phenomena. I have not intentionally ignored a body of observation simply because it failed to conform to some preconceived notion; rather, such observations form the starting point for theoretical development. The resulting model is straightforward, and accounts for the major structural, chemical, and mechanical phenomena of the field.

Whether the new model is adequate remains to be seen. The present model, however, seems beset with difficulties of a fundamental nature. I have little doubt that a patchwork approach could ameliorate some of the difficulties, but one wonders about the value of adhering too tenaciously to an idea chiefly because of its familiarity. A breath of fresh air could impart new life into a static field.

DISCUSSION

Dr. Sipkema: Concerning the tension increase after shortening, what happens when you wait a long time? Does that cause the tension increase to return to its original level?

Dr. Pollack: I can not say for sure because we have never extended the experiments beyond about 5 s tetanus. For 5 s tetani it tends to remain at a plateau. Whether tension remains flat beyond 5 s remains to be seen.

Dr. Marban: I would have expected step–wise shortening to be reflected most prominently at the level of a single sarcomere. It would be much more evident in the single myocardial cell preparation then in the enormous muscle. Yet it is hard for me to see the step–wise events in the smaller preparations.

Dr. Pollack: One of the difficulties is that we are dealing with a region that is so extremely small that it places severe limits on our ability to make measurements. The single myofibril is such a tiny preparation that it is difficult to make those measurements, so what you saw is a lot of noise, and superimposed on this noise there are fluctuations that we think are steps, but perhaps are not convincing. We are now working on improving our methodology to check and see if they are really there. We think they are, because the characteristics of those steps and pauses are rather similar to those measured with a host of other methods that we used in bigger preparations.

Dr. Fozzard: Back to this 1–2% shortening that produces extra shortening. Did you make the stiffness measurements in the two situations before and after. Did stiffness increase after the shortening?

Dr. Pollack: We have not made any stiffness measurements. One would perhaps like to measure stiffness to assess the number of cross–bridges. On the other hand, I am not so sure that stiffness reflects that exclusively.

Dr. Westerhof: I would like to ask how you explain that there is a descending limb in the force–length relation as based on thick filament shortening. Do you have a hypothesis that can be tested to explain this phenomena?

Dr. Pollack: Yes, the hypothesis is indeed developed. I spent about 6 years writing a book on the subject *[35]*. The hypothesis is well described in the book.

Dr. Noble: A strong argument with respect to the thick filament shortening is that if you take living muscle and with X–ray diffraction measure the repeat distance of the myosin projections, 43 nm, this does not change. How do you get around that one?

Dr. Pollack: What you are saying is that you have 43 nm projections coming out of a thick filament and if the thick filament shrinks, then these projections should get closer together. Therefore the X–ray diffraction pattern ought to reflect this diminution of spacing. How is it possible to have thick filament shortening? The kind of a mechanism that I am suggesting is this: the thick filament does not shorten uniformly; it occurs step by step, giving you step–wise sarcomere shortening. What may be happening is that the first molecule along the filament shortens, then the second, then the third, and so on, giving you step–wise shortening of the thick filament and therefore step–wise shortening of the sarcomere. As you can see from this model, only several of these molecules will shorten during an isometric contraction. This will be enough to give you the maximum tension. If you look at the relaxed phase of X–ray diffraction pattern, you see predominantly a 43 nm layer line, reflecting the spacing of these projections. If you go to the contracting state, what you would see is two components of the X–ray diffraction pattern: One represents the shortening region, which probably will not contribute much, since few molecules have shortened. The other represents the rest of the thick filament, and will contribute in exactly the same as it did in the resting state – again the 43 nm component. What you expect, then, in the activated state is a persistence of this 43 nm reflection, which is consistent with what is observed.

Dr. Bassingthwaighte: The implication of your diagram is that the thick filament shortens only in the region of no actiomyosin overlap. That means that maximal thick filament shortening should occur when the sarcomere is longest and overlap is least. Is that so?

Dr. Pollack: No. In the model, the entire thick filament can shorten. The model is elaborated in detail in my forthcoming book. I am sorry that I did not have an opportunity to discuss the details of the model here.

Dr. Bassingthwaighte: If thick filament shortening can occur in the overlap zone, it would imply that the actin either slides or also shortens.

Dr. Pollack: Yes, that is right. And there is evidence to support both possibilities. Most people think that thin filament does not change length at all. But the evidence is accumulating that thin filaments undergo very important length changes. The Oosawa/Yanagida group *[36]* has consistently advocated conformational changes in actin, and most recently there is evidence from Clarence Schutt *[37]* who has actually proposed a contraction model based on thin filament shortening.

REFERENCES

1. Huxley AF, Niedergerke R: Structural changes in muscle during contraction. Interference microscopy of living muscle fibres. *Nature (Lond)* 1954; 173:971–973.
2. Huxley HE, Hanson J: Changes in the cross–striations of muscle during contraction and stretch and their structural interpretation. *Nature (Lond)* 1954;173:973–976.
3. Pollack GH: The cross–bridge theory. *Physiological Reviews* 1983;63:1049–1113.
4. Huxley HE, Faruqi AR, Whitock S: Correlation between X–ray diffraction studies on contracting muscle and electron microscope observations using rapid freezing techniques. *Biophysics J* 1987;51:5a.
5. Hoyle G: *Muscles and Their Neural Control*. New York, John Wiley & Sons, Inc., 1983.
6. Pollack GH: Quantal mechanisms in cardiac contraction. *Circ Res* 1986;59:1–8.
7. Huxley AF: Comments on : Quantal mechanisms in cardiac contraction. *Circ Res* 1986;59:9–14.
8. Baba S: Regular steps in bending cilia during the effective stroke. *Nature* 1979;282:717–720.

9. Gelles J, Schnapp BJ, Sheetz MP: Kinesin–driven movement on the nanometer scale. *Biophys J* 1988;53:29a.
10. Bartoo ML, Tameyasu T, Burns DH, Pollack GH: Stepwise shortening in single myofibrils. *Biophys J* 1988;53:370a.
11. Granzier HLM, Myers JA, Pollack GH: Stepwise shortening of muscle fiber segments. *J Musc Res & Cell Motil* 1987;8:242–251.
12. Granzier HLM, Mattiazzi A, Pollack GH: Isotonic velocity transients and step–wise shortening in single frog muscle fibers. Submitted.
13. Housmans P: Discussion. In: *Contractile Mechanisms in Muscle*, Pollack GH, Sugi H (eds), New York, Plenum Press, 1984, pp 782–784.
14. Sugi H, Pollack GH (eds): *Molecular Mechanism of Muscle Contraction*. New York, Plenum Press, 1988.
15. Tameyasu T, Toyoki T, Sugi H: Nonsteady motion in unloaded contractions of single frog cardiac cells. *Biophys J* 1985;48:461–465.
16. Huxley AF: Muscle structure and theories of contraction. *Prog Biophys Chem* 1957;7:255–318.
17. Gordon AM, Huxley AF, Julian FJ: The variation in isometric tension with sarcomere length in vertebrate muscle fibres. *J Physiol (Lond)* 1966;184: 170–192.
18. ter Keurs HEDJ, Iwazumi T, Pollack GH: The sarcomere length–tension relation in skeletal muscle. *J Gen Physiol* 1978;72:565–592.
19. Pollack GH: *Muscles and Molecules: Uncovering the Principles of Biological Motion.* Ebner & Sons, Seattle, 1990.
20. Horowitz A, Caljouw CJ, Pollack GH: Effect of small sarcomere–length releases on force level during sarcomere–isometric tetanit. Submitted.
21. Horowitz A, Wussling M, Pollack GH: Force increase caused by small sarcomere–length release: dependence on size, velocity and timing of release. Submitted.
22. Gordon AM, Huxley AF, Julian FJ: Tension development in highly stretched vertebrate muscle fibres. *J Physiol (Lond)* 1966;184:143–169.
23. Huxley AF, Peachey LD: The maximum length for contraction in vertebrate striated muscle. *J Physiol (Lond)* 1961;156:150–165.
24. Podolsky RJ: The maximum sarcomere length for contraction of isolated myofibrils. *J Physiol (Lond)* 1964;170:110–123.
25. Carlsen F, Fuchs F, Knappeis GG: Contractility and ultrastructure in glycerol–extracted muscle fibers. II Ultrastructure in resting and shortened fibers. *J Cell Biol* 1965;27:35–46.
26. Carlsen F, Knappeis GG, Buchthal F: Ultrastructure of the resting and contracted striated muscle fiber at different degrees of stretch. *J Biophys Biochem Cytol* 1961;11:95–117.
27. Walcott B, Dewey MM: Length–tension relation in Limulus striated muscle. *J Cell Biology* 1980;87:204–208.
28. De Beer EL, Gründeman RLF, Wilhelm AJ, Caljouw CJ, Klepper D, Schiereck P: Caffeine suppresses length dependency of calcium–activation curves in skinned skeletal and cardiac muscle preparations of the rabbit. *Am J Physiol.* 1988;254:C491–C497.
29. Wang K: Cytoskeletal matrix in striated muscle: The role of titin, nebulin and intermediate filaments. In: *Contractile Mechanisms in Muscle*, Pollack GH, Sugi H (eds) New York, Plenum Press, pp 285–306, 1984.
30. Dewey MM, Brink P, Colflesh DE, Gaylinn B, Fan S–F, Anapol F: Limulus striated muscle provides an unusual model for muscle contraction. In: *Contractile Mechanisms in Muscle*, Pollack GH, Sugi H (eds) New York, Plenum Press, pp 67–88, 1984.

31. Levine RJC, Kensler RW: Structure of short thick filaments from Limulus muscle. *J Mol Biol* 1985;182:347–352.
32. Dewey MM, Colflesh D, Gaylinn B: Mass measurement of isolated thick filaments from Limulus striated muscle by STEM microscopy. *Biophys J* 1985;47:468a.
33. Edelmann L: The cell water problem posed by electron microscopic studies of ion binding in muscle. *Scanning Electron Microscopy* 1988;2(2):851–865.
34. Periasamy A, Burns DH, Holdren DN, Pollack GH, Trombitás K: A-band shortening in single fibers of frog skeletal muscle. *Biophys J* 1990;57:815–828.
35. Pollack GH: *Muscles and Molecules: Uncovering the Principles of Biological Motion.* Ebner & Sons, Seattle, 1990.
36. Yanigida T, Oosawa F: Polarized flouresence from E-ADP incorporated into F-actin in a myosin-free single fiber: conformation on ϕF-actin and changes induced in it by heavy meromyosin. *J Mol Biol* 1978;126:507–524.
37. Schutt CE, Lindberg U: Reconsideration of the role of actin in the generation of tension during muscle contraction. Submitted.

Myocardial Mechanics

CHAPTER 23:

MODELING THE EFFECT OF CONTRACTILITY ON DISTRIBUTION OF CORONARY FLOW

Pieter Sipkema and Nicolaas Westerhof

ABSTRACT

The effect of contractility in the isolated, maximally vasodilated, blood perfused cat heart on coronary flow amplitude is considerable. The flow distribution across the wall (the endo/epi ratio) depends also on the contractile state. We developed a model for the non–beating myocardium to predict the endo/epi ratio as a function of the level of contractility. This ratio is constant if all layers undergo the same change in contractile state. The endo/epi flow ratio decreases with increased contractility if the transmural vessels are relatively more affected than vessels parallel to the myocardial surface or when the effect of contraction is stronger on endocardial vessels than on epicardial vessels.

INTRODUCTION

Coronary flow is affected by the vasoactive state of the smooth muscle and by contraction of the heart muscle [1]. Cardiac contraction impedes coronary blood flow, so that systolic arterial inflow is small compared to diastolic arterial inflow. The flow distribution depends on the contractile state [2, 3]. To study the effects of contraction alone, changes in vasoactive state should be avoided. When perfusion pressure does not vary over the cardiac cycle, the variations in the vasculature are directly mirrored in flow. We studied diastolic to systolic coronary arterial flow differences (the oscillatory flow amplitude) at constant perfusion pressure during maximal vasodilation and found that there is only a weak coupling between systolic left ventricular pressure (LVP) and coronary flow in isovolumic and isobaric beats [4]. Downey et al. [5] also reported that systolic flow is inhibited in empty beating hearts.

As an explanation for these findings we suggested that the time varying elastance concept, originally proposed by Suga et al. [6] for the ventricular lumen, holds true, in adapted form, for the coronary vascular compartment. The time varying elastance results from the time varying stiffness of the cardiac muscle. For the coronary system the time varying elastic properties CE(t) (i.e. coronary elastance) can be represented by a varying capacitor (inverse of elastance). Because the vascular lumen is not in isovolumic conditions both vascular compliance (inverse of elastance) and vascular volume decrease. A decrease in volume results in an increase in coronary resistance (Poiseuille's law, resistance inversely proportional to volume squared). Therefore, vascular compliance decreases and resistance increases during these changes in elastance. Both the ventricular lumen and the coronary vasculature depend on their varying elastance and their load so that coronary flow impediment in systole is not dependent on LVP.

Flow distribution in the non beating left ventricular wall was measured with microspheres by Goto et al. [2] and in the beating heart by Van Winkle et al. [3]. Goto et al. studied the impeding effect of myocardial contraction on regional myocardial blood flow in the rabbit heart that was maximally vasodilated and perfused at a perfusion pressure of 100 mmHg. During diastolic arrest they found an endo/epi ratio of 2.17, at a LVP equal to zero. During systolic arrest they studied endo/epi ratios at a Plv of 0 and 60 mmHg and found ratios equal to 0.29 and to 0.49, respectively. Their conclusion is that myocardial contraction impedes subendocardial flow more than subepicardial flow. Van Winkle et al. [3] found that abolishing myocardial contraction reduced the subendocardial resistance, unloading did not affect it. They concluded that compression of the subendocardial vessels is unrelated to ventricular lumenal pressure.

The aim of the present study is to model the endo/epi ratio in non-beating myocardium as a function of the level of contractility.

METHODS

The aim of the model is the prediction of the pressure–flow relation from the given pressure– cross–sectional area relation. The computer program is tested on measurements done on a latex microtube [7], where the calculated pressure–flow relations using length and the measured pressure–cross–sectional area relation, Poiseuille's law, and accounting for diameter– and shear dependent viscosity compared well with the experimental pressure–flow relations.

The model consists of seven layers and the maximal diameters and lengths are taken from a study on the anatomy of the pig heart performed by van Bavel [8]. The relation between length and diameter was obtained from a

Figure 1: Pressure vs. cross–sectional area.
A = as in Eq. (1); A0 = Amplitude; B = Shift; P0 = Pressure of symmetry point; C = Slope (elastance).

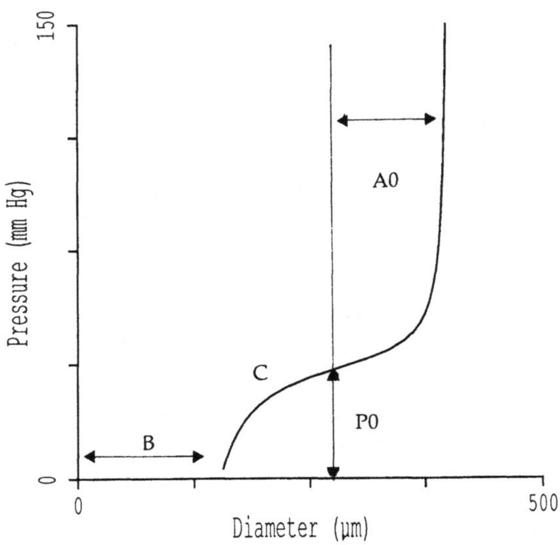

plot of 685 vascular segments. The relation between proximal and distal segment diameters is obtained at the nodes of these segments. Veins and venules were not measured in that study and we have assumed their cross–sectional area to be 3 times as large as arteries and arterioles.

The pressure–cross–sectional area relation (Fig. 1) is described with a shifted tangent function [7]:

$$A = A0 \cdot (B + 2/\pi \cdot \arctan[(P - P0)/C]) \tag{1}$$

where A0 is the amplitude, B the shift of the function, P0 the pressure at the point of symmetry (related to intramyocardial pressure) and C is related to the slope of the midportion of the graph (and the slope is related to elastance).

Note that A0 is the cross–sectional area, while diameter is plotted. For large values of pressure we find: $A = A0 \cdot (B + 1)$; this is the area calculated from the maximal diameter in the study of van Bavel [8]. The parameters for the model are given in Table 1. Note that the value of the parameter C at the venous side is smaller than at the arteriolar side. All the P0 and B values were assumed to be the same. The vessels marked with a5 and v5 (Table 1) are the transmural vessels, i.e. running from endocardium to epicardium and perpendicular to all other vessels. All other vessels are parallel to the muscle fibers. The transmural vessels are divided into seven equal parts (seven layers are used) and the ends of the segments (the nodes) are connected to the vessels from the appropriate layer. The model calculation starts with the endocardial layer, then the segment of the transmural vessel is added. The input pressure at this node is the input pressure for the next layer. The flow through this layer is found with an iteration procedure. The flow that gives the correct pressure is then added to the flow through the endocardial layer. This summated flow is the flow through the next segment of the transmural vessel. This procedure is repeated until the epicardial layer is reached.

Table 1: Diastolic parameters used in the model.

VG	n	L (cm)	P0 (kPa)	A0 (mm^2)	C (kPa)	B
v5	1	2	1	0.171	0.3	1.2
v4	2	0.715	1	0.0514	0.3	1.2
v3	4	0.481	1	0.0174	0.3	1.2
v2	8	0.339	1	0.00668	0.3	1.2
v1	16	0.243	1	0.00270	0.3	1.2
cap	5000	0.050	1	0.0000357	0.3	1.2
a1	16	0.162	1	0.00089	1.0	1.2
a2	8	0.226	1	0.00223	1.0	1.2
a3	4	0.322	1	0.00586	1.0	1.2
a4	2	0.478	1	0.0171	1.0	1.2
a5	1	2	1	0.0571	1.0	1.2

cap – capillaries; L – vessel length; n – number of vessels; VG – vessel generation.

Endo/epi ratios were calculated in the following situations:

1) In the diastolic state (see Table 1): all P0 = 1 (kPa) and B = 1.2. The values for the parameter C at the venous side are smaller than at the arterial side.

2) In the systolic state where it was assumed that all generations have the same values for the parameters P0 = 10(kPa) and C = 5(kPa). All other parameters keep the values as listed in Table 1.

3) The transmural vessel is assumed to be more sensitive to increase in tissue stiffness than the parallel vessels; A0 of the transmural vessel is reduced to 0.02. All other values are as given under 2.

4) The endocardial vessels are more sensitive than the epicardial vessels. P0 and C vary linear from P0 = 10 and C = 5 (as given under 2)) for the endocardial layer to P0 = 1 and C = 1 for the epicardial layer.

RESULTS

In the diastolic state, with all layers identical, an endo/epi ratio of 0.9 is calculated. This is not according to the literature where values in the order of 1.5 - 2.0 are reported. We have therefore assumed that the number of arterioles increases from epicardium to endocardium, with 80% more vessels in the endocardial layer.

Table 2 shows, for the situations mentioned in the methods, total flow, epicardial flow and the endo/epi ratio after the incorporation of the increasing number of vessels per layer in the model:

1) The diastolic endo/epi ratio is: 1.58.

2) In the systolic state with P0 = 10 and C = 5 for all generations. The systolic endo/epi ratio is 1.59. This result is not in agreement with the literature where values smaller than 1 have been reported.

3) The transmural vessel is more sensitive to increase in tissue stiffness than parallel vessels. The endo/epi ratio is now equal to 0.78. This value depends on the difference in sensitivity and can be further reduced.

4) The endocardial vessels are more sensitive than the epicardial vessels. P0 and C vary linearly across the wall. The systolic endo/epi ratio is now equal to 0.20, this is too low when compared to experimental data. This value depends on the range for P0 and C and can be increased if the range is decreased.

Table 2: Flow values in ml/s obtained for various situations model with number of vessels linearly increasing from endocardium to epicardium. Perfusion pressures were approximately 10 kPa (75 mmHg).

Situation	Total Flow	Epi Flow	Endo/Epi Ratio
1. Diastolic	$14.3\ 10^{-3}$	$1.58\ 10^{-3}$	1.58
2. Systolic (fixed)	$0.57\ 10^{-3}$	$0.063\ 10^{-3}$	1.59
3. Transmural vessel	$0.77\ 10^{-3}$	$0.127\ 10^{-3}$	·0.78
4. Systolic (variable)	$2.55\ 10^{-3}$	$0.74\ 10^{-3}$	0.20

DISCUSSION

LVP under maximal vasodilation has a poor predictive value for coronary flow impediment, but contractility does affect coronary flow [1, 4]. Maximal vasodilation is used to avoid the interference of autoregulation with direct mechanical effects. At all levels of contractility it was found during isovolumic beats that phasic arterial inflow is virtually unaffected by the magnitude of LVP, but the amplitude of phasic coronary inflow decreases as contractility decreases. We suggested earlier [1, 4] that the intramyocardial vessels are directly affected by the properties of the surrounding muscle fibers i.e. the time varying elastance concept, as suggested by Suga et al. [6] for the ventricular lumen, is applicable to the coronary vascular compartment as well. As was pointed out in the introduction the elastance concept does predict similar oscillatory flow amplitudes for different LVPs in isovolumic beats and for low loaded, isobaric beats, at a constant perfusion pressure. For a multi layered time varying elastance concept the predictions would be similar, because for each layer the oscillatory flow amplitude would be independent of developed pressure and total phasic flow would be the addition of all these layers. Exact quantitative predictions are up to now impossible because knowledge on a transmural distribution of elastances is still lacking and instantaneous phasic flow measurements in different layers has not been reported yet.

Calculations for a 7 layer model of the ventricular wall were performed in different conditions. The diastolic endo/epi ratio was made equal to 1.58, by a choice of the number of vessels. In the systolic state we assumed P0 = 10 and C = 5 for all generations and found an endo/epi ratio equal to 1.59. This result is not in agreement with the literature where values smaller than 1 are reported. We therefore assumed that the transmural vessels are more sensitive to increase in tissue stiffness than parallel vessels. The endo/epi ratio is now equal to 0.78; thus, due to a significant pressure drop across the transmural vessel in systole, we find endo/epi ratios in the right order of magnitude. Another possibility is that the endocardial vessels are more sensitive than the epicardial vessels. P0 and C vary linear across the wall. The systolic endo/epi ratio is then equal to 0.20.

CONCLUSION

The time varying elastic properties of the cardiac muscle surrounding the intramyocardial vessels affect the compliant (capacitive) and resistive properties of the coronary vessels and thus coronary inflow. We have performed calculations, under 'maximal vasodilation', of the endo/epi ratio in non–beating myocardium as a function of the level of contractility. This ratio is constant if all layers undergo the same change unless the transmural vessel behaves differently or the parameters vary linearly across the wall. The second alternative is probably more realistic, because, as seen in Table 2, epicardial flow is closer to diastolic epicardial flow than in the first alternative.

Further work is necessary to study in detail how the mechanical events alter the hemodynamics. Reliable techniques for instantaneous flow measurement in different layers would advance our understanding of coronary hemodynamics considerably.

DISCUSSION

Dr. Reneman: When you are uncoupling LVP and coronary perfusion pressure in your biological model, you are showing that phasic flow is determined to some extent by the LVP; the total flow is determined only by the LVP and you are dealing here with a pressure dependent flow model. This is indicating that you are working with a non–autoregulating heart. Do you expect a difference when you have an autoregulating heart?

Dr. Sipkema: Yes, but then it gets more complicated: you have two simultaneous effects and it is very difficult to separate them. We are thinking about doing those experiments. You have to do the experiments over very short periods of time, so after you make a change in the system, you have, say, two seconds before regulation starts. In that way we can separate the effects of vasoactivity and contraction.

Dr. Reneman: But then you would expect different results.

Dr. Sipkema: When the bed in the dilated state is compared with the constricted state, then the same contraction may have less effect and the bed may be "protected." But we have to further study this point.

Dr. Beyar: You have shown that intramyocardial compression of coronary flow is independent of LVP at constant contractility. As we know, based on fluid–fiber models, intramyocardial pressure drops from LV cavity pressure at the endocardium to 0 pressure at the epicardium. Therefore, based on your data, it seems that the intramyocardial pressure does not affect coronary compression. To explain this observation, we have modified our earlier fluid–fiber model, which does not take into account the fibrous cross–linkage between layers, to include collagen struts in the radial direction. When you include these elements you get a subendocardial intramyocardial pressure in the fluid which is much higher than the LVP, in cases where significant shortening occurs. Such a model predicts significant intramyocardial pressure for an empty beating heart. One gets considerable intramyocardial pressure due to the stretching of this collagen fiber in the radial direction. Such a model keeps the intramyocardial pressure high in spite of a reduction in afterload and therefore maintains the magnitude of coronary compression. Therefore, our view is that the coronary compression is still mediated via the intramyocardial pressure in the fluid around the coronary vessels, which is higher than the left ventricular cavity pressure. This difference between subendocardial pressure and LVP depends on fiber shortening leading to radial thickening of the subendocardial myocardium.

Dr. Sipkema: We have done these experiments and we find the same difference as in the isovolumically contracting heart.

Dr. Noble: Supposing you had a very simple assumption that the layers of the ventricle move differently. One could then imagine that the penetrating vessels would just shut off. If you make that sort of assumption, would you then get differences between systole and diastole in endocardial flow being recorded?

Dr. Sipkema: What you are suggesting is that we increase the effect on the transmural vessel. We have done a couple of those calculations and the more you shift the pressure–diameter relation to the left, the more this vessel is squeezed by the fibers and the smaller the endo/epi ratio that you can get.

Dr. Beyar: A comment to Dr. Noble's remark on the shear between layers. At Johns Hopkins, we have measured the shearing between layers to be less than 8°, and it is very unlikely that his amount of shearing can shut off the flow.

Dr. Pollack: There is one possible way of looking at the generation of these high forces that devise intermyocardial pressure, which perhaps has not been considered before. Consider the myocardial cell and the thick filaments being interconnected by these structures. Imagine that as the myocardial cell shortens, the thick filaments are shortening and getting fatter as they get shorter. If they are all interconnected by this run like connections, then you can understand perhaps why there would be side force and side pressure generated during the time that these myocardial cells are shortening. None of this might be related to the genesis of intermyocardial pressure.

Dr. Sipkema: But are you suggesting now that the muscle fiber changes in volume?

Dr. Pollack: No, it changes in diameter into tensile structures.

Dr. Sipkema: Yes. Side force is indeed a possibility for direct interaction.

REFERENCES

1. Krams R, Sipkema P, Westerhof N: Contractility is the main determinant of coronary systolic flow impediment. *Am J Physiol* 1989;257:H1936–H1944.
2. Goto M, Flynn AE, Doucette JW, Muehrcke D, Jansen CMA, Husseini W, Hoffman JIE: Impeding effect of myocardial contraction on regional myocardial bloodflow. *Faseb Journal A404,* 1990.
3. Van Winkle DM, Swafford AN, Downey JM: The subendocardial extravascular resistance in dog hearts is independent of ventricular lumenal pressure. *Faseb Journal A946,* 1990.
4. Krams R, Sipkema P, Westerhof N: The time varying elastance concept may explain coronary systolic flow impediment. *Am J Physiol* 1989; 257:H1471–H1479.
5. Downey JM, Downey HF, Kirk ES: Effects of myocardial strains on coronary blood flow. *Circ Res* 1974;34:286–292.
6. Suga H, Sagawa K, Shoukas AA: Load independence of the instantaneous pressure–volume ratio of the canine left ventricle and effects of epinephrine and heart rate on the ratio. *Circ Res* 1973;32:314–322.
7. Sipkema P, Westerhof N: Mechanics of a thin walled collapsible microtube. *Ann Biomed Eng* 1989;17:203–217.
8. Van Bavel ET: Metabolic and myogenic control of blood flow studied on isolated small arteries. Thesis University of Amsterdam, 1989.

CHAPTER 24:

INTRAMYOCARDIAL PRESSURE REVISITED

Nicolaas Westerhof

ABSTRACT

It is reasoned that intramyocardial pressure (IMP) and ventricular pressure result from the varying stiffness of the cardiac muscle. This varying elastance concept, well known for the ventricular lumen, should also be applied to interstitial spaces. Since pressure depends on the momentary value of the time–varying elastance and the momentary volume (ventricular lumen and interstitial volume, respectively), ventricular and IMP ventricular pressure are not necessarily related. Therefore, IMP should not be viewed as the result of pressure in the ventricular lumen but as a direct result of cardiac muscle contraction.

INTRODUCTION

Intramyocardial pressure (IMP) is universally assumed to be equal to the ventricular lumen pressure at the endocardium, and to decrease linearly, to negligible values at the epicardium. Since IMP is considered to be an extravascular pressure, and thus affects transmural pressure of the blood vessels, it is supposed to play a major role in the explanation of coronary flow impediment in systole. These generally accepted notions consequently imply that the left ventricular pressure (LVP) can be used as a predictor of coronary flow. This would mean that during contractions without development of LVP, i.e. during isobaric contractions [1], the coronary flow would not decrease during the cardiac contraction. We have shown recently that this is not the case [1, 2]. Coronary flow was found to decrease strongly in systole in the isolated blood perfused isobarically beating cat heart [2]. It is therefore doubted that these generally held views are correct.

It is suggested here that any "cavity" in the myocardium, e.g. the ventricular lumen, the interstitial spaces and the coronary vessels, are subject to the action of the surrounding cardiac muscle. The pressure in these "cavities" arises from the varying stiffness of the muscles surrounding them. This leads to a time varying elastance, i.e. pressure–volume ratio. The concept is shown in Fig. 1. The varying elastance concept as applied to the ventricular cavity is well known [3]. When the time varying elastance is a given characterization of the ventricle, the pressure in the ventricle is high in systole when isovolumic conditions are approached and low when isobaric conditions are present. Thus pressure in the ventricular lumen depends on the varying elastance in combination with the load on the ventricle [3].

It is suggested that the interstitial "cavity" is also subject to a varying elastance and thus its IMP depends directly on the varying muscle properties and the load on the cavity. The load on an interstitial cavity depends on how difficult, or easy, it is to displace liquid from the cavity. Measurement of IMP requires the introduction of an artificial "cavity". This may affect the conditions of the loading conditions of the "cavity" and thus the IMP. When

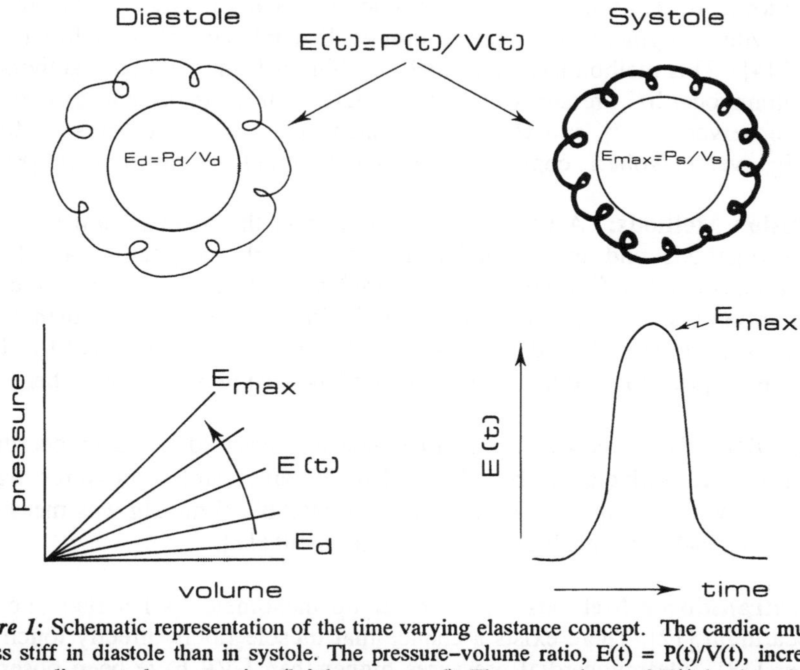

Figure 1: Schematic representation of the time varying elastance concept. The cardiac muscle is less stiff in diastole than in systole. The pressure–volume ratio, E(t) = P(t)/V(t), increases during cardiac muscle contraction (left bottom panel). The wave shape of E(t) is given in the right bottom panel. When the elastance (E) of any intramyocardial cavity is small, as in diastole (E_d), the pressure P_d and volume V_d are related as indicated. When E is large (high elastance, as in systole, E_{max}) any cavity will in general be subject to an increase in pressure and/or decrease in volume. Pressure change is maximal if volume cannot change isovolumic conditions. If the cavity is in connection with the environment its volume can change strongly but pressure will not change isobaric conditions.

the "cavity" which is created for IMP measurement is under isovolumic conditions IMP will be high, while under isobaric conditions IMP will be low.

Thus LVP and IMP can both be explained from the varying stiffness of cardiac muscle during contraction, but their respective values depend on the load on the cavities and therefore these pressures are in general not related, except when the conditions are exactly defined. It is therefore not surprising that IMPs higher and lower than ventricular pressure have been observed.

Here, we will try to explain the IMP data reported in the literature on the basis of the time varying elastance concept.

METHODS

Intramyocardial pressure measurements have been carried out for a long time mainly in order to gain insight into the effects of cardiac contraction on coronary flow.

Isovolumic or Closed Methods: An intramyocardial cavity was created, either with a segment of vessel or thin–walled tube which was filled with liquid [4]. This method has been used in different forms by many authors [5, 6]. Intramyocardial pressures higher than left ventricular (or aortic) pressure were observed [4, 7], and IMP was found to exist with pressures up to 60 mmHg, even in isobaric contractions with negligible LVP development [8].

Perfusion Methods: A vessel was embedded in the myocardium as in the closed methods and the perfusion pressure to the segment was slowly decreased [4]. When flow ceased, the perfusion pressure was assumed to equal IMP. Intramyocardial pressure was usually found to be smaller than LVP. However, systolic endocardial IMPs up to 120 mmHg were observed in iso–baric beats [9]. More extensive information has been presented elsewhere [6].

Open Methods: A needle or small cannula was inserted in the myocardium and, if desired, saline could be infused [10]. Intramyocardial pressures higher than the LVP were found [10] and a systolic IMP of 60 mmHg was measured in isobaric beats with negligible LVP development [11].

Microtransducer Methods: A catheter tip manometer is inserted into the myocardium [12]. There exists a chance that muscle fibers directly touch the sensor [13]. Intramyocardial pressures higher than LVP have been observed [12]. This method has recently been discussed in detail [6, 13].

Servo–Null Method: With a micropipette and a servo–null system Heineman and Grayson [14] recently measured IMPs and found that the IMP never exceeded LVP.

RESULTS AND DISCUSSION

The results by a great many authors show that in all methods, except the servo–null method, IMPs were observed to be higher than LVP. In most methods isobaric beats without pressure development in the LV demonstrated considerable IMPs [6]. Almost all methods find that IMPs at the endocardium are higher than at the epicardium [6].

The Varying Elastance Concept

The concept brought forward here is based on the similarities between "cavities" in the myocardium, the lumen of the LV and the interstitial cavities. The pressures in these "cavities" are LVP and IMPs, respectively. The surroundings of the "cavities" (i.e. the myocardium) becomes stiff in systole as compared to diastole. Ventricular lumen elastance was first introduced by Suga et al. [3], and the time varying elastance, E(t), of a cavity was defined as the ratio of pressure, P(t), and volume, V(t) [3]. For the sake of simplicity we use $E(t) = P(t)/V(t)$ and omit the "rest volume" [3]. Possible curvilinearity and dependence on the load of the pressure–volume relation are also not considered. These approximations are not essential for the qualitative reasoning used in the following sections. The elastance is small in diastole and large in systole; the so–called end–systolic pressure–volume relation or E_{max} is denoted. This relationship has been extensively discussed [3] for the ventricular lumen. Since E(t) varies in time, but is approximately independent of the load on the ventricle, it follows that the pressure as a function of time depends on the E(t) and on the volume change. For isovolumic contractions (volume constant) pressure will be high in systole. When volume is increased in diastole and then kept constant again during the contraction, the pressure in systole will be higher. For a decrease in volume during contraction (ejection), the pressure in systole will be lower than for isovolumic beats. When the ventricle is made to eject against a very low load pressure, it will have hardly, if any, increase in systole, but the change in volume will be large. Increased contractility results in an increase in the E_{max} [3], and thus, for a similar volume and in isovolumic conditions, a higher pressure is found in systole with increased contractility.

We now consider a hypothetical interstitial volume and the pressure within it, the IMP. Pressure and volume of that cavity are also subject to a time varying elastance relation $E(t) = P(t)/V(t)$. If volume can not leave this "cavity" then pressure will increase along with E(t) in systole (isovolumic conditions). When liquid can be displaced (for instance to the environment) volume decreases and pressure will be lower than during isovolumic conditions. When contractility increases, the E_{max} will be larger [3] and, for similar loading conditions, IMP will increase.

For constant conditions of loading of the interstitial "cavity", but changing load on the LV IMP remains the same but LVP will change. Since the pressure in the ventricular lumen and in an interstitial "cavity" both depend on their loading conditions they may be different and are not proportional. Only when the conditions are chosen such that the ventricular lumen and the interstitial "cavity" remain under the same conditions, will an increase in contractility result in a similar change in the two pressures. It is therefore not surprising that many researchers observe higher IMPs than LVP, a finding that cannot be explained when IMP is assumed to be derived from the LVP.

We have recently reported data on coronary flow impediment in systole that can be explained only on the basis of the time varying elastance concept [1, 2]. The coronary vessels also form a "cavity" in the myocardium and are therefore subject to the time varying properties of the cardiac muscle. LVP is therefore a poor predictor of coronary flow impediment.

Interpretation of Intramyocardial Pressure Findings

In all measurement techniques to obtain IMP, relatively large (with respect to interstitial spaces), cavities are created. If damage is small the methods to measure IMP (except the open method) may approach isovolumic conditions so that high IMPs are to be expected. If damage is large, liquid may be squeezed out to the environment when the muscle contracts and IMPs are expected to be low. Thus reports where high IMPs are reported may point to experimental results based on little damage. During subendocardial pressure measurement a longer channel to the environment is present than in subepicardial measurements. This long channel forms a large resistance to the environment and may keep the endocardial "cavity" under more isovolumic conditions than the epicardial "cavity" so that the endocardial pressures are higher than the epicardial pressures. When the volume of the created "cavity" is increased by infusion of fluid, the varying elastance concept predicts an increase in IMP. This is indeed observed for the isovolumic method [7], the open method [15], and the microtransducer method [16].

The measured IMP in ischemic areas is less than in normally contracting areas with obviously the same LVP [17]. If IMP is an outcome of the LVP, the same IMPs would be expected in both of these areas.

Post extra–systolic potentiation [17] or increased contractility ("strength of contraction" [4]) increase the E_{max} [3] and are predicted to produce a higher IMP; this was indeed found to be the case [4, 17].

We therefore conclude that IMP, and coronary flow impediment in systole, should not be viewed as the result of pressure in the ventricular lumen but rather, as a direct result of cardiac muscle contraction.

DISCUSSION

Dr. Reneman: I like that concept very much. People who have been working in open heart surgery, 20–25 years ago, saw the phenomenon of an empty beating left ventricle and when blood perfusion is at 70 mmHg you can see back-flow from the myocardium into the aorta. In other words, it is generating pressure in the wall as the full ventricle. Basically what you are saying is that in normal filling in normal function there is a relation between left ventricular pressure and IMP but in empty beating hearts there is not. How do you explain in your concept the gradient from endocardium to epicardium, is an empty beating left ventricle. Are you thinking about differences in force in epicardial or endocardial layers? But then I have a problem that when you take into account torsion of the left ventricle there is a homogeneous distribution of fiber forces. Or, are you thinking that the influences on the elastance are different for the endocardial to epicardial layers, which will explain a difference in the intramyocardial pressure in the wall.

Dr. Westerhof: What I am going to say is purely hypothetical. First, let me suggest that people who found very high intramyocardial pressures probably did a good experiment because they made little damage. I am not saying that it is a true IMP, but it may be. A very crucial experiment to do, therefore, is to approach the sensor from the endocardium. All measurements up till now were done by pushing a needle on some tube segment in from the epicardium. When you do this you make a hole so that you have more likelihood that fluid will leak out in systole and you have a lesser isovolumic condition. It may thus be that the IMP is found higher endocardially because the conditions there are more isovolumic. If approached from the endocardium, we may find an intramyocardial pressure that is lower endocardially than epicardially.

All methods show that IMP can be higher than left ventricular pressure, except Heineman's technique *[14]*. He first made a hole and then put this micropippete in. I do not know how much damage he did, but it may well be that the leak was considerable and the site of the measurement far from isovolumic. The thing to worry about is not the size of the sensor. Rather, it is the way you put it in. If you have a large leak here you get lower intramyocardial pressures since conditions are less isovolumic.

REFERENCES

1. Krams R, Sipkema P, Westerhof N: Can coronary systolic–diastolic flow differences be predicted by left ventricular pressure or time–varying intramyocardial elastance? *Basic Res Cardiol* 1989;84:149–159.
2. Krams R, Sipkema P, Westerhof N: Varying elastance concept may explain coronary systolic flow impediment. *Am J Physiol* 1989;257:H1471–H1479.
3. Suga H, Sagawa K, Shoukas AA: Load independence of the instantaneous pressure–volume ratio of the canine left ventricle and effects of epinephrine and heart rate on the ratio. *Circ Res* 1973;32:314–322.
4. Johnson JR, DiPalma JR: Intramyocardial pressure and its relation to aortic blood pressure. *Am J Physiol* 1939;125:234–243.
5. Nematzadeh D, Rose JC, Schryver TH, Huang HK, Kot PA: Analysis of methodology for measurement of intramyocardial pressure. *Basic Res Cardiol* 1984;79:86–97.

6. Westerhof N: Intramyocardial pressure. *Basic Res Cardiol* 1990;85:105–119
7. Gregg DE, Eckstein RW: Measurements of intramyocardial pressure. *Am J Physiol* 1941;132:781–790.
8. Salisbury PF, Cross CE, Rieben PA: Intramyocardial pressure and strength of left ventricular contraction. *Circ Res* 1962;10:608–623.
9. Baird RJ, Manktelow RT, Shah PA, Ameli FM: Intramyocardial pressure. The persistence of its transmural gradient in the empty heart and its relationship to myocardial oxygen consumption. *J Thoracic and Cardiovasc Surg* 1972; 64:635–646.
10. Laszt L, Mueller A: Der Myokardiale Druck. *Helv Physiol Acta* 1958; 16:88–106.
11. Kreuzer H, Schoeppe W: Zur Entstehung der Differenz zwischen systolischem Myokard–und Ventrikeldruck. *Pflügers Arch* 1963;278:199–208.
12. Armour JA, Randall WC: Canine left ventricular intramyocardial pressures. *Am J Physiol* 1971;220:1833–1839.
13. Rabbany SY, Kresch JY, Noordergraaf N: Intramyocardial pressure: interaction of myocardial fluid pressure and fibre stress. *Am J Physiol* 1989;257:H357–H364.
14. Heineman FW, Grayson J: Transmural distribution of intramyocardial pressure measured by micropipette technique. *Am J Physiol* 1985;249:H1216–H1223.
15. Brandi G,McGregor M: Intramural pressure in the left ventricle of the dog. *Cardiovasc Res* 1969;3:472–475.
16. Kreuzer H, Schoeppe W: Das Verhalten des Druckes in der Herzwand. *Pflügers Arch* 1963;278:181–198.
17. Kreuzer H, Schoeppe W: Der Myokarddruck bei vernederter Coronar-durchblutung und bei Ischemie. *Pflügers Arch* 1963;278:209–220.

CHAPTER 25:

ON THE MECHANISM OF TRANSMURAL MYOCARDIAL COMPRESSION AND PERFUSION

Rafael Beyar, Reuven Kamminker, Dan Manor, Riva ben Ari and Samuel Sideman

ABSTRACT

While the waterfall (WF) model predicts coronary arterial inflow reasonably well, it fails to predict venous flow. This is adequately predicted here by a pressure dependent (PD) model of transmural myocardial perfusion to the intramyocardial pressure (IMP) distribution, assuming pressure dependence of the different intramyocardial compartments. The vessel area is allowed to vary with transmural pressure. Under ischemic conditions two phenomena are predicted: (1) an alternating flow between subendocardium and subepicardium, and (2) a collapse of the intramyocardial arterial system at the subendocardium. It is also shown that when the left ventricle (LV) operates under conditions involving high shortening, such as decreased afterload and increased contractility, a considerable difference between the IMP at the subendocardium and the LV cavity pressure can develop. These effects are related to the stiffness in the radial direction due to the collagen fibers interconnecting the muscle fibers, which need be included in the analysis of myocardial compression over a wide range of load and contractility.

INTRODUCTION

The mechanisms involved in the interaction between myocardial function and transmyocardial perfusion are highly complex and are the subject of extensive research. However, in spite of these efforts, access into the subendocardial layers has been difficult and measurements of pressure and flow in these locations are scanty and unreliable. Measurements are more available for the entry (arterial) and exit (venous) parts of the coronary circulation and for the subepicardial layers [1–5]. Models are therefore required to extrapolate the available data to the unreachable compartments. Consequently, all present models are limited by both data and assumptions. For instance, our earlier coronary perfusion model [6], which employs the WF concept of the coronary circulation, does not account for the coronary veins, and a new concept is therefore needed. Recent models have suggested that a relationship between the local transmural pressure and the local cross–sectional area can be used to describe the coronary circulation [7–9]. Here we present a model which accounts for the pressure dependent (PD) cross–sectional area of the coronary intramyocardial vessels, including the coronary venous system as well as the collateral circulation, and avoids the need to use the coronary backpressure which is different from the atrial pressure. The present PD model is tied to LV mechanics via an approximate linear transmural IMP distribution. Possible mechanisms that may affect the correct development of the IMP are discussed in light of the known fibrous structure and function of the myocardium, based on a modification of our earlier fluid–fiber model of transmural ventricular mechanics [10].

THE CORONARY MODEL

General Structure

The coronary model is presented in Fig. 1 in an electrical analog form. The intramyocardial bed is structured of a normal and an ischemic vascular bed, each comprised of 3 layers: subepicardial, midwall and subendocardial. The ischemic bed is interconnected to the normal bed by arterial collaterals at each layer. The bed can be subjected to ischemia by either occluding the artery or imposing a local stress characterized by simple resistance. Each layer of the model is represented by 4 compartments: large vessels, small vessels, small veins and large veins. Each compartment is characterized by a PD resistance and capacitance. An epicardial capacitance is also assumed in the model. The three layers are connected by the penetrating arteries which are simulated as simple resistors.

The Pressure–Dependent Resistance and Capacitance

The vessels are assumed to have a nonlinear relationship between their local cross–sectional area, A, and transmural pressure, P_{tr}. Although data in the literature is not available for all types of vessels, we adopt the relationship given by Spaan et al. [11] with certain modification (Fig. 2). The positive pressure part of the relationship is given by a fourth order polynomial

$$A = K_0 + K_1 \cdot P_{tr} + K_2 \cdot P_{tr}^2 + K_3 \cdot P_{tr}^3 + K_4 \cdot P_{tr}^4 \qquad (1)$$

where K_i are the polynomial constants.

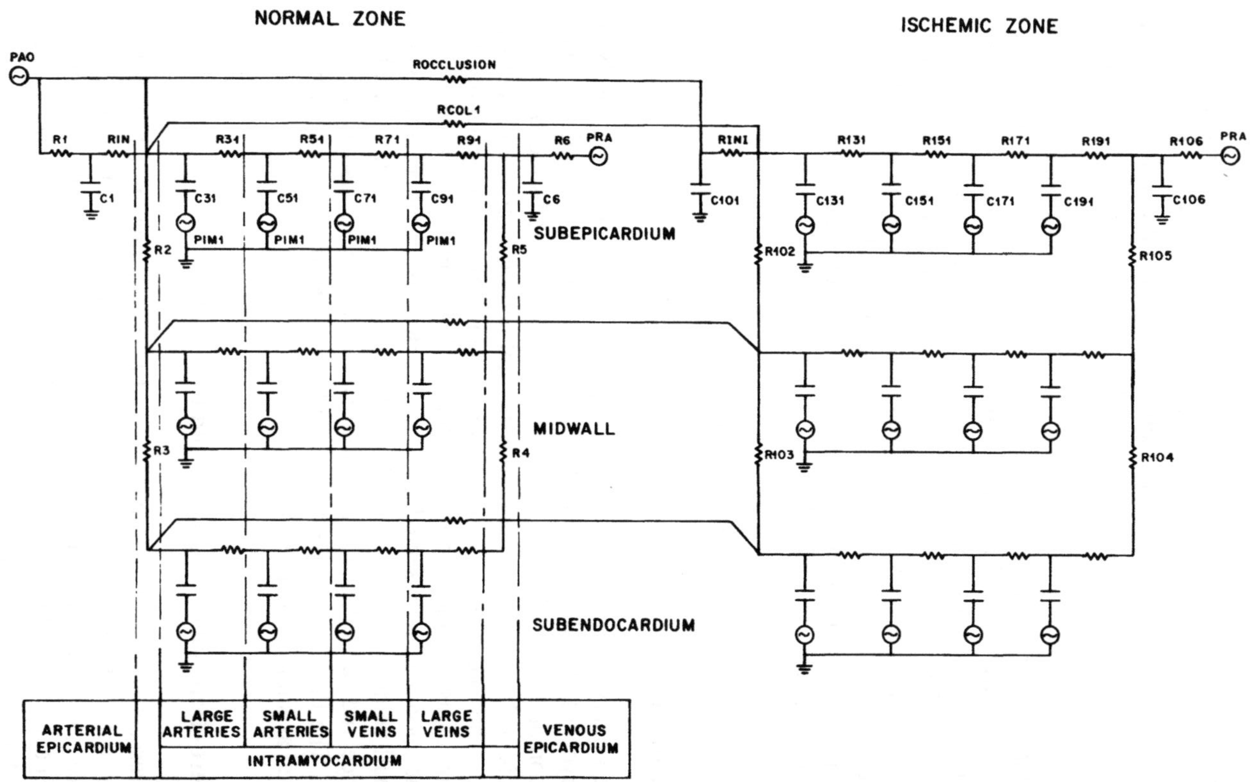

Figure 1: An analog presentation of the intramyocardial bed. The normal zone is connected to the ischemic zone by collaterals, at all layers. Each zone is divided to 3 layers: the subendocardium, mid–layer and the subepicardium. Four capacitors and four resistors account for large and small intramyocardial arteries and veins in each layer. The capillaries are included within the small venular compartment. Capacitors are used to describe the arterial and venous epicardial vessels.

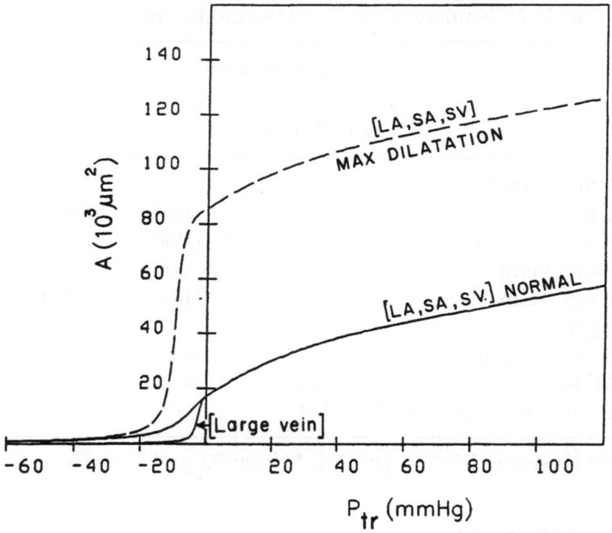

Figure 2: The relationship between vessel transmural area, A, and its transmural pressure P_{tr}. Note that maximum vasodilatation is represented by an upward parallel shift of this relationship for the large arteries (LA), small arteries (SA) and small veins (SV). Also note that the large veins are assumed to be more sensitive to collapse, as reflected by the steeper relationship at transmural pressure below zero.

The negative part is given by:

$$A = A_o \cdot [1 + 2/\pi \cdot atan\,(P_{tr} - P_o/C) + E] \qquad (2)$$

where A_o, P_o C and E are constants. Note that veins are considered to collapse at smaller transmural pressures than arteries. Also note that, for lack of actual data, the degree of vasodilation is simulated as a parallel upward shift of the curve.

The resistance and capacitance of each compartment are related to the instantaneous transmural pressure by:

$$R_{(i,j)} = K_{r1}(i) \cdot K_{r2}(j) \cdot 1/A^2$$
$$C_{(i,j)} = K_{c1}(i) \cdot K_{c2}(j) \cdot dA/dP_{tr} \qquad (3)$$

where $K_{r1}(i)$, $K_{c1}(i)$ and are constants related to layer (i), and $K_{r2}(j)$ and $K_{c2}(j)$ are related to compartment (j). The numerical data of the model coefficients are given in Table 1.

Table 1: Numerical values of the model coefficients.

	Coefficients			
	K_{r1} (*Unit)	K_{c1} (**Unit)	K_{r2}	K_{c2}
Compartment				
Large arteries	–	–	70,000	0.003
Small arteries	–	–	100,000	0.0125
Small veins	–	–	30,000	0.024
Large veins	–	–	10,000	0.009
Layer				
Subepicardial	1.6	1.0	–	–
Midway	1.1	1.0	–	–
Subendocardial	0.75	1.0	–	–

*[mmHg·s·cm·10^{-10}·33 gmLV]; **[(cm/33 gmLV)·10^{8}]

Intramyocardial Pressure

The extensive research on the transmural IMP distribution [12–15] is reviewed by Hoffman and Spaan [16]. As a first approximation, and for simplicity of comparison between the WF and PD models, the IMP of the LV is assumed here to decrease linearly from the LV cavity pressure at the endocardium to zero pressure at the epicardium. However, the assumption that IMP depends on the cavity pressure at the endocardium is contrasted by various measurements that show that the IMP is higher than the LV cavity pressure in the subendocardial layers. While this information has commonly been considered to be an artifact, attributed to the local distortion caused by the measurement device, recent theoretical studies which account for the collagen fibers in the transverse stress analysis, have shown that the IMP may indeed be higher than the LV cavity pressure. Also, evidence that significant intramyocardial compression may occur even at zero cavity pressure [17] or at low afterloads with normal contractility [18], suggest that relating the IMP only to the LV cavity pressure may considerably underestimate the IMP at the subendocardial layer at certain conditions. Therefore, while it is assumed here that the LV cavity pressure is dominating the IMP, it is noted that modification of this assumption may be necessary under certain conditions.

Numerical Calculations

The model was solved by the Runge Kutta Verner solution of the first order. A Vax 780 was used for the simulation with a run time of approximately 10 minutes. Time dependent graphic and numerical output is produced by the model.

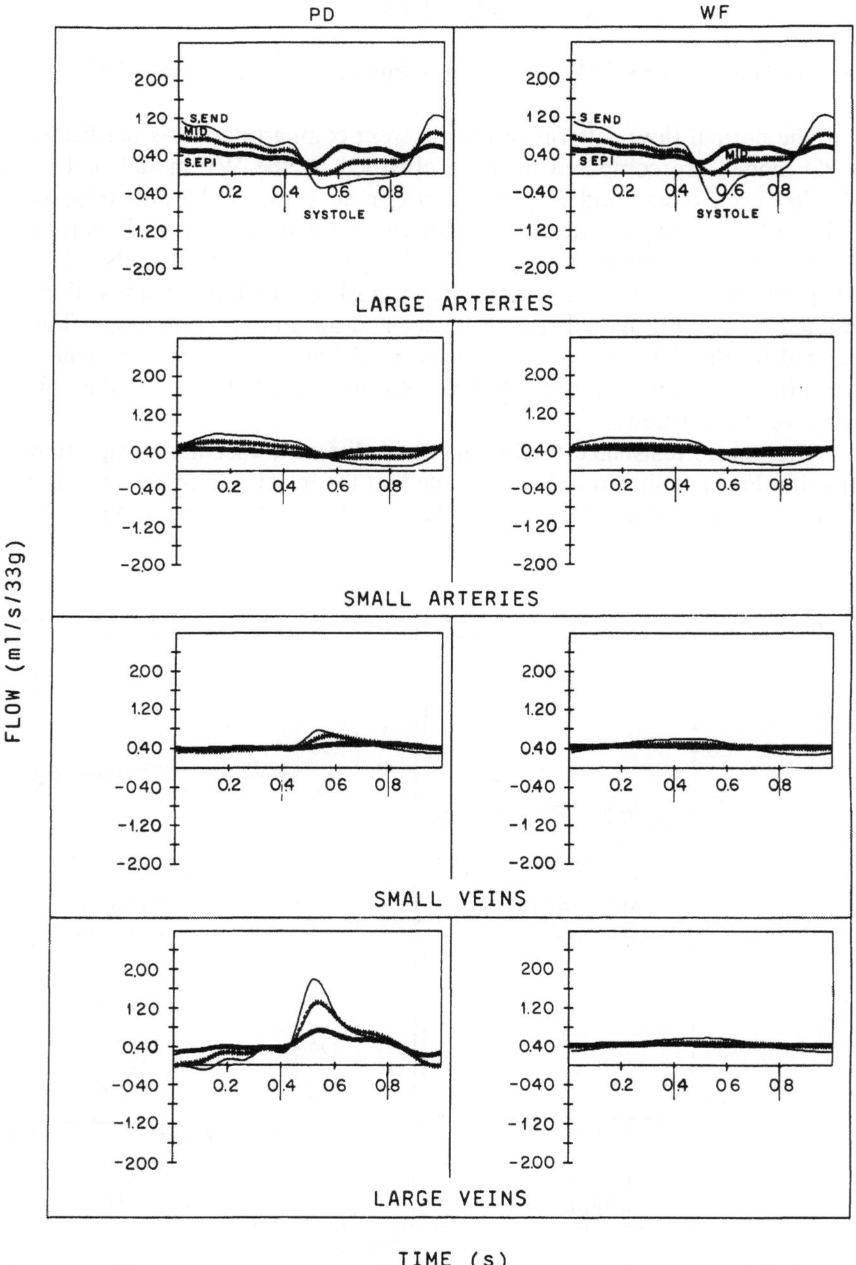

***Figure 3*:** Normal transmural flow distribution in the three layers for the pressure dependent (PD) vs. the waterfall model (WF). Note the similarity in arterial flows for both large and small vessels. In contrast, note the dissimilarity in venous flow which starts at the smaller vessels and is most obvious for the larger subendocardial veins.

RESULTS

Pressure Dependent vs. Waterfall Flow Patterns in the Normal LV

The normal flow patterns in the different compartments, as predicted by the PD model, are compared to those obtained by the WF model in Fig. 3. Note that both these models predict retrograde flow in the subendocardial arterial inflow during systole and the patterns of the inflows, as well as flow in the small arterial system, are quite similar. However, the models differ in their predictions of the venous outflow; positive systolic venous flow is predicted by the PD model for all layers and an almost continuous flow is predicted by the WF model. In view of available experimental evidence on venous flows [19], it is apparent that the WF model fails to predict the typical cardiac venous outflows.

The phasic resistance to flow in the 4 different vascular compartments defined in Fig. 1 is shown in Fig. 4. Note that the resistance of the venous bed increases during systole, reflecting the large decrease in venous volume which

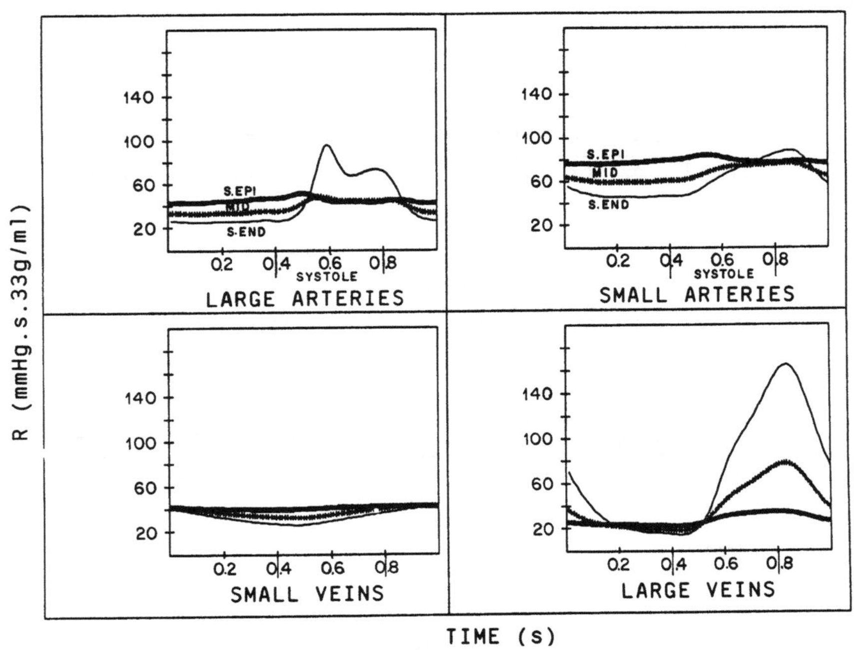

Figure 4: Resistance as a function of time in the different compartments of the 3 layers. Note the large change in resistance during systole for the large veins, particularly in the subendocardium. Also note that the subendocardial larte arteries compartments show a relatively large change in resistance.

Figure 5: Flow into the 3 myocardial layers in the ischemic LV for the pressure dependent model, with complete coronary obstruction, and with collateral compensation supplying the baseline flow under maximum vasodilatation. Note that the negative subendocardial flow and positive subepicardial flow at the beginning of systole reverses in diastole.

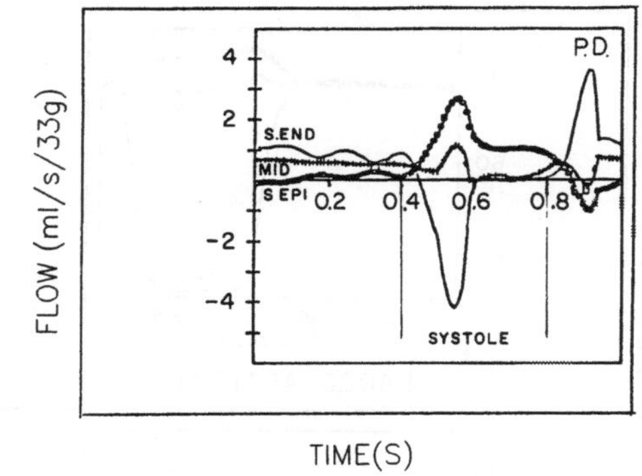

occurs mostly in the larger intramyocardial veins. Also noteworthy is the significant increase in the arteriolar resistance of the subendocardial layers during systole.

The inflow to the different layers during ischemia is particularly interesting. As seen in Fig. 5, at the beginning of systole there is retrograde subendocardial flow and positive subepicardial flow. This pattern reverses itself during the isovolumic relaxation phase. This suggests the existence of an alternating flow between the subendocardium and subepicardium during ischemia, which can be attributed to a gradient in the IMP. This behavior may partially explain the "hidden flow" during ischemia, which is the difference between flow measurements using tracer washout techniques and microspheres measured flow [20, 21].

A challenging prediction is shown in Fig. 6: the intramyocardial arteries will collapse during systole at the subendocardium due to the large local negative transmural pressures there. This prediction, which appears only during the ischemic process due to the low arterial pressure distal to the obstruction, has yet to be validated experimentally.

Intramyocardial Pressure and Myocardial Compression

As stated above, it is commonly accepted that the IMP increases from the zero epicardial pressure to the LV cavity pressure at the endocardium. This concept is consistent with predictions of fluid–fiber based models of myocardial mechanics [10, 22] as well as some experimental measurements [16]. However, Krams et al. [18] have recently shown that myocardial compression is a function of the LV elastance and is therefore dissociated from the LV cavity pressure when the LV encounters a low afterload. Obviously, a model that uses LV cavity pressure as a sole predictor of the IMP will underestimate the compression effects in those conditions.

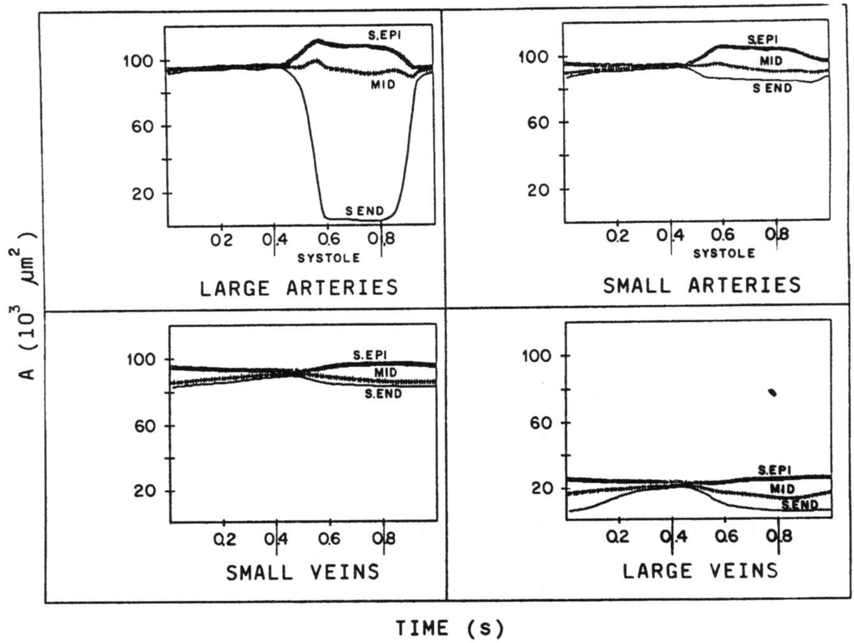

Figure 6: The cross–sectional area as a function of time in the different compartments for the ischemic case. Note that the subendocardial arteries collapse here during systole, and that the venous compartments also undergo large changes in the cross–sectional area during systole.

To examine the relationship between the LV cavity pressure and the IMP, we have to extend our earlier model of LV mechanics [10]. Structured as a highly tethered muscle fiber architecture, the model does not include any transverse fibrous stiffness, and therefore has no stress component in the radial direction other than the hydrostatic fluid pressure. Adding the collagen connections between muscle fibers [23, 24] introduces stiffness in the cross fiber direction in the model.

As shown in Fig. 7, the modified model predicts that the IMP remains high at decreasing afterload conditions, in spite of the decrease in LV cavity pressure. The difference between subendocardial IMP and LV cavity pressure is found to be related to fiber shortening; the higher the fiber shortening, the higher the difference. Therefore, myocardial compression by virtue of the IMP alone may explain the observations of Krams et al. [18]. This mechanism is further detailed in Fig. 8.

When the myocardium contracts, it shortens in all in–plane directions but increases in thickness in the radial direction, thus stretching the collagen fibers which connect between the layers of muscle fibers. The stretching causes a high stress in the radial direction which generates a significant

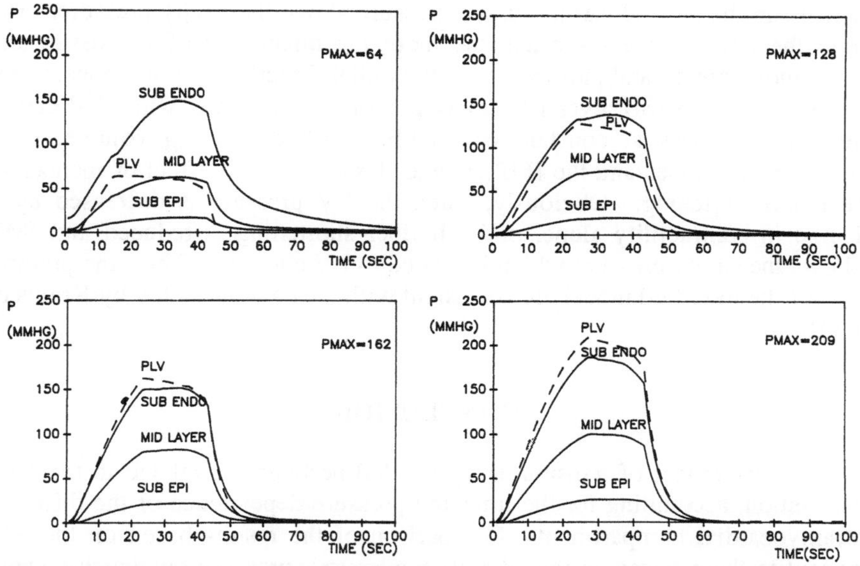

Figure 7: The intramyocardial pressure (IMP) for the 3 layers of the myocardium under conditions of varying afterload pressure with constant contractility. Note that while the LV cavity pressure decreases considerably, the IMP at the subendocardial layer remains high and exceeds the LV cavity pressure.

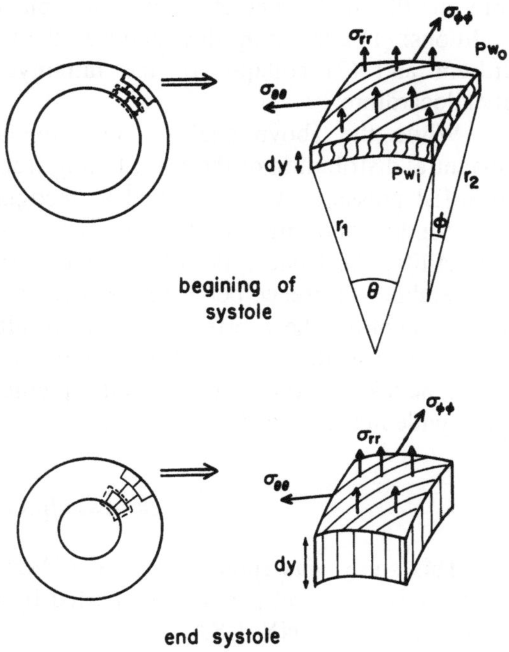

Figure 8: A shell of muscle at ED *(top panel)* of thickness *dy* undergoes in–plane shortening and transverse thickening as it deforms to ES *(bottom panel)*. As a result of the thickening, the radial stiffness components stretch and generate stress in the radial direction. This generated stress component causes an increase in the intramyocardial pressure above LV cavity pressure.

increase in the subendocardial tissue pressure above the cavity pressure. The more the muscle fibers shorten, the more the thickening of the tissue and hence more stretch and more stress in the radial directions, with a consequent larger increase in IMP over LV cavity pressure. Therefore, when LV cavity pressure decreases at constant contractility with decreasing volume, fiber shortening increases and the IMP changes less due to the effect of increasing the radial stretching. Obviously, when the LV pressure is decreased by a change in contractility alone, midwall fiber shortening is reduced and IMP follows the cavity pressure which is related to LV elastance. Thus, the predictions of the modified model are consistent with the data presented by Krams et al. [18].

CONCLUSION

A new model of transmural myocardial perfusion is related to the IMP distribution, accounting for the time and pressure dependence of the different intramyocardial compartments. Comparison of the results based on the WF concept to those based on vessel area–transmural pressure relationship show that while the WF model predicts arterial inflow reasonably well, it fails to predict the venous flow, which is adequately predicted by the PD model. The PD model predicts that a marked decrease in cross–sectional area of the large intramyocardial veins occurs during systole at the subendocardium. Furthermore, the PD model predicts 2 additional phenomena to occur under ischemic conditions: (1) alternating flow between the subendocardium and the subepicardium and (2) collapse of the intramyocardial arterial system at the subendocardium.

While the above analysis and conclusions are based on a linear transmural distribution of the IMP, falling from the LV cavity pressure to zero epicardial pressure, it is important to recognize that when the LV operates under conditions of high shortening, such as decreased afterload and increased contractility, the basic fluid–fiber model [10] can be modified to predict a considerable difference between the IMP at the subendocardium and the LV cavity pressure. These effects are the result of the stiffness in the radial direction due to the collagen fibers interconnecting the muscle fibers which need to be taken into account for better prediction of myocardial compression over a wide range of loads and contractility.

Acknowledgment

This study was sponsored by the Women's Division of the American Technion Society and partially supported by the Pearl Milch and H. Kaplan Endowment Funds, NY, USA.

DISCUSSION

Dr. Strauss: Do you know what happens to the myocardial pressure in a tamponade model.

Dr. Beyar: No. We have not related to that specifically. I assume that an increase in the IMP will occur due to the increase in the external pressure around the LV. In addition, increased fiber shortening due to the external pressure should increase IMP due to the stretch of fibers in the radial direction. This should cause a significant impairment in subendocardial flow under these conditions.

Dr. Spaan: It is all getting more complicated every time I listen to you. Why do you predict the collapse of arterioles in the endocardium during ischemia?

Dr. Beyar: What happens is that a subendocardial pressure of about 175 mmHg is generated while the entrance pressure after the coronary stenosis is in the range of 40 to 60 mmHg. Therefore, the pressure external to the vessel is higher than the pressure in the vessel, which leads to zero cross-sectional area corresponding to a total collapse at the levels of the arterioles, as well as in the venous system.

Dr. Spaan: And that is during normal contraction? during ischemia? Looking at the model, I see no blood flow so there is low contractility and there would be restoration of flow.

Dr. Beyar: We use the same IMP distribution for normal and ischemic cases. This may be incorrect since myocardial shortening may alter the IMP. Under the condition of this simulation, this is what is predicted.

Dr. Westerhof: Do I understand you correctly that in your model the arterial resistance hardly changes so that the arterial inflow does not change due to arterial resistance change but rather due to venous changes?

Dr. Beyar: Correct. The arterial resistance is only slightly changed throughout the cycle, therefore, the changes in arterial inflow are affected to a large extent by discharge of the arterial capacitance and to a lesser extent by the venous capacitance.

Dr. Reneman: How do you define collapse in your model?

Dr. Beyar: I tried not to define collapse. We have seen that on the venous side there is a marked decrease of the venous diameter accompanied by a marked increase in venous resistance which we can attribute to partial collapse and not to total collapse. On the arterial side, in acute ischemia, we got a reduction of the cross-sectional area almost to zero, which is practically a complete collapse.

REFERENCES

1. Kajiya F: Evaluation of coronary inflow and outflow from myocardium by laser Doppler method. *Med Biol Eng Comput* 1985;23(Supp 2):1315–1316.

2. Tillmanns H, Ikeda S, Hansen H, Sarma JSM, Fauvel JM, Bing RJ: Microcirculation in the ventricle of the dog and turtle. *Circ Res* 1974;34:561–569.
3. Ashikawa K, Kanatsuka H, Suzuki T, Takashima T: Phasic blood flow velocity pattern in epimyocardial microvessels in beating canine left ventricle. *Circ Res* 1986;59:704–711.
4. Chilian WM, Marcus ML: Phasic coronary flow velocity in intramural and epicardial coronary arteries. *Circ Res* 1982;50:775–781.
5. Chillian WM, Eastham CL, Marcus ML: Microvascular distribution of coronary vascular resistance in beating left ventricle. *Am J Physiol (Heart and Circulation Physiol)* 1986;20:H779–H788.
6. Beyar R, Sideman S: Time dependent coronary blood flow distribution in the left ventricular wall. *Am J Physiol (Heart and Circ Physiol 21)* 1987;252:H417–H433.
7. Bruinsima P, Arts T, Dankelman J, Spaan JAE: Model of the coronary circulation based on pressure dependence of the coronary resistance and compliance. *Basic Res Cardiol* 1988;83:510–524.
8. Kresh JY, Fox M, Brockman S, Noordergraaf A: Model based analysis of transmural vessel impedance and myocardial circulation dynamics. *Am J Physiol (Heart Circ Physiol 27)* 1990;258:H262–H276.
9. Chadwick RS, Tedgui A, Michel JB, Ohayon J, Levy BI: Phasic regional myocardial inflow and outflow: comparison of theory and experiments. *Am J Physiol (Heart Circ Physiol 27)* 1990;258:H1687–H1698.
10. Beyar R, Sideman S: A computer study of left ventricular performance based on fiber structure, sarcomere dynamics, and transmural electrical propagation velocity. *Circ Res* 1984;55:358–375.
11. Spaan JAE: Coronary diastolic pressure flow relation and zero flow pressure explained on the basis of intramyocardial compliance. *Circ Res* 1985; 56:293–309.
12. Van Der Meer JJ, Reneman RS, Schneider H, Wieberdink J: Technique for estimation of intramyocardial pressure in acute and chronic experiments. *Cardiovasc Res* 1970;4:132–140.
13. Brandi G, McGregor M: Intramural pressure in the left ventricle of the dog. *Cardiovasc Res* 1969;3:472–475.
14. Baird RJ, Manktelow RT, Shah PA, Ameli FM: Intramyocardial pressure: a study of its regional variations and its relationship to intraventricular pressure. *J Thorac Cardiovasc Surg* 1970;59:810–823.
15. Salisbury PF, Cross CE, Rieben PA: Intramyocardial pressure and strength of left ventricular contraction. *Circ Res* 1962;10:608–623.
16. Hoffman JIE, Spaan JAE: Pressure flow relation in the coronary circulation. *Physiol Rev* 1990;70:331–390.
17. Baird RJ, Goldbach MM, De La Rocha A: Intramyocardial pressure: The persistence of its transmural gradients in the empty heart and its relationship to myocardial oxygen consumption. *J Thorac Cardiovasc Surg* 1970;59:810–823.
18. Krams R, Sipkema P, Zegers J, Westerhof N: Contractility is the main determinant of coronary systolic flow impediment. *Amer J Physiol (Heart Circ Physiol 26)* 1989;:H1936–H1944.
19. Armour JA, Klassen GA: Pressures and flows in the epicardial coronary veins of the dog heart – responses to positive inotropism. *Can J Physiol Pharmacol* 1983;62:38–48.
20. Maxwell MP, Hearse DJ, Yellon DM: Is there a component of coronary collateral flow which cannot be detected by radiolabelled microspheres? *Cardiovasc Res* 1987;21:747–754.

21. Yoshida S, Akizuki S, Gowski D, Downey JM: Discrepancy between microspheres and diffusible tracer estimates of perfusion to ischemic myocardium. *Am J Physiol (Heart Circ Physiol 18)* 1985;248:H255–H264.
22. Arts T, Reneman TS, Veenstra PC: A model of the mechanics of the left ventricle. *Annals Biomed Eng* 1979;7:299–318.
23. Caulfield JB, Borg TK: The collagen network of the heart. *J Lab Invest* 1979; 40:364–372.
24. Nevo E, Lanir Y: Structural finite deformation model of the left ventricle during diastole and systole. *Trans of the AAME* 1989;111:342–349.

Circulation and Coronary Flow

CHAPTER 26:

A MODEL FOR DISTURBED CORONARY ARTERY FLOW WITH PHONOCARDIOGRAPHIC VERIFICATION

Walter Welkowitz, Metin Akay, Jin–zhao Wang, John Semmlow, and John Kostis

ABSTRACT

A basic problem exists of detecting constricted coronary arteries and the resultant coronary flow. A measurement technique is described which detects diastolic murmurs and extracts them from noise using time windowing and parametric signal processing methods. The experimental results can be detailed by two theoretical models; an incremental network model of the coronary artery tree, and a sound generation model based upon turbulent flow. Measurements on a group of 23 pre– and post–angioplasty patients confirms the underlining assumptions and the efficacy of the procedure. Analysis of 100 records (63 coronary artery disease (CAD), 37 normal) confirms the utility of the approach for screening patients.

INTRODUCTION

One of the major causes of death is the development of blockages in the coronary artery system with the subsequent infarct development and loss of contractility in regions of the myocardium. Therefore, there exists an important diagnostic problem of detecting partially constricted coronary arteries and the resultant reduced coronary blood flow. Auditory characteristics have been used to locate stenoses in several blood vessels, especially in the carotid artery [1]. Audible murmurs from the chest wall during diastole have been correlated with coronary artery stenoses [2]. Semmlow, et al. [3] developed a procedure for noninvasive diagnosis of coronary artery disease based upon the assumption that partially occluded coronary arteries generate unique auditory features. Computer–based signal processing techniques isolate the diastolic portions of these acoustic signals, when the coronary blood flow, and the associated turbulence generated signals, have maximum values. Modern signal processing techniques are used to estimate the spectral characteristics of these isolated diastolic sounds [4, 5, 6]. Measurements over limited patient populations suggest that these spectral characteristics can be used to detect partially occluded coronary arteries [7, 8]. The measurements can be explained using two theoretical models: (1) an incremental network model of the coronary artery tree, and (2) a sound generation model based upon turbulent flow and the excitation of arterial resonances.

THEORETICAL MODEL

In 1970, Lees and Dewey [9] suggested that quantitative diagnosis of coronary artery stenosis would be possible if the fundamental mechanisms of vascular murmur generation and transmission were known. Boughner and Roach [10] showed that isolated arteries resonate when excited by a sound stimulus at a characteristic frequency. Kim and Corcoran [11] showed that turbulent flow in latex tubes produced two superimposed spectra: a turbulent spectrum caused by the disturbed flow, and a resonant spectrum resulting from the natural frequency of the tube.

The first step in developing a model of vascular murmurs is to generate a model of the coronary artery tree [16]. The model for the left coronary artery shown in Fig. 1 is derived from the topological model of Sapoznikov et al. [12]. The 68 elements of that model have been reduced here to 14. However, these can be subdivided to describe stenoses. The physical parameters for the model can be obtained from angiograms. The terminal branches representing the arteriole system are considered resistive with the resistance determined by the areas they serve. As shown in Fig. 2, each segment is described by an

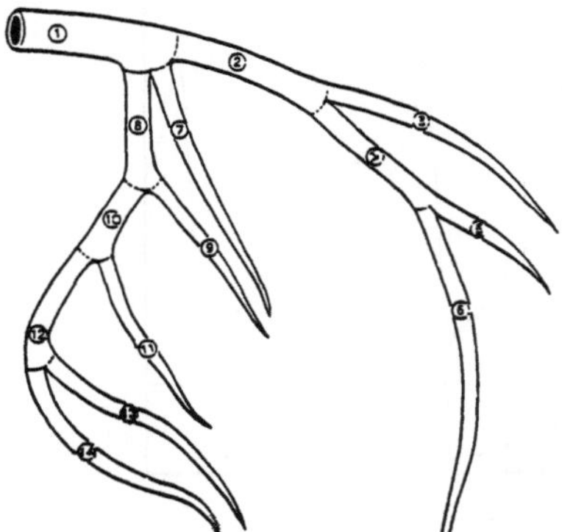

Figure 1: A segmental coronary artery model.

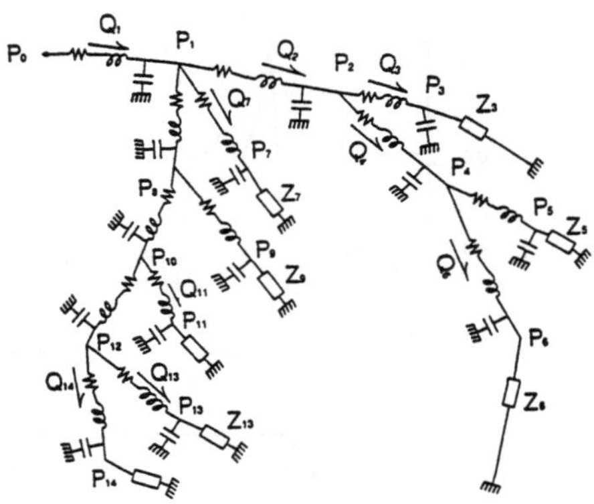

Figure 2: The coronary artery model with each branch replaced by a circuit analog.

analog circuit with parameters calculated from geometry, elasticity of the walls, and viscosity of the blood. Figure 3 shows a typical predicted coronary flow using measured ascending aorta pressure waveforms and appropriate area variations. Figure 4 shows the average coronary flow (in relative units), in a particular segment, as a function of stenosis size with and without regulation. For the case of regulation, the resistance of the arterioles is adjusted to increase flow to blood starved areas.

The coronary artery model described can be used to calculate flow in any stenosed coronary segment. It is proposed that the experimentally ob–served generated sounds are due to turbulent sounds inducing resonant vibrations in a short length of coronary artery, either in the stenosed region or in the following dilated region, as there is usually a widening of a region just beyond the stenosis. This region may be where the sound from the turbulence is propagated with its resonance producing the observed sounds. Subsegments are used to describe these regions. The stenosed region and its four subsegments is shown in Fig. 5. The resonance characteristics of a subsegment can be described by its transfer function. These resonance amplitudes and locations are very sensitive to parameter variations. These include length changes, diameter changes and viscosity variations (including viscoelastic effects of the coronary artery walls.

The characterization of the important features of the turbulent pressure field at the arterial wall has been well documented [13, 14, 15]. Fredberg [13] described the RMS pressure fluctuations as a function of blood velocity and diameter of the stenotic segment. When combined with Fredberg's wide band

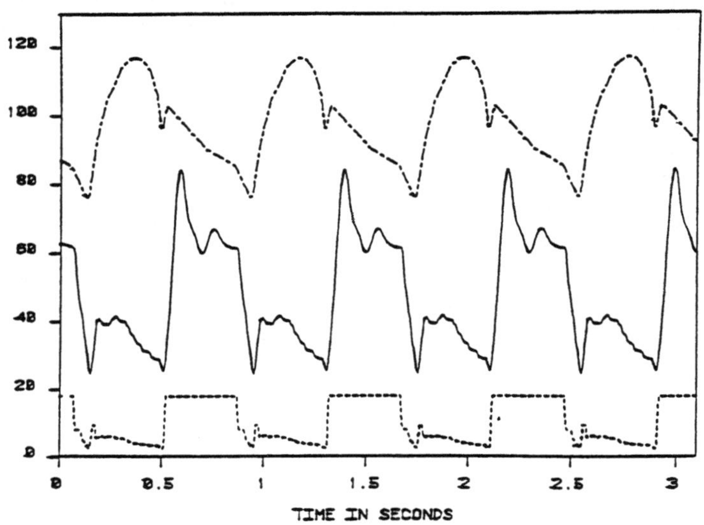

Figure 3: Predicted coronary artery flow waveforms (——) with measured ascending aorta pressure (—·—·—) and appropriate area variations (---).

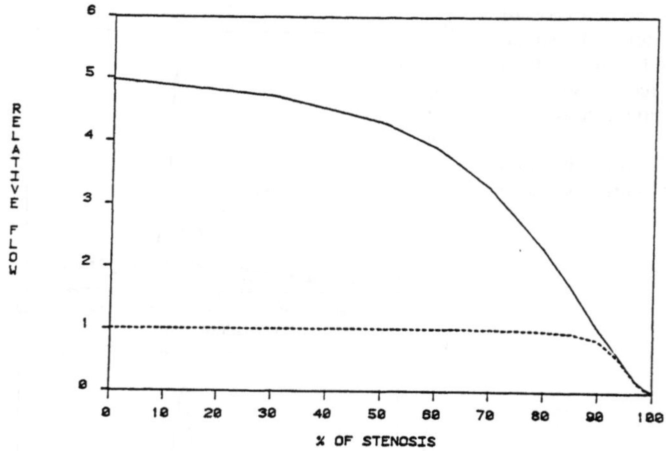

Figure 4: Relative average coronary flow in a particular segment as a function of stenosis size: (---) without regulation; (——) with maximum regulation.

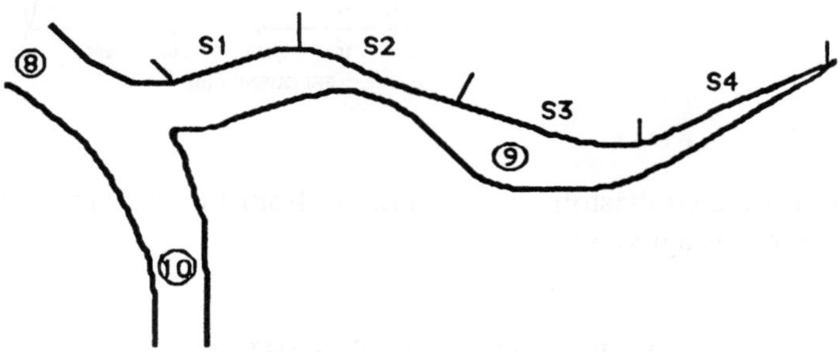

Figure 5: A stenosed region of the artery and its subsections. S1: Proximal to stenosis segment; S2: stenotic segment; S3: poststenotic dilation segment; S4: distal to dilation segment.

spectrum equation [13] our coronary artery network model can be used to describe the source of sound excitation power which produces the resonant signals at any segment. Figure 6 compares results obtained from recordings on a patient with the model prediction for a typical stenosis. This patient had a 50% occlusion in the proximal third of the left coronary artery. The solid line in Fig. 6A is the transfer function for the stenotic section. The turbulent sound spectrum calculated using Fredberg's formula is shown in the dashed line of Fig. 6A. The multiplication of these two spectra produces the predicted output sound spectrum (short/long dashed line). The spectrum

Figure 6: A comparison of model prediction of sound generation and the patient diastolic sound recordings.
(A) Transfer function for the stenotic section.
(B) Spectrum from isolated diastolic heart murmurs, same patient.

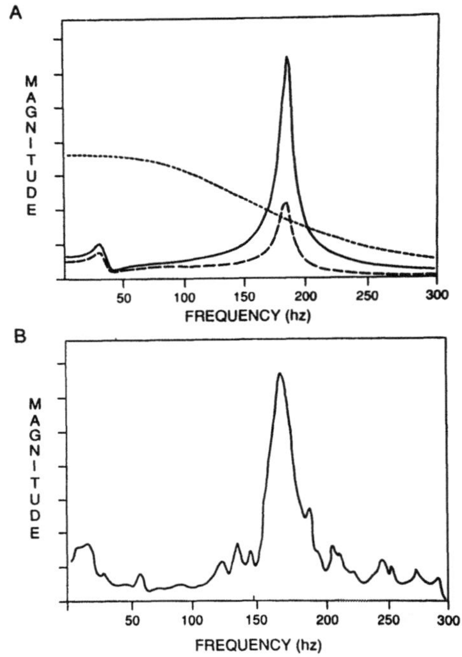

obtained from isolated diastolic heart murmurs from this patient is shown in Fig. 6B. The results agree well.

EXPERIMENTAL STUDIES

Two experimental studies were carried out to evaluate the approach described [17]. The first of these had as its goal the validation of the basic concept of the existence of an acoustic correlate of the coronary artery stenosis. Twenty–five patients undergoing a balloon angioplastic procedure for partially unblocking stenosed coronary arteries had chest sound measurements made before and after the procedure. Parametric modeling techniques were utilized for signal processing [18]. Isolated diastolic heart sounds were analyzed using the autoregressive (AR) method. Model parameters were used to estimate if the recordings were made before or after the coronary lesions were enlarged through the angioplastic surgery. Using spectral characteristics as well as changes in zero location and reflection coefficient amplitudes, a blind judgment was made of whether a record was made before or after angioplasty. In the blind procedure, the persons making the decisions had no knowledge of which record of each pair was taken before or after the angioplasty. Table 1 shows the patient condition (in terms of coronary artery stenoses) before and after angioplasty. Table 2 shows the parameter values obtained from the AR analysis. Fig. 7 shows the average AR spectra obtained before and after angioplasty on a typical patient. Note that the dotted post angio line of Fig. 7 corresponds nicely with a typical spectrum of a normal heart. In the blind evaluation, the estimate of which record was made before angioplasty was correct in all but two (patients 9 and 17) of the 23 angioplasty patients. In addition, both normal patients were correctly assessed. This blind assessment agrees well with a detailed retrospective review of Table 2.

The second study was an investigation of the diagnostic capability of the procedure described in the first study but applied to a mixed group of abnormal and normal patients. One hundred sound recordings were obtained from 57 patients (63 abnormal and 37 normal records) and the isolated diastolic sounds were analyzed using the AR procedure. This analysis was again done in a blind fashion. Patients with occlusions less than 30% were considered normal. Table 3 shows the condition of some of the abnormal patients as well as the derived acoustic parameters. Records marked with the

Table 1: Patient condition in terms of coronary artery stenoses, before and after angioplasty.

#	Pt. ID	Pre-Angio Condition	Pre # Lesion	Post-Angio Condition	Post Lesion
1	JP34	LAD 90%, RCA 30% 30%, 35%	4	LAD 20%, RCA 30% RCA 35%	3
2	TC27	CFX 99%, CFX 90%, 90% LAD 55%	4	CFX 40%, CFX 20%, LAD 55%	3
3	DB32	LAD 95%	1	LAD 50%	1
4	NJ31	CFX 90%	1	CFX 20%	1
5	AC24	RCA 90%, 55%, 40% LAD 40%,30% CFX 35%, 35%	7	RCA 20%, 70%, 20% LAD 35% 25% CFX 35%, 35%	7
6	RB26	CFX 95%, 40%, RCA 100%, 95%	4	CFX 40%, 40%, RCA 100% RCA 95%	4
7	GV21	LAD 99%, RCA 95% RCA 60% LAD 70%, RCA 25%	5	LAD 20%, 20% RCA 60%, 95% RCA 25%	5
8	HD20	Normal	-	Normal	-
9	RD11x	RCA 50%, 75%, LAD 20%, 40% CFX 30%, 30%	6	RCA 50%, 35%, LAD 20% 40% CFX 30%, 30%	6
10	FG10	RCA 90%, RCA 90%, RCA 35% LAD 40%, CFX 20%	5	RCA 10%, RCA 35%, CFX 20% LAD 40%	4
11	RH18	RCA 95%, CFX 95%, 95%	3	RCA 10%, CFX 95%, 95%	3
12	EC01	LAD 80%	1	LAD 5%	1
13	RK03	CFX 70%	1	CFX 20%,	1
14	TD05	LAD 95%, RCA 50%, 60%	3	LAD 25%, 45%, RCA 50%, 60%	4
15	AP02	LAD 90%	1	LAD 20%	1
16	WL04	LAD 95%	1	LAD 30%	1
17	RR13x	LAD 20%, 20%, CFX 75%, 99%	4	LAD 20%, 20%, CFX 75%, 30%	4
18	WF06	LAD 35%, RCA 85%, 85% 95% CFX 100%, 99%, 99%	7	RCA 85%, 85%, 30% LAD 100%, 99%, 99%	6
19	PB14	LAD 95%, 20%, 20% RCA 100%, RCA 60%	5	LAD 10%, RCA 100%, RCA 60% LAD 20% 20%	5
20	RA08	LAD 95%, 20%, 20% RCA 25% RCA 25%, 25%,	6	LAD 20%, 20%, 20% RCA 25% RCA 25%, 25%	6
21	RF78+	Normal	-	Normal	-
22	LW33	LAD 90%, CFX 50%	2	LAD 10%, CFX 50%	2
23	GN35	CFX 85%, LAD 40%, 40%	3	CFX 30%, LAD 40%, 40%	3
24	LG36	LAD 50%, 20%, CFX 90%	3	LAD 50%, 20%, CFX 30%	3
25	WW25	RCA 100%, RCA 80% LAD 35%, CFX 50%	4	RCA 100%, LAD 25%, CFX 50% LAD 35%	4

LCA – left coronary artery; LAD – left anterior descending artery; CFX – circumflex arteries; RCA – right coronary artery; + – pseudo-normal; x – patient was incorrectly diagnosed.

Table 2: Parameter values obtained before and after angioplasty.

Patient #	Prea-Angi ap	Prea-Angi P2	Post Angi ap	Post Angi P2	apr	r	Pre Angio ar	Post Angio ar
#1	770.00	18.15	576.00	6.31	1.337	2.88	12.00	20.00
#2	1447.98	24.21	485.22	4.48	2.99	5.40	15.50	15.40
#3	2893.50	69.50	802.00	16.21	3.60	4.2	16.00	9.00
#4	4978.00	363.10	3146.00	147.23	1.58	2.46	1.67	1.18
#5	2342.00	125.90	632.00	6.31	3.71	19.95	2.73	27.96
#6	2902.78	211.34	2467.00	84.33	1.18	2.50	2.35	2.73
#7	665.70	28.90	526.00	16.35	1.27	2.34	1.06	5.53
#8	528.00	5.01	612.00	3.98			16.00	31.00
#9	2172.20	100.00	10759.20	1000.00	0.20	0.1	0.78	0.26
#10	10403.70	570.16	4587.00	260.61	2.27	2.19	0.68	0.78
#11	1281.98	39.81	763.00	10.77	1.67	2.89	0.43	6.95
#12	2673.81	100.00	283.40	3.16	9.25	31.64	1.42	11.59
#13	4370.00	251.18	1530.00	79.43	2.86	3.16	0.2	1.23
#14	2227.11	125.89	478.30	5.01	4.66	25.17	1.59	8.98
#15	3264.48	125.89	4825.00	125.89	0.68	1.00	0.80	0.45
#16	1417.00	50.11	826.00	19.95	1.72	2.51	2.76	4.08
#17	2264.90	135.83	8558.00	316.22	0.26	0.43	0.63	0.20
#18	838.01	22.38	654.21	11.22	1.27	2.00	4.30	9.43
#19	5476.00	316.22	3615.00	177.82	1.51	1.78	0.41	1.83
#20	2638.00	100.00	1082.00	25.11	2.44	3.98	8.01	3.00
#21	304.00	3.98	364.00	2.51			15.19	6.09
#22	993.92	28.57	661.69	13.18	1.50	2.17	2.32	8.21
#23	3426.00	173.22	2377.11	75.06	2.93	2.51	8.27	16.94
#24	2381.48	100.22	1693.60	59.12	1.40	1.69	19.95	22.58
#25	845.45	21.60	426.257	2.41	1.99	10.75	2.45	24.00

ap – absolute power between 400–800 Hz; P2 – amplitude of the second peak; apr – ratio of ap before surgery/ap after surgery; r – ratio of P2 before surgery/P2 after surgery; ar – ratio of absolute power below 400 Hz/after 400 Hz. Additional information for patients 1–25, on amplitude phases, is available from the authors.

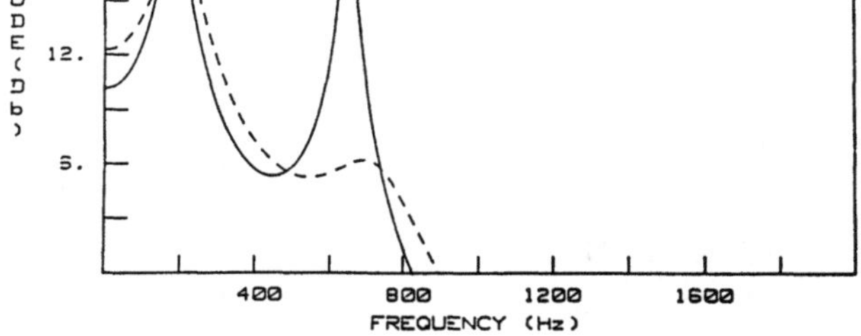

Figure 7: Average spectra obtained from a typical patient (#14) before and after angioplasty. (——) preangioplasty; (---) post-angioplasty.

Table 3: Patient condition as well as derived acoustic parameters for part of the group of abnormal (AD) patients used as data base. For legend, see Table 3.

#	Pt.ID	Cond.(%)	P2	ap	apr	r	g3	z_1	z_2	z_3
1	RD11A	LAD 20,40, RCA 50,35, CFX 30,30	1000.00	10759.20	.26	.050	.70	.56/±.12	.83/±.14	.98/±.31
2	SJ19A	CFX 30,30	641.00	6320.00	.30	.300	.70	.51/±.16	.71/±.20	.98/±.33
3	FG10B	LAD 40, RCA 90 RCA 40,90, CFX 20	570.16	10403.00	.68	.270	.82	.91/±.11	.96/±.31	.77/±.32
4	SJ19B	CFX 80,30	547.00	6177.00	.50	-	.80	.50/±.15	.64/±.19	.97/±.32
5	JN31B	CFX 90	363.10	4978.00	1.67	.690	.79	.96/±.08	.97/±.32	.50/±.42
6	RR13A	LAD 20,20 CFX 75,30	316.20	8558.00	.20	-	.54	.88/±.10	.96/±.32	.83/±.40
7	PB14B	LAD 95,20,20 RCA 100,55	316.20	5476.00	.41	.070	.69	.92/±.07	.98/±.32	.67/±.35
8	FG10A	LAD 40, RCA 40,20, CFX 20	260.61	4587.00	.78	.960	.78	.85/±.09	.57/±.33	.96/±.37
9	PK03B	CFX 70	251.18	4370.00	.20	.005	.59	.83/±.14	.96/±.33	.71/±.40
10	PB26B	CFX 95,40 RCA 100,95	211.34	2902.00	1.06	1.190	.74	.94/±.08	.97/±.32	.52/±.30
11	PB14A	RCA 100,55 LAD 10,20,20	177.82	3615.00	1.83	.890	.62	.93/±.08	.49/±.29	.97/±.32
12	GH35B	CFX 85 LAD 40,40	173.22	3426.00	.66	.200	.57	.97/±.08	.89/±.24	.90/±.40
13	RR13B	CFX 75,99,LAD 20,20	135.83	2264.90	.63	.730	.68	.75/±.17	.70/±.20	.97/±.31
14	AC24B	LAD 40,30 RCA 90,55,40 CFX 40	125.90	2342.00	2.73	1.580	.58	.93/±.69	.96/±.31	.64/±.24
15	AP02B	LAD 90	125.89	3264.80	.80	.310	.62	.91/±.09	.96/±.33	.61/±.40
16	TD05B	LAD 95,RCA 50,60	125.89	2227.11	1.58	.830	.44	.94/±.09	.97/±.32	.73/±.35
17	RA08B	LAD 95,20,20 RCA 25,25,25	100.00	2638.00	8.01	10.000	.69	.97/±.07	.96/±.30	.79/±.44
18	RD11B	LAD 20,40 RCA 60,75,CFX, 30,30	100.00	2172.00	.78	.250	.63	.85/±.10	.67/±.23	.96/±.33
19	AK30B	LAD 45,85 RCA 45,50,70	85.07	1839.60	2.14	1.590	.58	.94/±.09	.95/±.31	.67/±.33
20	RB26A	RCA 100	84.33	2467.00	5.53	8.090	.58	.94/±.10	.53/±.25	.95/±.32
21	RJ29B	LAD 40	77.66	1817.19	2.54	2.090	.36	.94/±.09	.95/±.30	.65/±.39
22	JZ22B	LAD 90,35,40	73.45	1529.00	.78	.340	.70	.89/±.11	.38/±.32	.96/±.41
50	WB52B2x	CFX 20,95	4.12	552.70	20.61	95.80	.19	.98/±.07	.73/±.22	.94/±.33
51	SK63A x	LAD 90,60	4.00	457.60	6.52	21.82	.28	.92/±.10	.79/±.37	.58/±.39
52	DS57B2	LAD 25, CFX 100	3.98	304.00	15.19	25.12	.01	.94/±.09	.76/±.33	.64/±.34
53	AF64A x	RCA 55	3.62	392.52	9.82	38.97	.31	.94/±.10	.79/±.35	.60/±.37
54	JE46B2x	LAD 50, CFX 100	3.56	389.05	5.76	12.21	.14	.91/±.08	.62/±.17	.82/±.39
55	WW25A x	RCA 25, 100 CFX 50 LAD 35	3.51	426.20	17.52	.85	.23	.94/±.10	.69/±.34	.53/±.41
56	SK63B	LAD 60 LAD 35,90	2.90	308.61	7.60	20.90	.16	.92/±.12	.42/±.34	.82/±.49
57	HS79B1x	CFX 80,85, RCA 100 LAD 35,90	2.63	310.34	13.10	71.43	.01	.95/±.10	.77/±.35	.68/±.36
58	CZ59B2x	LAD 50,25 RCA 25	2.23	334.07	8.82	38.66	.01	.92/±.11	.67/±.35	.63/±.40
59	WB52B1*x	CFX 20,95	2.12	273.22	18.11	31.23	.28	.92/±.09	.60/±.27	.82/±.33
60	WC71B1*	LAD 30, RCA 100, CFX 100	1.93	199.42	5.20	6.20	.00	.86/±.10	.63/±.15	.78/±.45
61	DS57B1*	LAD 25, CFX 100	1.92	253.84	12.02	30.96	.20	.88/±.08	.22/±.37	.78/±.49
62	WC71B2	LAD 30, RCA 100, CFX 100	1.62	295.79	8.62	43.85	.33	.92/±.11	.71/±.35	.59/±.37
63	ED61B x	RCA 80 LAD 90	1.45	201.36	11.36	58.76	.24	.94/±.10	.60/±.31	.73/±.39

P2 – amplitude of the second peak; ap – absolute power between 400–800 Hz; apr – ratio of ap below 400 Hz/ap above 400 Hz; r – ratio of first peak/second peak; g3 – third reflection coefficient; z1 – first pair zeros pair amplitude/phase; z2 – second pair zeros pair amplitude/phase; z3 – third pair zeros pair amplitude/phase. Sampling frequency is 4 kHz; + – pseudo–normals; * – parameters of these patients were not used for statistical analysis; x – patients incorrectly diagnosed.

Figure 8: Sensitivity vs specificity obtained from this study using the AR second peak as the decision parameter.

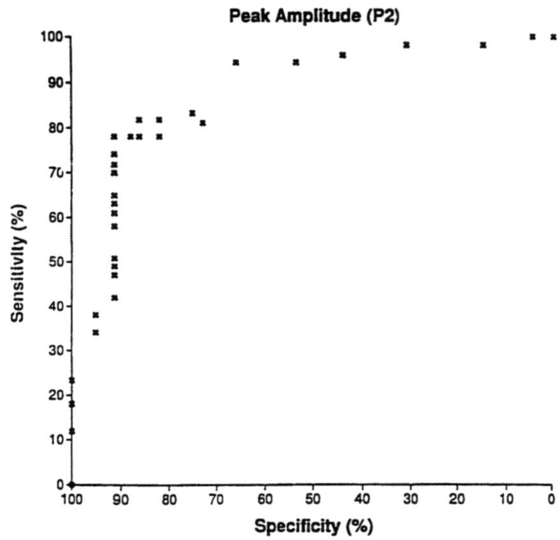

Table 4: Acoustic parameters obtained from 37 normal patients. For legend, see Table 3.

#	Pt.ID	p2	ap	apr	r	g3	z1	z2	z3
1	JN31A x	147.23	3146.00	1.18	.53	.58	.90/±.11	.95/±.32	.40/±.32
2	AP02A x	125.89	4825.00	.45	.19	.20	.79/±.16	.75/±.28	.96/±.31
3	RK03A x	79.43	1530.00	1.23	.50	.62	.91/±.09	.96/±.33	.61/±.40
4	VP37N1*+	8.67	739.70	18.58	61.70	.39	.94/±.08	.82/±.32	.39/±.33
5	AS40N +	7.52	720.20	15.20	69.16	.26	.94/±.08	.82/±.32	.45/±.35
6	WC62N1*	5.01	628.87	8.77	40.10	.15	.94/±.10	.75/±.36	.74/±.49
7	TC65N	4.65	514.14	1.00	3.83	.22	.74/±.16	.79/±.17	.87/±.43
8	WK47N1*x	4.60	470.27	9.34	34.47	.16	.93/±.14	.56/±.21	.81/±.38
9	VP37N2+	4.40	396.90	16.71	70.70	.14	.95/±.09	.58/±.26	.81/±.32
10	RF78N1*+	3.98	304.00	15.19	25.12	.01	.94/±.09	.76/±.33	.64/±.34
11	CD55N1*	3.50	388.86	6.26	18.36	.13	.92/±.10	.53/±.39	.79/±.36
12	GT50N	3.25	355.33	7.96	29.00	.08	.93/±.10	.78/±.39	.68/±.36
13	JS70N +	3.17	331.00	7.50	30.90	.11	.94/±.11	.64/±.33	.81/±.37
14	FT75N1*	3.16	380.04	5.68	16.78	.22	.91/±.10	.45/±.39	.79/±.39
15	EK56N1*	3.12	370.22	6.10	18.33	.21	.91/±.10	.71/±.37	.83/±.43
16	CP68N1*x	3.05	367.59	6.97	27.67	.41	.93/±.10	.77/±.37	.78/±.41
17	VY49N1*	3.01	335.38	6.16	17.90	.13	.91/±.10	.50/±.16	.81/±.40
18	WC62N2	2.96	220.52	6.74	12.18	.28	.92/±.10	.79/±.37	.58/±.39
19	MA39N +	2.95	304.00	3.16	6.05	.10	.88/±.10	.48/±.32	.81/±.41
20	FT75N2	2.75	562.44	15.51	137.90	.11	.97/±.09	.62/±.36	.79/±.40
21	RF78N2+	2.51	364.00	6.04	12.59	.01	.90/±.09	.78/±.33	.68/±.34
22	DV69N1*	2.29	300.40	2.18	4.36	.04	.89/±.12	.64/±.41	.76/±.43
23	CR53N1*	2.20	244.00	10.82	36.81	.24	.93/±.16	.73/±.32	.80/±.36
24	WK47N2	2.13	249.32	13.74	57.49	.27	.94/±.14	.57/±.24	.76/±.35
25	EK56N2	2.07	272.44	15.51	137.90	.05	.97/±.11	.67/±.34	.76/±.36
26	RD76N1*	2.02	294.88	8.54	34.17	.09	.92/±.10	.47/±.25	.74/±.39
27	CP68N2	1.77	168.20	10.26	24.28	.03	.91/±.09	.64/±.35	.82/±.37
28	VY49N2	1.77	297.40	10.84	56.85	.34	.93/±.11	.52/±.22	.70/±.42
29	EL54N1*	1.73	316.60	9.22	50.89	.00	.93/±.11	.71/±.35	.59/±.37
30	JB73N	1.54	358.33	2.69	12.38	.01	.85/±.12	.41/±.21	.83/±.44
31	RD76N2	1.25	308.23	7.01	42.36	.20	.91/±.10	.67/±.35	.77/±.43
32	OZ72N	1.21	370.22	7.51	68.37	.16	.93/±.10	.50/±.33	.81/±.35
33	CD55N2	1.19	404.11	4.62	26.73	.02	.89/±.13	.68/±.29	.83/±.38
34	EL54N2	1.05	367.59	6.97	27.67	.41	.93/±.10	.67/±.37	.78/±.41
35	CR53N2	1.03	363.40	10.00	84.98	.13	.90/±.12	.69/±.25	.73/±.34
36	OM74N	.85	255.07	20.00	261.11	.03	.92/±.12	.69/±.32	.72/±.36
37	DV69N2	.52	426.21	18.65	629.20	.02	.95/±.09	.77/±.27	.78/±.39

symbol (x) on these tables shows which records were incorrectly diagnosed in the blind evaluation. Table 4 shows the parameters from the normal patients. In this study 80 out of the 100 records were correctly diagnosed. A curve of sensitivity vs. specificity from this study using the parameter (P2) as the variable, the second peak in the AR spectrum, is shown in Fig. 8.

CONCLUSION

Two experimental studies were evaluated using two theoretical models: an incremental network model of the coronary artery tree, and a sound generation model based upon turbulent flow. The underlying assumptions and the efficacy of the procedure were confirmed by the experimental analysis of a group of 23 pre– and post–angioplasty patient records and 100 normal/diseased records (63 coronary artery disease (CAD), 37 normal) using advanced signal processing techniques such as the autoregressive (AR) and autoregressive moving average (ARMA) methods.

DISCUSSION

Dr. Pollack: I wonder whether you could be certain of the etiology of the sound. We published a paper about 8 or 9 years ago *[Brozovich & Pollack, Biophys J, 1983;41:35–40]*] in which we look at isolated skeletal muscles. During shortening or lengthening, we could record sounds of a substantial magnitude. So it occurred to us that during diastole, the myocardium is not static; you are getting some length change. I wonder, therefore, if it is possible that some of the sound occurs because the muscle itself is generating it?

Dr. Welkowitz: It is certainly possible. What we have done is as follows. When we approached this subject there was little theory. We measured sounds so we knew sounds exist. Why they exist is more difficult to determine. We have done more than I have shown here. We have taken, for example, multiple blockage cases and matched theory with measurements. That is harder to explain from your approach, because you get strange spectra when you have two blockages. They merge under certain circumstances or they have multiple peaks. I would say we have compared 6 or 7 cases where we laboriously measured everything and in all cases we obtained reasonable agreement. While it does not prove the theory it gives one a feeling of confidence.

Dr. Pollack: When you have stenosis, you are going to get some local change in muscle, shortening or lengthening too.

Dr. Welkowitz: That is true. But we looked at the difference in sounds before and after angioplasty. The differences seemed to confirm our theory.

Dr. Reneman: The idea of flow induced sounds has been worked out for the carotids very nicely by Lees *[1]* 10 years ago. He indicated clearly that you needed a 60% diameter reduction before you could pick of turbulent flow induced sounds and that is in a large sized artery. What is the degree of narrowing before you get turbulent flow?

Dr. Welkowitz: We find we can pick up noise at about 25% narrowing.

Dr. Reneman: Are you picking up sounds due to turbulence or is it sound induced by flow separation?

Dr. Welkowitz: We measured sounds with a 25% blockage. Let's say we do not know why this noise occurs. We referred to Fredberg's *[13]* work with Lees and we referred to Fredberg's equations. There is a drop-off with reduced blockage, but if you examine the equations for low velocity when there is not as much constriction, they do predict the presence of some sounds. They just predict smaller sounds. The question then is: are the lower sounds measurable? If you look at the system as a circuit, then we are amplifying the sounds with the resonances. The resonance gives us measurable output even from a lower sound. Lees and Dewey *[9]* essentially did not look very much at the resonances. They used the turbulent breakover point to do calculations, for example, on the size of the vessel or the size of the constriction. We believe that what we see in our measurements would not be see just from the turbulence and in fact we do not see a broad band spectrum. We always see resonances. So we concentrated on the resonances that we see and in the second half of the model, we are trying to explain why we get the resonances that we do see.

Dr. Reneman: That is an important question. But do you see the resonance only when you have enough energy to excite resonances in the wall.

Dr. Welkowitz: Absolutely. In a normal subject (cath proven), we do not see a high frequency peak. We also find some students in the laboratory (we do not have angiograms on them) show even less output in that frequency region, they show a very clean drop-off. Thus, we seem to have evidence that a normal subject does not have this frequency peak and that a diseased subject does. Clearly as the amount of constriction gets smaller, we get into a region which is questionable. Besides the "blind" analysis, we did decision making from the spectral numbers. There, we used 30% blockage as a cut-off between normal and abnormal.

Dr. Noble: Some of your data seem to show a substantial amplitude of sound at low frequencies. Can you explain to me what this sound is?

Dr. Welkowitz: Yes. This is due to all sorts of noise in our system. We have a high pass filter of 170 Hz and anything observed below that is probably just a mixture of filtered electrical and mechanical noise, not necessarily due to any stenosis.

Dr. Noble: Some of those amplitudes were as large as the abnormal peaks.

Dr. Welkowitz: Because of this we are looking into the AR processing in detail now. AR processing is a very non-linear technique, since it is a way of getting rid of some of the noise and keeping the signal. It assumes certain spectral characteristics of the signal. When one applies only noise to our system, one gets spectra out of the AR processor with the low frequency characteristics we have observed from patient data.

REFERENCES

1. Lees RS, Kistler J: Carotid phonoangiography. In: *Noninvasive Diagnostic Techniques in Vascular Disease*, Bernstein A (ed), C.V. Mosby Co, St. Louis, 1978, pp 187–194.
2. Sangster JF, Oakley CM: Diastolic murmur of coronary artery stenosis. *British Heart J* 1973;352:840–844.
3. Semmlow J, Welkowitz W, Kostis JB, Mackenzie J: Coronary artery disease–correlates between diastolic auditory characteristics and coronary artery stenoses. *IEEE Trans Biomed Eng* 1983;30(2):136–139.
4. Akay M, Semmlow J, Welkowitz W, Bauer M, Kostis J: Detection of coronary occlusions using AR modelling of diastolic heart sounds. *IEEE Trans Biomed Eng* 1990; in press.
5. Akay M, Semmlow J, Welkowitz W, Bauer M, Kostis J: Noninvasive detection of coronary occlusions using Eigenvector methods before and after angioplasty. *IEEE Trans Biomed Eng* 1990; in press.
6. Semmlow JL, Akay M, Welkowitz W: Noninvasive detection of CAD using parametric analysis methods. *IEEE, EMBS Magazine* 1990;9:33–36.
7. Akay M, Bauer M, Semmlow JL, Welkowitz W, and Kostis J: Autoregressive modelling of diastolic heart sounds. *IEEE EMBS 10th Ann Intl Conf*, New Orleans, 1988.
8. Akay M, Bauer M, Semmlow JL, Welkowitz W, Kostis J: Analysis of diastolic heart sounds before and after angioplasty. *IEEE EMBS 10th Ann Intl Conf*, New Orleans, 1988.
9. Lees RS, Dewey CF: Phonoangiography. *Proc Natl Acad Sci* 1970;67(2):935–942.
10. Boughner DR, Roach MR: Effect of low frequency vibration on the arterial wall. *Circ Res* 1971;39:136–144.
11. Kim BM, Corcoran WK: Experimental measurement of turbulence spectra distal to stenosis. *J Biomechanics* 1974;7:335–342.
12. Sapoznikov D, Halon DA, Lewis BS, Gotsman MS: A graphic computerized system for reporting and analysis of coronary angiograms. *Comp and Biomed Res* 1983;16:334–339.
13. Fredberg JJ: Origin and character of vascular murmurs: model studies. *J Acoust Soc Am* 1977;66:1077–1085.
14. Clark C: Turbulence velocity measurements in a model of aortic stenosis. *J Biomechanics* 1976;9: 677–687.
15. Tobin R: Wall pressure spectra scaling downstream of stenosis in steady tube flow. *J Biomech* 1976;9:633–640.
16. Wang J–Z, Tie B, Welkowitz W, Kostis J, Semmlow J: Incremental network analogue model of the coronary artery. *Med and Biol Eng and Comp* 1989;27:416–422.
17. Akay M: Noninvasive detection of coronary artery disease using advanced signal processing techniques. *Ph.D. Thesis.* Dept of Biomedical Eng, Rutgers University, New Brunswick, NJ, 1990.
18. Semmlow JL, Akay M, Welkowitz W: Noninvasive detection of CAD using parametric analysis methods. *IEEE, EMBS Magazine* 1990;9:33–37.

CHAPTER 27:

LEFT VENTRICULAR PRESSURE VERSUS CONTRACTILITY AS CAUSES FOR PULSATILE CORONARY FLOW

Erik Kouwenhoven and Jos A.E. Spaan

ABSTRACT

The effects of cardiac contraction on coronary flow have been attributed to an intramyocardial pressure (IMP) generated by the left ventricular pressure (LVP), and, more recently, to the time varying elastance of the myocardium. Experiments on open chest anesthetized goats show an early systolic transmission of LVP on coronary flow. It is suggested that time varying elastance protects the intramyocardial vessels from squeezing by LVP in systole.

INTRODUCTION

Cardiac contraction reduces net coronary blood flow, which is evidenced in the maximally dilated heart when pressure–flow relations, during cardiac arrest and in the beating state, are compared. At equal perfusion pressure the mean flow is lower in the beating heart [1]. Many mechanisms have been proposed explaining this flow impediment.

Initially, systole was viewed as a period of reduced coronary flow due to an increase of vascular resistance or increased back pressure. The latter mechanism was called an intramyocardial waterfall, as venous pressure does not determine flow during systole according to this model. The validity of a waterfall concept may be questioned, however, as it is not compatible with two basic observations. First, it is observed that, when perfusion pressure is kept constant in an animal preparation, negative coronary arterial flows may appear in systole [2]. Arterial back flow can only be obtained from a shift of blood volume into the perfusion system. Second, measurement of venous flow [3] show that this increases in systole and decreases in diastole, i.e., arterial and venous flow are out of phase. These facts imply that considerable volume displacements occur in a heart beat: a difference in arterial and venous flow must be due to a changing intramyocardial blood volume.

Emptying and filling of coronary vessels was the basis of the intramyocardial pump concept [2–4]. This model states that an intramyocardial pressure (IMP), caused by the left ventricular pressure (LVP), induces volume variations through a large compliance. An analysis of arterial and venous flow changes induced by arterial occlusion or cardiac arrest showed that the time in which volume variations take place, expressed in the time constant, is in between 1.5 and 2 seconds long. This large time constant makes occurrence of vascular collapse – a waterfall mechanism – unlikely, as the duration of a heart beat is too short to displace the necessary amount of volume.

The basic observation of flow reduction at vasodilation in a beating heart is not explained by this concept. An extended intramyocardial pump model [5], however, considering the volume dependency of vascular resistance as well, can give a realistic description of the relation between pressures and flows in the coronary circulation.

The mechanisms mentioned so far commonly assume that the LVP is transmitted to the myocardium, giving rise to an IMP. This compresses the microvessels, whether it be local (waterfall) or global (intramyocardial pump). Theoretical considerations show that the IMP decreases from LVP at the subendocardium to atmospheric at the subepicardium. This is in accordance with the distribution of microspheres over the myocardium, showing that reduction of coronary flow due to cardiac contraction is larger in the subendocardium than in the subepicardium. Another argument in favor of the idea of cavity pressure being the mediator between cardiac contraction and coronary flow concerns measurement of right coronary arterial flow. This appears to be primarily determined by perfusion pressure (aortic pressure) and is not significantly reduced by cardiac contraction [6]. As right ventricular cavity pressure is much lower than LVP, this is in agreement with the above mentioned results.

TIME VARYING ELASTANCE CONCEPT

Convincing evidence has recently been provided that coronary arterial flow is related to changing elastic properties of the myocardium [7–9], rather than ventricular cavity pressure. This idea is closely related to the Suga-Sagawa concept [10] for description of heart function, stating that the ventricle as a pump can be described by its time varying elastance. An increasing elastance or stiffness of the ventricular muscle causes either increasing pressure in the isovolumetric phase or a decreasing volume in the ejection phase. This macroscopic notion of the ventricle was generalized to a microscopic level, postulating that the microvessels are situated in a continuum – the myocardium – with a variation of the elasticity modulus over the heart beat. As a first approximation, macroscopic and microscopic elastance were taken as equal.

Krams et al. [7] performed experiments in the isolated, blood perfused and maximally vasodilated cat heart. An intraventricular balloon enabled control of cavity pressure and/or volume. Perfusion was through a cannula in the aorta, pressure at the tip of the cannula was maintained constant. The difference of systolic and diastolic coronary arterial flow was approximately equal in isovolumetric and isobaric beats, revealing its independency from LVP. In contrast, the systolic diastolic flow difference was highly dependent on the contractile state of the myocardium. As elastance is closely linked with contractility – in the Suga–Sagawa concept contractility is defined as the maximal elastance in a beat – this was provided as evidence for the time varying elastance concept. These experiments leaded to a controversy about the mechanism responsible for coronary flow impediment. On the one hand, the experimental results of Krams et al. [7, 8] are incompatible with the LVP as the quantity influencing coronary flow. On the other hand, the observation of non–impeded right coronary arterial flow, and the endocardial/epicardial differences are in contrast with the time varying elastance concept. A contribution to the solution of this controversy is presented in this study.

At this point it is important to notice that the mechanism responsible for affecting coronary flow due to cardiac contraction can only be understood if considerable volume displacements are taken into account. The essential contribution of these intramyocardial blood volume shifts on the coronary arterial flow profile is illustrated by the observation of retrograde arterial flow. As an example of its significance, data obtained in our laboratory are presented in Fig. 1. These signals were measured in the open chest,

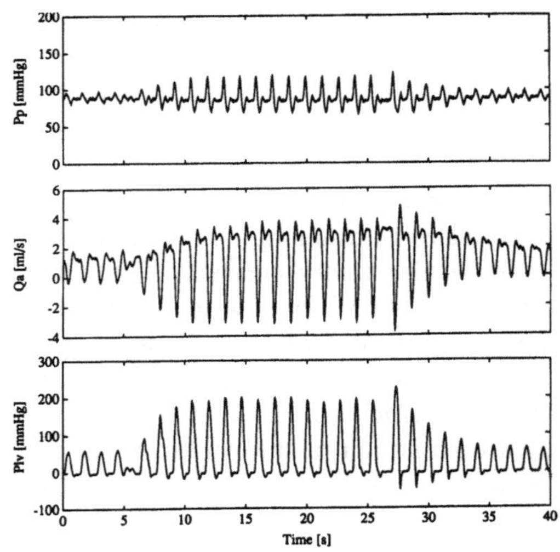

Figure 1: Measured perfusion pressure (P_p), coronary arterial flow (Q_a), and LVP (P_{lv}), in the anesthetized open thorax dog with cannulated and artificially perfused left main coronary artery. Paired pacing of the LV was used to enhance contractility.

anesthetized dog with cannulated and artificially perfused left main coronary artery. Contractility was enhanced by the paired pacing technique [11]. This leads to an increase of developed LVP, and a concomitant large effect on flow. It is seen that arterial flow is retrograde, practically during the whole of systole. This negative component of flow can only be due to emptying of coronary vessels into the perfusion system. Therefore, whatever the mechanism proposed for understanding coronary flow impediment, it should explain these capacitive phenomena.

INTERACTION OF ELASTANCE AND INTRAMYOCARDIAL PRESSURE

In our laboratory experiments were performed on the anesthetized open thorax goat. The left coronary descending artery (LAD) was cannulated and connected to a perfusion system, enabling control of perfusion pressure. The experimental protocol consisted of occluding the aorta descendens. The response of coronary arterial flow on the subsequent rise in LVP was studied. A typical tracing of the phasic ventricular pressure, arterial flow and perfusion pressure is given in Fig. 2.

It is noticed that these results are hard to interpret with an intramyocardial pump concept: a twofold increase of LVP pulse does not result in a flow pulse increase of the same magnitude. Apparently the changes in LVP are only transmitted to the early systolic peak in flow; the higher

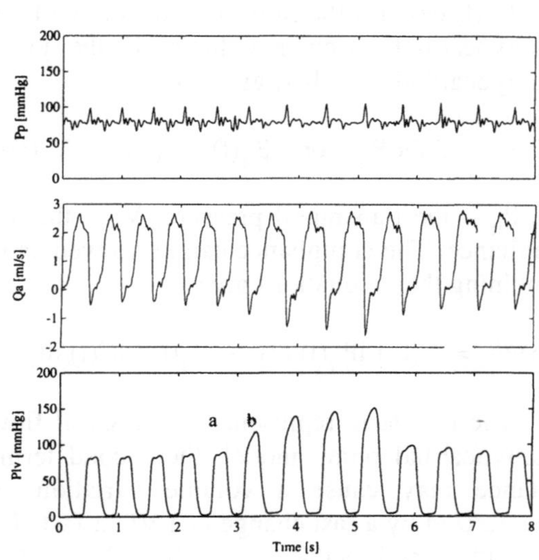

Figure 2: Measured signals in the anesthetized open thorax goat. Given are coronary perfusion pressure (P_p), coronary arterial flow (Q_a), and LVP (P_{lv}). The aorta was occluded during a few beats.

pressure later in systole barely affects arterial flow. Hence, ventricular pressure influences the rapid initial peak systolic flow, but later in systole the arterial flow profile cannot be ascribed to the ventricular pressure alone. Then why did the results of Krams et al. diverge from ours in that they did not found a relation between ventricular pressure pulse and arterial flow pulse? In some of our experiments such a relation was absent as well, while analyzing the early systolic part of the signals revealed a clear relation between instantaneous LVP and arterial flow. Moreover, the nature of the signals obtained by Krams et al. differ from ours especially in the early systolic phase. This may be due to the difference in experimental preparation.

Apparently, the transmission of LVP to the coronary bed is higher in early systole than in mid–systole, i.e. is varying over the cardiac cycle. An explanation is that the systolic increase of elastance impedes the transmission of cavity pressure to the coronary circulation. This implies that the observed flow signal is caused by (1) a time varying transmission of LVP and (2) a volume variation of intramyocardial vessels due to time varying elastic properties of their surroundings. The unification of both lines of thought is thus not a superposition of the two effects, but involves an interaction through time varying parameters. A simple way to formulate this interaction is by analyzing the arterial flow due to volume variations. As was discussed earlier, these volume variations contribute significantly to the observed flow. The flow component due to capacitive effects is assumed to be proportional to the first time derivative of volume according to the intramyocardial pump

$$Q_c(t) \propto dV(t)/dt \qquad (1)$$

$Q_c(t)$ denotes the flow component due to volume variations and $V(t)$ the intramyocardial vascular volume. If the pressure–volume relation of the intramyocardial vessels is given by

$$C = dV/dP_{tr} \quad \text{or} \quad P_{tr}(t) = (V(t) - V_0(t))/C(t) \qquad (2)$$

with P_{tr} – the transmural pressure, V_0 – the unstressed volume and C – the compliance. The compliance can be viewed as the reciprocal of the elastance. Combining these equations gives

$$dV(t)/dt = C(t)\,[dP_{tr}(t)/dt] + P_{tr}(t)\,[dC(t)/dt] \qquad (3)$$

The first term represents the classical flow component according to the intramyocardial pump model, the second term demonstrates that a varying elastance may cause a volume variation as well. If early systole is characterized by a fast change of LVP, a fast change of transmural pressure as well and a high value of C then the first term is expected to have a considerable contribution to the volume changes. Later in systole, the first term is expected to be smaller, both as a result of the decrease in C and of the reduced rate of change in P_{tr}. This simple analysis can thus explain the basic experimental findings. In reality the situation is of course much more complex. In this analysis resistance in the coronary circulation was neglected, let alone the dependency of resistance on vascular volume [5]. The concept of a time constant, determining the rate with which vessels fill and empty, is therefore not an item in the analysis. Besides, an aspect that was implicitly assumed in the above mentioned derivation is the unstretched volume being constant. From the Suga–Sagawa concept for the whole ventricle it is known, however, that the end–diastolic and the end–systolic pressure–volume relationships have different volume intercepts.

Another consequence of the time varying transmission of LVP is that the time constant is subject to variation as well. In systole its value is smaller than in diastole. This results in a reduced systolic transmission as mentioned earlier, as well as a faster flow response on ventricular pressure changes.

CONCLUSIONS

The discussion about interaction of cardiac contraction and coronary flow was revived by the introduction of the time varying elastance concept. The present study discusses the possibility of time varying elastance protecting the microcirculation from squeezing by the LVP. Experiments performed in our laboratory support this idea. However, more data as well as mathematical models, are necessary to gain profound understanding about the complex mechanisms that determine the coronary flow profile.

Acknowledgements

The authors wish to thank Drs. S.Y.A. Han, Dr. I. Vergroesen and Ing. A. Boekee for their contribution to the experimental work.

DISCUSSION

Dr. Noble: The main problem is that with the elastance model, the results depend on knowing the pressure–volume relationship or the stiffness relationship of each intra-myocardial space within which you have vessels and knowing what the load is. You do not know what those are because the measurements you made are irrelevant, they pertain to completely different artificial spaces. Those are unknown and therefore you can not disprove the theory, simply because it is, with the present methods, untestable. That does not mean that it is right of course, but the problem is that it is untestable.

Dr. Spaan: I did not reject the elastance model; I said that I do not know what to do with it. We did some calculations in a model with changing pressure–volume relation and that generated phasic flow. Hence, the hypothesis is testable. The problem is: why has the LVP no effect? As you can see, the pressure transmission is only in the early phase or systole and this is consistent in all observations.

Dr. Westerhof: We have done experiments where the heart beat isobarically, so no development of LVP took place and yet there was impediment of coronary flow. I am not saying that the varying elastance concept explains it all, but you need it since LVP can not explain it all.

Dr. Spaan: What happens to dP/dt (Fig. 2)? It may be that contractility changes.

Dr. Westerhof: Can it change from one beat to the other?

Dr. Spaan: I don't know.

Dr. Noble: I think the answer is no. Between beat a and beat b (in Fig. 2) there would be no change in dP/dt but you have a change in the flow.

Dr. Westerhof: The question is what happens to contractility. The assumption that contractility and dP/dt are always related may not be correct.

Dr. Spaan: This negative peak has been shown by Ashikawa *[12]* when he measured velocity in epicardial vessels.

Dr. Westerhof: His results may be just compounding the problem because he sees the effect in the epicardium, where the IMP is assumed to be negligible.

Dr. Spaan: I am not sure that varying elastance explains it all, but LVP does not explain it all either.

Dr. Westerhof: I was surprised to find that LVP had so little effect in some beats of occlusion and a clear effect in other beats.

Dr. Beyar: The intramyocardial pressure may be disassociated from the LV pressure, especially when the afterload level is low. Whether it is elastance that causes intramyocardial compression, or the elastance generates IMP, dissociated from the ventricular pressure at low afterloads, both are possibilities for coronary compression. You describe the elastance concept based on changing the volume of the interstitial space. From what I understand, there is no significant change during one cycle in this space and it should remain constant. Could you comment on that?

Dr. Westerhof: I do not know. You have interstitial spaces and no change in volume may point to isovolumic conditions while a volume change may relate to isobaric or anxotomic conditions. You can not say what the IMP is because you do not know whether the interstitial space is isovolumic or isobaric or somewhere in between. I wish we could go into that.

Dr. Spaan: If, in the theoretical analysis, the parameters of the arterial compartment are changing to account for varying elastance, you can predict pulsatile flow. However, compliance changes, resistance changes, etc. How are we going to make a distinction between what is doing what? You can explain certain observations by either compliance or time varying resistance.

Dr. Westerhof: What we are missing is data on the small scale in the peripheral and the very small vessels. What happens there to velocity, diameters and pressures?

Dr. Beyar: Especially in the subendocardium.

Dr. Sideman: What we heard today indicates that we will maintain this activity for many more years! Whatever we thought we knew was today questioned and a lot of new speculations came in to be discussed and studied. I challenge those having the vision to propose new suppositions and assumptions. I expect that we shall meet again to discuss them in the future.

REFERENCES

1. Downey JM, Kirk ES: Inhibition of coronary blood flow by a vascular waterfall mechanism. *Circ Res* 1975;36: 753–760.
2. Spaan JAE, Breuls NPW, Laird JD: Forward coronary flow normally seen in systole is the result of both forward and concealed back flow. *Basic Res Cardiol* 1981;76: 582–586.
3. Spaan JAE, Breuls NPW, Laird JD: Diastolic–systolic coronary flow differences are caused by intramyocardial pump action in the anesthetized dog. *Circ Res* 1981;49: 584–593.
4. Hoffman JIE, Spaan JAE: Pressure–flow relations in coronary circulation. *Physiol Rev* 1990;70: 331–390.
5. Bruinsma P, Arts T, Dankelman J, Spaan JAE: Model of the coronary circulation based on pressure dependence of coronary resistance and compliance. *Basic Res Cardiol* 1988;83:510–524.
6. Lowensohn HS, Khourim EM, Gregg DE, Pyle RL and Patterson RE: Phasic right coronary arterial flow in conscious dogs with normal and elevated right ventricular pressures. *Circ Res* 1976;39: 760–766.
7. Krams R, Sipkema P, Zegers J, Westerhof N: Contractility is the main determinant of coronary systolic flow impediment. *Amer J Physiol (Heart Circ Physiol 26)* 1989;257:H1936–H1944.
8. Krams R, Haelst ACTA van, Sipkema P, Westerhof N: Can coronary systolic–diastolic flow differences be predicted by left ventricular pressure or time–varying intramyocardial elastance? *Basic Res Cardiol* 1989;84: 149–159.
9. Krams R: The effect of cardiac contraction on coronary flow. *Ph.D. Thesis*, Amsterdam: Free University of Amsterdam, 1988.
10. Suga H, Sagawa K, Shoukas AA: Load independence of the instantaneous pressure–volume ratio of the canine left ventricle and effects of epinephrine and heart rate on the ratio. *Circ Res* 1973;32:314–322.
11. Hoffman BF, Cranefield PF: Paired–pulse stimulation of the heart. *Amer Heart J* 1966;71:143.
12. Ashikawa K, Kanatsuka H, Suzuki T, Takishama T: Phasic blood flow velocity pattern in epimyocardial microvessels in the beating canine left ventricle. *Circ Res* 1986;59:704–711.

CHAPTER 28:

ARE FLOW TRANSPORT CAPACITIES AND METABOLIC REQUIREMENTS MATCHED REGIONALLY IN THE MYOCARDIUM?

James B. Bassingthwaighte, James H. Caldwell, Gary Raymond, Keith Kroll and Gary V. Martin

ABSTRACT

Data were acquired on the uptake of metabolic substrates and on well-retained ions in small regions of the myocardium, 1 to 2 min after injection into the left atrium, and compared with estimates of local myocardial blood flows in each of the 250 to 500 regions, per heart. The data showed approximately linear relationships between the deposition of these solutes and the local flows. Models for blood–tissue exchange kinetics based on an assumption of the constancy of permeability–surface area products for capillary and cellular uptake in all regions being the same were not compatible with the data. Using axially distributed multiregional models appropriate to the solutes, the data could be fitted only when the permeability–surface area products were proportional to the local flow in each region. Within myocardial pieces of 0.1 to 0.2 G, the range of flows was about six–fold; if within the pieces having the lowest flows the regional metabolism was at the average rate for the whole heart, tissue viability could not be supported by flows available. These regions must therefore have less than average metabolic requirements. Likewise high flow regions would be extravagantly perfused if their metabolic rates were at the average rate. We conclude that although the link between transport capacities and flow is well established, that between local metabolism and flows is not proven, but it is highly likely that local flows transport capacities and metabolic requirements are well matched.

INTRODUCTION

In 1909 Christian Bohr [1], in examining the loss of CO_2 from the blood and the uptake of oxygen from the alveolar into the blood during passage through alveolar capillaries, summarized the data by the following equation:

$$PS = -F \ln(1 - E) \tag{1}$$

where the PS is the capillary permeability–surface area product in ml/(g.min), F is regional blood flow in the same units, and E is the transcapillary extraction of the solute. For CO_2, the extraction is the fractional loss of CO_2 relative to the concentration in the pulmonary artery; for oxygen, E is the increment in oxygen concentration from the pulmonary artery to the pulmonary vein divided by the pulmonary venous concentration. The basic assumption in the application of the formula was that flows F and PS's were everywhere constant, that is, all capillaries have the same flows and other properties.

Equation (1) was redeveloped independently in later years. Renkin [2] applied it to the extraction of tracer labeled potassium during passage through the capillaries of skeletal muscle. Here the PS represented the composite permeability–surface area product of the capillary wall, the interstitium, and of the membrane of the muscle cells. The potassium extraction was a net pseudo steady state extraction, maintained for many minutes by an almost constant uptake of tracer potassium into the intracellular potassium pool within the muscle cells.

Christian Crone [3] gave Eq. (1) its modern meaning in pioneering the development of the use of the multiple indicator dilution technique for the estimation of PS. In his study, the extraction E was the fraction of the permeating solute taken up relative to an intravascular solute such as plasma albumin, during the first few seconds following a bolus injection of the pair of solutes into the vascular inflow, while collecting a sequence of samples from the outflow from which the extraction was measured. This did not require that there be a large extravascular sink for the permeant solute since the measurement was made during the first moments when the concentration of tracer in the extravascular region was so small that there is essentially no reflux from tissue back into the blood. Both Renkin's and Crone's approaches made the implicit assumption that all regions of the organ under study had identical flows and capillary permeability–surface area products. Crone's instantaneous extraction technique was specific for estimating PS from the barrier closest to the blood, namely the capillary wall. Renkin's estimate of PS was for a composite barrier between blood and myocyte. Sheehan and Renkin [4] later improved the analysis of these pseudo steady state experiments by accounting for 2 barriers.

METHODS AND RESULTS

Spatial Heterogeneities in Local Blood Flows

Data of 3 different sorts lead to the hypothesis that local flows, transport capacities, and metabolic rates are matched. The initial stimulus was the observation of persisting stable spatial heterogeneities of local blood flows in the left ventricular myocardium. The data of King et al. [5] demonstrated that myocardial blood flows in sections of the myocardium amounting to about 1/250 of the mass of the heart showed about a six-fold range of flows. Observations were made at 4 to 6 different times in awake baboons and it was observed that there was remarkable spatial stability. High flow regions were observed to have high flows at all times, and low flow regions to have low flows at all times. King and Bassingthwaighte [6] analyzed these data further showing that the contribution of temporal fluctuations to the overall variation was very small. The relative dispersion (standard deviation/ mean flow for the whole heart) averaged around 25 to 30, of which the temporal fluctuation contribution was only 2 to 3 of the variation.

The observed broad heterogeneity of flow distributions are not artifacts of the microsphere technique. By using a reference "molecular" microsphere to test the adequacy of the standard 15 or 16 micron–microsphere technique, Bassingthwaighte, Malone et al. [7] in rabbit hearts and Bassingthwaighte,

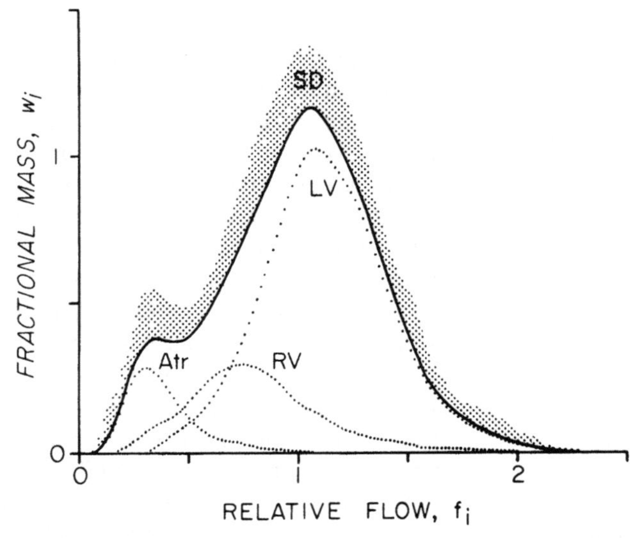

Figure 1: The distribution of relative flows in the hearts of 13 awake baboons, from 13,114 estimates at 4 to 6 times in 2,706 tissue pieces. The solid line represents the distribution for the whole heart, while the dotted lines are its LV, right ventricular (RV), and atrial (Atr) components. The standard deviation (SD) for the whole heart curve is shown by the shaded region. The area under each curve represents its fraction of the total heart weight, and the mean is its flow relative to the total heart flow. These are, by definition, unity for the total heart curve. For the regional curves the values are: LV, 0.70 and 1.14; RV, 0.20 and 0.81, and atria, 0.10 and 0.41. The relative dispersion of each curve is a measure of the spatial heterogeneity of the flow. The values are: whole heart, 0.38; LV, 0.30; RV, 0.32; and atria, 0.17. Reprinted from King et al. [5] with permission from the American Heart Association.

Malone et al. [8] in sheep hearts demonstrated that although the error in flow estimation was slightly greater with microspheres than with the reference marker, the overall variation in regional flow estimates was essentially the same. These studies affirmed the results of many earlier studies.

The implications of relatively broad heterogeneities of flows are interesting. Systematic deviations due to the particulate nature of the microspheres contribute very little to the variation [8, 9]. A six to seven fold range of flows observable at the size of pieces sampled in these recent studies implies heterogeneity of flows would be broader yet if the sizes have been further reduced. A ten or twenty–fold range of flows would be expected at the size of pieces representing the flow region supplied by a terminal arteriole, something of the order of 100 micrograms of tissue [10–12].

With such variation there is, therefore, a substantial fraction of the ventricular myocardium with flows less than 40% of the mean. In the heart the overall arterio–venous extraction for oxygen is about 70%. This would imply that if regions having flows less than that 40% had normal or average oxygen requirements, that the oxygen extraction would have to be greater than 100% locally; to get the same oxygen usage in such a region the extraction would have to be 0.71/0.4 or 175%, an impossibility. Thus, for long term survival, such persistently low flow regions cannot have average oxygen requirements, but must require less than the average in order to survive.

A similar but not quite so decisive statement might be made regarding the highest flow regions. If they had oxygen requirements only equal to the average for the heart, then they would have a two–fold excess of flow, something which would be obviously inefficient for an organ whose oxygen requirements are so high that over two thirds of the oxygen is normally extracted from the blood. The conclusion then follows naturally that local regions must be metabolically low and high flow regions must be metabolically high.

Heterogeneities in Transmembrane Transport of Monovalent Cations

A second set of information extends the information on regional heterogeneities to encompass information on transmembrane transport. Pohost et al. [13] and Nielsen et al. [14] observed approximately linear relationships between thallium deposition and regional myocardial blood flow as measured by microspheres in a series of dog hearts. That the relationships were linear is doubly interesting. Firstly, as seen below, the linear relationship is not really compatible with a constant value for the permeability–surface area product for thallium penetration of the membranes throughout the heart. That the slope was less than 1.0, which it would be if thallium were 100 extracted and retained, was not surprising since it is known that thallium is only

75 to 90 extracted during single passage through the myocardial capillaries. The second interesting feature was that on all of the graphs of thallium deposition versus microsphere deposition there was a positive intercept on the thallium axis. This means that in the very lowest flow regions, as measured by the microspheres, there was excess deposition of thallium as compared with the particulate spheres. A similar result is shown in data from our laboratory for rubidium, Fig. 2.

Modeling analysis was applied using the axially–distributed blood–tissue exchange model of Kuikka et al. [15], composed of capillary interstitial space and cellular space. This model accounts for barriers at the capillary and sarcolemmal membranes and for gradients along the length of the capillary–tissue unit, which is important for indicators of high permeability. Two versions of the model were applied, one with constant PS in all regions and one with PS directly proportional to the local flow in each of the multiple pathways covering the range of flows. As was observed in the data of Pohost et al. [13], some of the data show rubidium deposition apparently out of proportion to flow.

Depositions above the line of identity imply local extractions greater than 100%. It has been thought that this might be due to recirculation with continued deposition into low flow regions while wash out from high flow

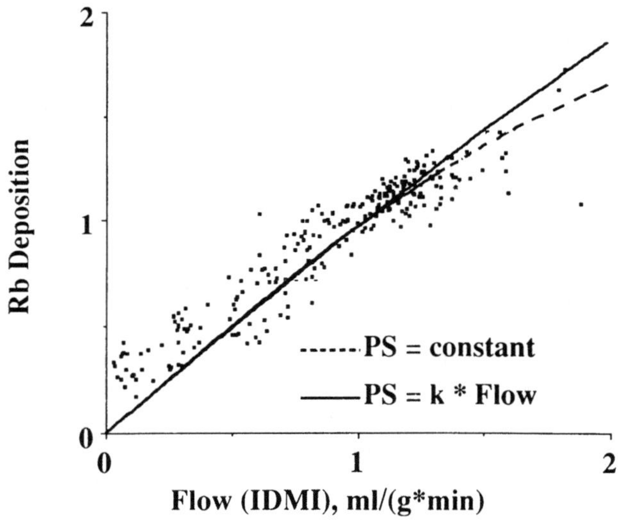

Figure 2: Deposition of rubidium locally versus local flow determined by the deposition of the molecular microsphere, IDMI. The linear regression gives a slope of about 0.8, compatible with a transorgan extraction of about 80%. Two model solutions have been fitted through the data, one with PS constant everywhere, and the second with PS proportional to flow in each region.

regions was beginning. However, this explanation should be regarded as speculative since it is not clear that the recirculated rubidium is at sufficiently high levels and the wash out from high flow region at a sufficiently fast rate for the observed results to be explained on this basis. To make the relationship linear would require a coincidence of such events.

An alternative explanation is that microspheres and IDMI are both extracted at the arteriolar ends of capillaries, and that some of these low flow regions are supplied by capillary networks that are distant from the supplying arterioles and are therefore receiving less of the "flow marker" than they deserve, and so their flows are underestimated. This will be true of any bed with a "portal" circulation, that is, a second capillary bed following the first capillary bed, or any region that is supplied through the capillary network after the flow markers have been extracted. However, this is an aside from the theme of this chapter.

The only conclusion that can be reached from these data on rubidium and thallium is that while the transport capacity for these ionic indicators may possibly be proportional to the local flow, a constant value of PS throughout the organ cannot be disproved from these data. The experiments failed to provide data at flows sufficiently high that the permeability barrier was a major obstruction to deposition, and so leaves the question still open.

The rubidium analyses suggested three model formulations for the local PS–to–flow relationships; these are diagramed in Fig. 3. The equations in the figure define the possibilities: (1) PS is constant; (2) the local value of PS in

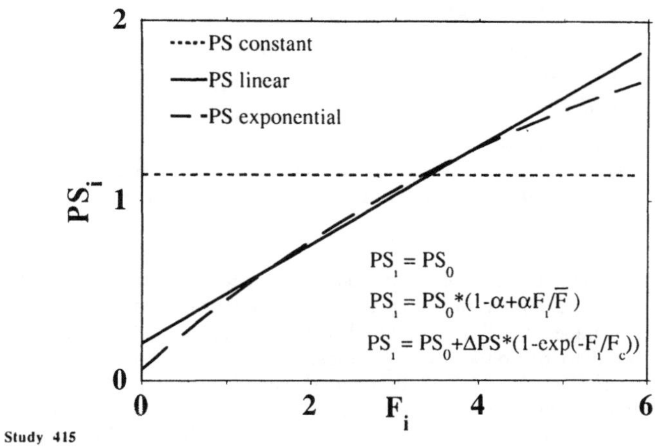

Figure 3: Three hypothesized relationships between the local permeability–surface area product, PS_i, and the local flow, F_i. Units of both axes are ml $g^{-1}min^{-1}$. $\bar{F}$ is the average flow.

the ith region is a linear function of flow with slope α and average value PS_o; and (3) a local PS_i rising exponentially from PS_o at $F_i = 0$ to a maximum at $PS_o + \Delta PS$, with a mean value dependent on the value of flow F_c at which PS $= PS_o + 0.63\ \Delta PS$.

Heterogeneities in Transmembrane Transport of Fatty Acids

A third set of data concerns the deposition of a fatty acid analog, iodophenylpentadecanoic acid, along with microspheres in the myocardium of dogs running on a treadmill at 10% grade [16]. Occlusion of the circumflex coronary artery about two centimeters below its origin contributed to the observed heterogeneity. Ninety seconds after left atrial injection of tracer microspheres and the fatty acid analog, the hearts were stopped and excised to obtain the regional depositions. The plot (Fig. 4) of fatty acid depositions versus the regional flows appears rather similar to those for rubidium and thallium. The near–linearity is striking.

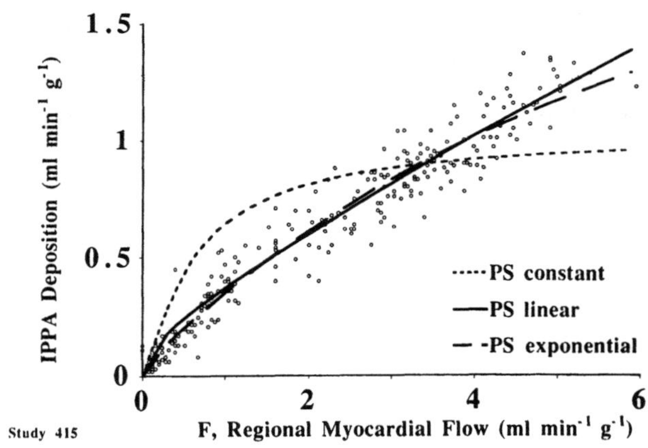

Figure 4: Deposition of 131–I–phenylpentadecanoic acid versus regional flow by the microsphere technique in a dog heart while performing exercise on a treadmill at 10% grade, 10 miles per hour. The extraction is about 30%. Three models are fitted to the data (for equations, see Fig. 3). For each, the coefficient of variation of model solution versus data is minimal. Model one with PS constant everywhere is an exceedingly poor fit. Model two with an exponentially increasing PS as a function of flow regionally is a better fit. Model three with PS proportional to flow provides the best fit. The coefficients of variation were 0.179 for PS constant, 0.088 for PS exponentially increasing, and 0.086 for PS proportional to flow. Note that the "exponential" model has been forced to be basically linear.

In these analyses, as is clear from Fig. 4, the constant PS model is quite unacceptable and that the model with PS proportional to flow gives a greatly improved fit to the data. This is a rather dramatic result and has much stronger implications than the data from rubidium and thallium. Rubidium and thallium both penetrate the clefts between endothelial cells by passive diffusion and may enter endothelial cells via the potassium channels which are normally open at the resting potential. Since open intercellular clefts and open potassium channels can be expected to be similar in all capillaries, then either the constant PS model might be suitable for these ions or there might be a flow–proportional PS, if there were large differences in capillary surface area throughout different regions. That is to say that the flow of proportionality in PS might be attributable to a constant P and flow proportionality in the surface area, S.

For fatty acid the story is quite different from that of rubidium and potassium because it appears that fatty acid transport is mediated through a special transport mechanism on the endothelial cell surfaces and probably on the sarcolemma as well. The data of van der Vusse et al. [17] shows that by increasing the albumin concentrations at constant albumin/fatty acid ratios, transcapillary transport is impeded, indicating that the transport, or at least the off–loading of fatty acids from the albumin which carries in the vascular space, is mediated by a saturable mechanism at the endothelial cell surface. Proportionality is therefore a strong indicator that the numbers of "receptor/translocator" sites on the endothelial cell surface is proportional to flow. Whether or not this flow proportionality is completely independent of capillary surface area is not known, since the capillarity, the numbers of capillaries per unit mass of tissue, is not known for high flow versus low flow regions in the myocardium. Wide ranges of capillary density in heart muscle have not been reported. It has been observed, however, by Gonzalez and Bassingthwaighte [18] that there is very little systematic dependence of blood volume content regionally upon flow regionally. We feel therefore that the most reasonable explanation for the flow proportionality of PS for the iodophenylpentadecanoic acid regionally is due not only to the number of binding sites available, but presumably also that the number of binding sites is regulated by the metabolic requirements of the tissue locally.

HYPOTHESIS AND DISCUSSION

Hypothesis

Impedance matching appears to be as desirable in nature as it is in designing electrical or mechanical networks. Our hypothesis is that there are matched capacities for the delivery of substrates by flow, by transport across the membranes, by intracellular translocation mechanisms, and by intracellular enzymic reactions. This would appear to be the cheapest way to build the system. The studies on the exercising dog are probably most relevant to the testing of this hypothesis since the impedance matching probably cannot hold for all states, including rest and modest activity, but would presumably be designed for the most demanding state that commonly occurs, as in dogs running over a significant distance. Likewise, it would make sense if training for increased human performance in endurance events would lead to an improvement in impedance matching in the various components of a system delivering substrates of energy metabolism to the sites in the cell where the ultimate reactions occur, namely intramitochondrial oxidative metabolism.

Implications of Intramitochondrial Heterogeneity with Impedance Matching

Heterogeneities in flow, transport, and metabolic rates which are locally matched for optimal performance provoke serious questions concerning how the heart functions. If metabolic requirements are so broadly heterogeneous as the flows which we measure, this implies that the mechanical force development in these regions is also heterogeneous. The preliminary analyses of Ritman et al. [19] would imply that there are heterogeneities in mechanical stress locally. This would appear to occur because of variations in the anatomical arrangements and imply more heterogeneity than will be expected in the mathematical models of stress development as proposed, for example, by Arts et al. [20].

The relationship between these observations and electrophysiological abnormalities is not known. However, it would seem likely that when changes in the physiological state result in deviations from the optimal matching of flow, transport, and metabolism that abnormal stresses must be most likely to occur. In untrained individuals stresses such as prolonged exercise might be more likely to create electrophysiological abnormalities, even leading to arrhythmias, in situations to which they are unaccustomed. Even so, the heart is probably designed to handle such situations by appropriate responses. For example, Upsher and Weiss [21] observed that iodopindalol binding sites were in greater abundance in high flow regions than in low flow regions. Since the local responses to beta adrenergic stimulus is increased metabolism, but also increased flow locally, it would appear that in their dog hearts at least, there was potential for the maintenance of the impedance matching over a range of physiological states. Our tentative conclusion is that there is impedance matching in flows, transport, and metabolism regionally in the myocardium.

CONCLUSION

This is a subject that we need considerably more investigation before conjecture can be considered proven. Since neither the measurement of local substrate concentrations (fatty acids, glucose, or ATP) nor the measurement of local enzyme concentrations such as lactate dehydrogenase or others, are measures of metabolic rate, but merely measures of metabolic capacity to some extent, the experiments that are needed are not simple. Firstly, we would like to know if nature subscribes to our simplifying hypothesis that there is an optimal matching of impedances, or if other factors such as structural maintenance are more important. Parsimony in energy and structural requirements requires some trade-off. Secondly, close matching of local impedances would lead to major simplifications in modeling blood-tissue exchange processes. Thirdly, as advancing technology reveals more and more of the fine-scale heterogeneities in the heart, we need to be able to distinguish normal heterogeneities from those brought on by pathological conditions.

Acknowledgements

This study was supported by NIH grants HL19139 and HL38736 from the National Heart, Lung and Blood Institute. The authors appreciate the help of Maureen Feeley in preparing the manuscript.

DISCUSSION

Dr. Strauss: What about using thallium?

Dr. Bassingthwaighte: In the experiment with rubidium, we also simultaneously injected thallium, iodo-desmethylimipramine (IDMI) and microspheres, a total of 4 tracers. Thallium looks like rubidium. It is interesting too. In our experiment, I think we made a mistake by stopping the heart at 60 s after injection, because there is recirculation of rubidium during that period. There are some provoking data, those data above the line of identity, that would seem to be physically impossible, since the deposition appears to be higher than can be explained by the observed "flows" estimated from microspheres. Thallium does the same thing. The molecular microsphere IDMI, unlike the ceramic or polystyrene spheres, does not suffer the rheological biases, which are in any case fairly small. So you would say how could this get here? Is recirculation enough to explain that. I think the answer is no. Dr. K. Tsuiki at Yamagata *[Personal Communication]*, showed me data he has on thallium vs microspheres, which shows the same kind of thing. If you go back to the published work of Pohost et al. *[13]* on thallium vs. microspheres, they all show this. In all of those earlier experiments, with the exception of Tsuiki's, there is recirculation present, so there is some doubt about why this might occur. Tsuiki's preliminary data were presented as regression lines and the individual points are needed to help figure out how rubidium could get into regions out of proportion to flow. There is a little mystery here, but it is possible that microspheres and IDMI are trapped prior to entering certain regions.

Dr. Schelbert: Do you think that the problem may be due to error in the microsphere method?

Dr. Bassingthwaighte: No. That is why I used the IDMI, the molecular microsphere, as the reference here *[8]*.

Dr. Schelbert: How large are they?

Dr. Bassingthwaighte: IDMI has a molecular weight of 400, so plasma skimming cannot reduce its deposition. The IDMI is a tricyclic antidepressant which gets bound to α_2 and 5HT receptors in the heart so it is virtually completely extracted. It gets into the red cells, so there could be some red cell contribution to this separation. Rubidium also gets into the red cells and so does thallium, so again, you would expect that to be self compensating. I do not think that is the explanation and I do not know what it really is at this point. We need to do a cleaner experiment with no recirculation. I suppose we will have to get rid of the red cells as well.

Dr. Spaan: What happens if the heart is vasodilated, does flow increase to all regions? Is the surface area for exchange, S, still proportional to flow or is the proportional relationship of PS to F broken down?

Dr. Bassingthwaighte: These are normal flows in this heart and our experiment only doubled the flows, which is not high enough to answer your question. One of the possible explanations is that high flow regions simply have much higher capillarity. But the problem with that idea is that we see a six–fold range of flows. No one to my knowledge has ever seen such variation in capillary densities; only 20–30% but not six–fold.

Dr. Spaan: But still you have a 3–fold range in both the regional flows and the rubidium depositions to the anatomical regions within the heart.

Dr. Bassingthwaighte: Yes, actually 6–fold, but those depositions do not directly give the PS distribution. They are strictly the raw data on the rubidium deposition. These 2 interpretations, PS constant versus PS/F constant, are indistinguishable from the data in Fig. 2. We need to do an experiment with adenosine or with exercise to reach much higher flows and see if that would distinguish the two ideas. That is why it is a failed experiment. But that is a different issue than trying to explain the points about the line of identity.

Dr. Westerhof: Would you like to speculate on mechanical forces or mechanical performance of these different locations, because if metabolism is fixed to flow, wouldn't you expect the local mechanics also to be related to flow?

Dr. Bassingthwaighte: Definitely. If I look at E. Ritman's and Y.C. Pao's *[19]* finite element analysis then you can see different force development or rather work levels in each of the individual regions. They have some calculations of that but I do not know the details enough to know whether that gives them as huge a range as this.

Dr. Schelbert: I was intrigued by your statement on myocardial regions with high and with low blood flows. You supported this statement by an illustration. Did you see such regions with high and low flows consistently in all of the animals which you examined? If so, were they associated with one particular anatomic region?

Dr. Bassingthwaighte: We sought extensively to determine whether the regions of high flow were reproducible from animal to animal, in these series of 13 baboons, and 11 sheep. Dr. Reneman has likewise sought for these consistencies in a long series of dog experiments that he and Dr. Prinzen and others performed. We could discover no systematic predictably high flow or low flow regions. They seem to be different in each heart. I think your dog data verify that, don't they Dr. Reneman?

Dr. Reneman: Yes, but they also showed differences in flows from the base to the apex, but with much inconsistency. From the literature, it appears different in closed chest preparations from open chest preparations, so you have to be fairly careful in interpreting the data.

Dr. Strauss: How do you relate ammonia trapping to FDG trapping? Do you have human data?

Dr. Bassingthwaighte: If you interpret ammonia trapping as a measure of flow locally and FDG trapping as a measure of a switching to glycolytic flux with ischemia, they go together. We do not have enough data on humans to make any comparable statement.

REFERENCES

1. Bohr C: Uber die spezifische Tätigkeit der Lungen bei der respiratorischen Gasaufnahme und ihr Verhalten zu der durch die Alveolarwand stattfindenden Gasdiffusion. *Skand Arch Physiol* 1909;22:221–280.
2. Renkin EM: Transport of potassium–42 from blood to tissue in isolated mammalian skeletal muscles. *Am J Physiol* 1959;197:1205–1210.
3. Crone C: The permeability of capillaries in various organs as determined by the use of the "indicator diffusion" method. *Acta Physiol Scand* 1963;58:292–305.
4. Sheehan RM, Renkin EM: Capillary, interstitial, and cell membrane barriers to blood–tissue transport of potassium and rubidium in mammalian skeletal muscle. *Circ Res* 1972;30:588–607.
5. King RB, Bassingthwaighte JB, Hales JRS, Rowell LB: Stability of heterogeneity of myocardial blood flow in normal awake baboons. *Circ Res* 1985;57:285–295.
6. King RB, Bassingthwaighte JB: Temporal fluctuations in regional myocardial flows. *Pflügers Arch (Eur J Physiol)* 1989;413(4):336–342.

7. Bassingthwaighte JB, Malone MA, Moffett TC, King RB, Little SE, Link JM, Krohn KA: Validity of microsphere depositions for regional myocardial flows. *Am J Physiol (Heart Circ Physiol 22)* 1987;253: H184–H193.
8. Bassingthwaighte JB, Malone MA, Moffett TC, King RB, Chan IS, Link JM, Krohn KA: Molecular and particulate depositions for regional myocardial flows in sheep. *Circ Res* 1990;66:1328–1344.
9. Yipintsoi T, Dobbs WA Jr., Scanlon PD, Knopp TK, Bassingthwaighte JB: Regional distribution of diffusible tracers and carbonized microspheres in the left ventricle of isolated dog hearts. *Circ Res* 1973;33:573–587.
10. Bassingthwaighte JB, King RB, Roger SA: Fractal nature of regional myocardial blood flow heterogeneity. *Circ Res* 1989;65:578–590.
11. van Beek JHGM, Bassingthwaighte JB, Roger SA: Fractal networks explain regional myocardial flow heterogeneity. In: *Oxygen Transport to Tissue XI. Adv. Exp. Med. Biol. 248,* (ed) K Rakusan. Plenum Press, New York, 1989, pp. 249–257.
12. Roger SA, van Beek JHGM, Bassingthwaighte JB: Estimating the functional myocardial microvascular unit size using fractal models. *Microvasc Res* 1990; in press.
13. Pohost GM, Zir LM , Moore RH , McKusick KA , Guiney TE , Beller GA: Differentiation of transiently ischemic from infarcted myocardium by serial imaging after a single dose of thallium–201. *Circulation* 1977;55:294–302.
14. Nielsen AP, Morris KG , Murdock R , Bruno FP , Cobb FR: Linear relationship between the distribution of thallium–201 and blood flow in ischemic and non-ischemic myocardium during exercise. *Circulation* 1980;61:797–801.
15. Kuikka J, Levin M , Bassingthwaighte JB: Multiple tracer dilution estimates of D– and 2–deoxy–D–glucose uptake by the heart. *Am J Physiol (Heart Circ Physiol 19)* 1986;250:H29–H42.
16. Caldwell JH, Martin GV, Link JM, Gronka M, Krohn KA, Bassingthwaighte JB: Iodophenylpentadecanoic acid – myocardial blood flow relationship during maximal exercise with coronary occlusion. *J Nucl Med* 1990;30:99–105.
17. van der Vusse GJ, Little SE, BAssingthwaighte JB: Transendothelial transport of arachidonic and palmitic acid in the isolated rabbit heart. *J Mol Cell Cardiol (Suppl.III)* 1987;19:S100 (abstract).
18. Gonzalez F, Bassingthwaighte JB: Heterogeneities in regional volumes of distribution and flows in the rabbit heart. *Am J Physiol (Heart Circ Physiol 27)* 1990;258:H1012–H1024.
19. Ritman EL, Heethaar RM , Robb RA , Pao YC: Finite element analysis of myocardial diastolic stress and strain relationships in the intact heart. *Eur J Cardiology* 1978;7:105–119.
20. Arts T, Reneman RS, Veenstra PC: A model of the mechanics of the left ventricle. *Ann Biomed Eng* 1979;7:299–318.
21. Upsher ME, Weiss HR: Relationship between beta–adrenoceptors and coronary blood flow heterogeneity. *Life Sci* 1989;44:1173–1184.

CHAPTER 29:

DIMINISHED TOLERANCE TO ISCHEMIA OF HYPERTROPHIED HEARTS

Luc H.E.H. Snoeckx, Will A. Coumans, Ger J. van der Vusse and Robert S. Reneman

ABSTRACT

The findings presented in this survey indicate that old, not fully compensated, hypertrophied hearts are more vulnerable to ischemia than non-hypertrophied hearts of comparable age. Also, young fully compensated hypertrophied hearts have a similar susceptibility to ischemia as non-hypertrophied hearts of comparable age, provided that they are perfused at a relatively high pressure, leading to a similar coronary flow rate per gram tissue.

INTRODUCTION

Protection of the hypertrophic heart during cardiac surgery is a continuing problem, because these hearts are more vulnerable to ischemia than non-hypertrophied hearts [1, 2, 3, 4]. Most of the data presently available have been derived from studies performed on severely hypertrophied, often decompensated, hearts. Less information is available about the ischemia tolerance of hearts with fully compensated left ventricular hypertrophy. In these hearts left ventricular pump function is believed to be normal or even enhanced [5, 6].

In this chapter we survey the results obtained in our laboratories in studies on the ischemia tolerance of hypertrophied hearts at different stages of the disease. The studies were performed on hearts of spontaneously hypertensive rats (SHR) of 8 and 18 months. Since in this type of left ventricular hypertrophy genetic factors may be involved, we also studied the ischemia tolerance of hypertrophied hearts in adult aorta-banded rats. In this model left ventricular hypertrophy is fully compensated [7]. The experiments were performed on isolated, ejecting, assisted Tyrode perfused hearts in which global ischemia was induced for 45 min and recovery following the ischemic insult was evaluated during a 45 min period of reperfusion. Since pressure levels are relatively high in these models of left ventricular hypertrophy, the hearts were perfused at high (100 mmHg) or low (60 mmHg) pressure. Elevation of coronary perfusion pressure has to be limited, because too high a pressure will impose an unacceptable load to the jeopardized heart following the ischemic insult. In the experiments on SHR, Wistar–Kyoto rats (WKY) served as controls. Banding of the aorta was performed on Lewis rats. In these experiments sham–operated rats of this strain were used as controls.

EXPERIMENTAL APPROACH AND PERFUSION SET–UP

The animals were anesthetized with ether, and after heparinization (1000 IU I.V.) the hearts were removed from the body and immersed in ice–chilled Tyrode's solution. Remnant lung tissue and fat were removed. The hearts were weighed and attached to a non–recirculating perfusion apparatus [8]. In short, the hearts were perfused with a Tyrode's solution, equilibrated at 37± 0.3°C and containing (in mmol/liter): NaCl (130.0), KCl (5.6), $CaCl_2$ (2.2), $MgCl_2$ (1.0), NaH_2PO_4 (1.2), $NaHCO_3$ (28.6). D (+) glucose (11.1) and pyruvate (5.0) were added as substrates. The pH of this solution ranged from 7.35 to 7.45. The perfusion system with the hearts was placed in a thermostabilized cage, in which the air temperature was also kept at 37°C. The isolated hearts were perfused in an assisted mode, which means that the perfusion pressure of the coronary arteries could be maintained at a preset level, irrespective of the mechanical performance of the heart. A second cannula was connected to the left atrium. The filling pressure was set at 10 mmHg. Aortic and left ventricular pressures were measured with external pressure transducers (CTC). The flow in the aortic cannula was measured distal to the offspring of the coronary arteries with an extracorporeal flowprobe, connected to an electromagnetic flowmeter (Skalar 601). During retrograde perfusion this signal equaled coronary flow, while during antegrade perfusion it represented cardiac output minus coronary flow. In this situation the coronary flow was measured by collecting the coronary effluent over 1 minute periods. During antegrade perfusion cardiac output was calculated by adding up aortic and coronary flow and expressed in ml/min/g heart tissue. All variables, including the electrogram, were fed into a microcomputer [9].

After attachment to the perfusion apparatus, retrograde perfusion was maintained during 10 min. Thereafter, the left atrial cannula was opened and antegrade perfusion was started during 20–30 min. Depending on the experimental group, diastolic aortic pressure was kept at either 60 or 100 mmHg throughout the experiments. Heart rate was generally kept constant at 300 beats per minute by pacing with the use of platinum electrodes attached to the right atrial wall. At the end of the pre–ischemic period, flow to the heart was completely stopped during 45 min by cross–clamping the atrial and aortic cannulas. Pacing of the hearts was discontinued. Thereafter reperfusion was started retrogradely for 15 min, whereupon electrical stimulation and antegrade perfusion was restarted for a period of 30 min under pre–ischemic conditions.

Coronary effluent samples were immediately stored on ice, and analyzed later for the content of purine nucleosides and oxypurines, such as adenosine, inosine, hypoxanthine, xanthine, and uric acid, using a Varian 5500 HPLC on a 0.4 x 30 cm prepacked microBondapak/C18 column [10]. The activity of lactate dehydrogenase (LHD) was determined using the technique described by Bergmeyer and Bernt [11]. At the end of the experiments the ventricles were immediately freeze–clamped. The tissue was freeze–dried overnight at a temperature below –30°C. Aliquots of the tissue were extracted at –15°C in a mixture, containing 3 $mol.l^{-1}$ perchloric acid and 5 $mmol.l^{-1}$ dithiotreithol. After 5 min centrifugaton at 1200 g and 4°C, a known amount of the supernatant was removed and neutralized with KH_2CO_3. The salt–precipitate of the neutralized extract was pelleted by centrifugation at 1200 g and 4°C during 60 min. Tissue contents of ATP, ADP and AMP were determined in 10 μl samples of the supernatant by HPLC on a VARIAN 5500 [12]. Tissue contents of creatine phosphate (CP) were determined as described before [13].

ISCHEMIA TOLERANCE OF HYPERTROPHIED HEARTS

Eight Months Old SHR and WKY Rats

Heart weights. The heart weights (2.06 g vs 1.64 g) and the heart weight to body weight ratios ($0.5 \cdot 10^{-2}$ vs $0.39 \cdot 10^{-2}$) were significantly higher in SHR than in WKY rats.

Perfusion pressure: 60 mmHg. At this pressure, cardiac output recovered completely following the 45 min of global ischemia in the WKY rats, but not in SHR. The values attained were significantly different for both strains (Table 1). One heart in the SHR did not regain normal electrical activity. In SHR the diminished recovery of cardiac output was associated with good recovery of coronary flow (Table 2), but with no restoration of myocardial ATP content and only limited restoration of myocardial CP content (Table 3). Restoration of coronary flow and myocardial CP content following the ischemic insult was practically complete in the WKY rats (Tables 2 and 3), but the myocardial ATP content recovered only partially. Recovery of the energy charge at the end of the reperfusion period was significantly better in WKY than in SHR [14]. The wash–out of LDH and lactate during the reperfusion period was similar in WKY and SHR [14], but that of degradation products of adenine nucleotides was significantly greater in WKY than in SHR.

Table 1: Recovery of cardiac output (% of control; median values) after 45 min of global ischemia in SHR and WKY rats.

Perfusion Pressure (mmHg)	Percent Recovery Cardiac Output			
	8 Months		18 Months	
	SHR	WKY	SHR	WKY
60	71*/**	100**	0*	38
100	94*/**	59**	8	12

* significantly different from WKY in same group.
** significantly different from corresponding hearts of 18 months old rats (Mann–Whitney U–test).

Table 2: Recovery of coronary flow (% of control; median values) after 45 min of reperfusion following the ischemic insult in SHR and WKY rats.

Perfusion Pressure (mmHg)	Percent Recovery Cardiac Output			
	8 Months		18 Months	
	SHR	WKY	SHR	WKY
60	88**	98	45*	84
100	95**	111**	68	87

*,** same as in Table 1

Perfusion pressure: 100 mmHg. At this pressure, recovery of cardiac output was significantly better in SHR than in WKY rats (Table 1), despite the fact that at the end of the reperfusion period the coronary flow (Table 2), the myocardial CP content (Table 3) and the energy charge were restored to pre-ischemic control values in both strains. All hearts regained normal electrical activity. The myocardial ATP content, however, remained low during the reperfusion period in both WKY and SHR. Despite similar myocardial flow values, the wash–out of LDH and degradation products of adenine nucleotides during the reperfusion period was greater in WKY than in SHR. The wash–out of lactate was similar in both strains [14].

Eighteen Months Old SHR and WKY Rats

Heart weights. Before perfusion the heart weights of the SHR were significantly higher than those of the WKY rats (median values: 2.55 vs 1.95 g in the 60 mmHg group and 2.78 vs 1.93 g in the 100 mmHg group). It is an interesting observation that the heart weights of WKY rats increased during

Table 3: Myocardial CP content (μmol/g dry weight) in 8 months old rats before and at the end of 45 min ischemia, and at the end of 45 min reperfusion (data at various moments derived from different hearts).

	Perfusion Pressure (mmHg)	Myocardial Creatine Phosphate (CP)		
		Pre–Ischemia	End–Ischemia	End–Reperfusion
SHR	60	41.5*	4.5	20.8*/***
WKY	60	39.3*	9.8	37.2*/**
SHR	100	——	——	36.3
WKY	100	——	——	47.0

* significantly different from end–ischemic values. ** significantly different from SHR hearts at end–reperfusion. *** significantly different from SHR hearts at end–reperfusion, when perfused at 100 mmHg (Mann–Whitney U–test).

ischemia and the following reperfusion period (from 1.95 to 2.30 g in the 60 mmHg and from 1.93 to 2.66 g in the 100 mmHg; median values), but not those of SHR. The increase was most pronounced in the WKY hearts perfused at 100 mmHg.

Perfusion pressure: 60 mmHg. At this pressure the recovery of cardiac output was very limited in WKY rats and absent in SHR (Table 1), despite the fact that there was good recovery of overall coronary flow in WKY and limited recovery of this variable in SHR. In the latter strain the overall coronary flow values at the end of reperfusion were significantly lower than those in younger SHR and in the comparable WKY (Table 2). In SHR the underperfusion of the myocardium, as measured with radioactive microspheres, was most pronounced in the endocardial layers leading to endo/epicardial flow ratios of 0.21 to 0.29 (control: 0.82). In WKY rats this ratio varied between 1.16 in the control situation and 1.07–1.37 during reperfusion (Fig. 1). At the end of the reperfusion period no rhythm disturbances were observed in the WKY heart, but none of the SHR heart regained normal electrical activity. The myocardial CP content was completely restored in WKY, but no recovery of the content of this high–energy phosphate was seen in SHR (Table 4). The energy charge showed good recovery in WKY, but not in SHR. In the latter strain the energy charge was even lower at the end of the reperfusion period than at the end of ischemia. The wash–out of degradation products of adenine nucleotides and LDH during the reperfusion period was significantly delayed in SHR as compared with WKY rats [15]. At the end of the reperfusion period the myocardial lactate content was still slightly increased in WKY and even higher than at the end of the ischemic period in SHR [15].

Table 4: Myocardial CP content (μmol/g dry weight; median values) in 18 months old rats before and at the end of 45 min ischemia, and at the end of 45 min reperfusion (data at various moments derived from different hearts).

	Perfusion Pressure (mmHg)	Myocardial Creatine Phosphate (CP)		
		Pre–Ischemia	End–Ischemia	End–Reperfusion
SHR	60	22.5*	4.4	6.4***
WKY	60	25.2*	4.1	33.9*/**
SHR	100	24.1*	3.1	13.1*
WKY	100	31.0*	4.4	22.9*

*, **, ***: See Table 3.

Perfusion pressure: 100 mmHg. At this perfusion pressure the recovery of cardiac output was practically zero in WKY and SHR (Table 1), despite the fact that overall coronary flow was restored rather well (Table 2) and normal electrical activity was regained. Although in the SHR endocardial blood flow was better restored at this perfusion pressure than when the hearts were perfused at 60 mmHg, these layers remained underperfused relative to the epicardial layers in the reperfusion phase. The left ventricular wall in WKY rats was perfused rather homogeneously at 100 mmHg (Fig. 1). In both WKY and SHR the myocardial CP content tended to restore itself; the recovery being relatively most pronounced in WKY rats (Table 4). Myocardial ATP content did not show any recovery, either in WKY or in SHR. In both WKY and SHR the energy charge tended to recover, but the control levels were not reached. Although in SHR the wash–out of degradation products of adenine nucleotides and LDH tended to increase at this perfusion pressure, it was still delayed as compared to the WKY rats perfused at the same pressure [15]. The myocardial lactate content was practically normalized in WKY and SHR hearts perfused at 100 mmHg [15].

ISCHEMIA TOLERANCE OF HYPERTROPHIED HEARTS DUE TO BANDING OF THE ABDOMINAL AORTA

Banding technique. Banding of the abdominal aorta was performed in 10 weeks old, male Lewis rats as previously described in detail [7]. In sham–operated animals the aorta was only isolated. The experiments were performed 10 weeks after abdominal surgery.

Heart weights. The heart weights (1.75 vs 1.46) and the heart weight to body weight ratios ($0.49 \cdot 10^{-2}$ vs $0.39 \cdot 10^{-2}$) were significantly higher in aorta–banded than in sham–operated rats.

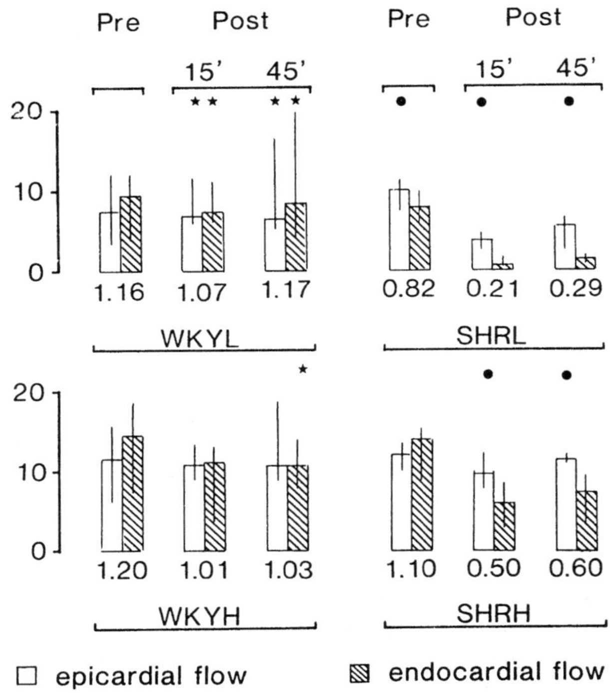

Figure 1: Endocardial and epicardial flow values in WKY and SHR of 18 months, perfused at 60 (WKYL and SHRL) and 100 mmHg (WKYH and SHRH) before (pre), and 15 and 45 min after ischemia (post). The endo/epicardial flow ratios are noted underneath the bars. Median values and 95% confidence limits are presented. Legend: * significantly different from the corresponding values in SHR (Mann–Whitney U–test); • significant difference between endo and epicardial flow (Wilcoxons matched–pairs signed–ranks test) (from [15] with permission).

Perfusion pressure: 60 mmHg. At this pressure only 4 out of 10 aorta–banded rats showed electrical and mechanical recovery. The recovery of cardiac output was minor in these hearts. The other 6 hearts did not eject any detectable amount of fluid due to ventricular fibrillation alternated with periods of ventricular tachycardia. In the sham–operated hearts cardiac output recovered to 70% of the control value. During the reperfusion phase overall coronary flow was significantly lower in aorta–banded than in sham–operated hearts. The wash–out of degradation products of adenine nucleotides and LDH was significantly delayed in the hearts of aorta–banded rats as compared to those of sham–operated ones. The myocardial ATP content at the end of reperfusion period was significantly lower in aorta–banded than in sham–

operated rats. The myocardial CP levels also tended to be lower in the aorta-banded rats, but these differences did not reach the level of significance.

Perfusion pressure: 100 mmHg. In contrast to the hearts perfused at 60 mmHg, the recovery at 100 mmHg of heart function in the aorta-banded hearts was good and cardiac output was restored to 70% of the control value. No arrhythmias were observed, in both the aorta-banded and the sham-operated rats. A similar recovery of cardiac output was found in the WKY rats (to 75% of the control value). The enhanced recovery of heart function in the aorta-banded rats was associated with increased overall coronary flow, to a level similar to the one in sham-operated rats, and improved wash-out of LDH. The early wash-out of degradation products of adenine nucleotides also tended to increase at a perfusion pressure of 100 mmHg, in both aorta-banded and sham-operated hearts, but interestingly enough the overall release of degradation products of energy-rich phosphates was smaller at 100 mmHg than at 60 mmHg. At the higher perfusion pressure the myocardial ATP and CP contents were similar in aorta-banded and sham-operated hearts and significantly higher than in the hearts perfused at 60 mmHg.

DISCUSSION

The findings as presented in this survey indicate that hypertrophied hearts of 18 months old SHR are more vulnerable to ischemia than non-hypertrophied hearts of WKY rats of comparable age, when the hearts are perfused at a pressure of 60 mmHg. Lack of functional recovery of the hypertrophied hearts following the ischemic insult is associated with impaired post-ischemic myocardial perfusion, mainly of the endocardial layers, resynthesis of CP and ATP and wash-out of degradation products. Despite the accumulation of these products, the myocardium is apparently unable to resynthesize CP and ATP, probably because of inadequate oxygen supply. This is supported by the observation that the myocardial ATP content further decreases during reperfusion and the myocardial lactate content is higher at the end of reperfusion than at the end of the ischemic period, indicating that myocardial ischemia is maintained. The further decrease in energy charge during the reperfusion period supports this notion. It is interesting to note that 60% of the hypertrophied hearts of old SHR did not regain normal electrical activity at a perfusion pressure of 60 mmHg. Although in old WKY rats the myocardial CP content was completely restored, and the energy charge and coronary flow, also in the endocardial layers, showed good recovery, the restoration of pump function was limited (median value: 38% of the control value). This may be explained by maintained underperfusion of some area's, because the myocardial lactate content was still significantly increased at the end of the reperfusion period, as compared to the control value. Alternatively one is dealing with regionally stunned myocardium.

When perfused at 100 mmHg myocardial ATP content does not show any recovery in old WKY and SHR. Although in SHR overall coronary flow and perfusion of the endocardial layers, resynthesis of CP and the wash–out of degradation products, including lactate, is significantly improved, as compared with perfusion at 60 mmHg, the recovery of pump function remains very limited. A similar observation is made in non–hypertrophied old WKY hearts. The absence of good functional recovery, despite these improvements and the restoration of normal electrical activity, can probably be explained by the inability of old hearts to eject against a higher load following the ischemic insult.

It is an interesting observation that unlike the non–hypertrophied old WKY hearts, the hypertrophied hearts of old SHR do not show an increase in heart weight at the end of the reperfusion period. The old WKY hearts increase their weight during reperfusion most significantly when they are perfused at 100 mmHg. These observations are in accordance with disturbed myocardial perfusion due to vasoconstriction in the old SHR hearts, preventing the formation of myocardial edema. The myocardial underperfusion in old SHR at both high and low perfusion pressure could be explained by excessive vasoconstriction due to enhanced reactivity of smooth muscle cells to such mediators as norepinephrine [16]. Since ischemia induces an enhanced release of intracellular norepinephrine [17, 18], the almost complete absence of subendocardial flow at low perfusion pressure and the depressed flow to these layers at high perfusion pressure might be caused by the activity of catecholamines released from endogenous sites during the ischemic insult.

The findings in the present study also suggest that hypertrophied hearts of relatively young SHR (8 months) have a good resistance to ischemia, provided that they are perfused at a relatively high pressure (100 mmHg), leading to a coronary flow rate per gram tissue similar to that in non–hypertrophied hearts. At this perfusion pressure cardiac output at the end of reperfusion is even significantly higher in SHR (median value: 94% of control) than in WKY (median value: 59% of control), indicating that the load imposed on the non–hypertrophied heart is too high for adequate pump function following the ischemic insult. On the contrary, the functional recovery of non–hypertrophied hearts is complete when they are perfused at a pressure of 60 mmHg. At this perfusion pressure cardiac output is only partially restored (median value: 71% of control) in SHR. Since hypertrophy is less pronounced in young than in old SHR, it is quite well possible that the hypertrophy in these relatively young rats is still fully compensated, explaining why the hearts are able to cope with the relatively high load imposed and to restore pump function after the ischemic period. The similar findings obtained in the young aorta–banded rats, a model believed to represent fully compensated hypertrophy [7], support this notion. In the hypertrophic hearts due to banding of the aorta, severe ventricular arrhythmias were seen when the hearts were reperfused at 60 mmHg. These arrhythmias could be avoided by perfusing the hypertrophied hearts at a pressure of 100 mmHg. This improvement was associated with normalization of coronary flow rate per gram tissue.

Acknowledgements

This project was supported by NWO–grant 900–516–91. The authors are indebted to Jos Heemskerk, Karin van Brussel and Bianca Lahoye for their help in preparing the manuscript.

DISCUSSION

Dr. Noble: You did not suggest potassium as a possible cause of increased no–flow phenomenon. Do you have some reason for ruling that out?

Dr. Reneman: We do not have a reason for ruling it out. Potassium induced depolarization of smooth muscle cells, and, hence, vasoconstriction can indeed not be excluded, but we have not looked at it very carefully.

Dr. Hudlicka: Do you think that some of your experiments can be explained by the lower capillary density in hypertrophic hearts?

Dr. Reneman: It is true that there is a lower capillary density in hypertrophied hearts, but I have some difficulty in using this as the sole explanation. At the higher perfusion level, after all, we were able to show that in the control situation the myocardial flow in ml/min/g was similar in hypertrophied and non–hypertrophied hearts of older rats, in both the endocardial and epicardial layers. In the post–ischemic period, coronary flow was homogeneously distributed across the myocardial wall in the non–hypertrophied hearts, but subendocardial underperfusion was found in the hypertrophied hearts. These observations indicate that reduced capillary density is not of great importance. If it does play a role, then we have to assume that the capillary density in old hypertrophied hearts is lower in the subendocardial than in the subepicardial layers.

Dr. Marban: Coronar and Dennings described a form of no–reflow during reperfusion which they attributed to leukocytes. In your case there are no white cells present so you would think it is more likely that it would be an endothelial cell swelling.

Dr. Reneman: Yes, the role of leukocytes can be excluded in our experimental set–up, but this does not imply that they are not involved in no–reflow under other circumstances. Endothelial cell swelling is indeed a possibility to explain the no–

reflow phenomenon. Endothelial cell swelling has also been shown to occur in skeletal muscle upon reperfusion, following an ischemic insult, by several investigators.

REFERENCES

1. Cooley DA, Reul GJ, Wukash DC: Ischemic contracture of the heart: "stone heart". *Am J Cardiol* 1972;29:575–577.
2. Buckberg GD: Left ventricular subendocardial necrosis. *Ann Thorac Surg* 1977;24:379–393.
3. Schaper J, Schwartz F, Flameng W, Hehrlein F: Tolerance to ischemia of hyper–trophied human hearts during valve replacement. *Bas Res Cardiol* 1978;73:171–187.
4. Beyersdorf F, Elert O, Satter P: Determination of maximal ischemic tolerance of the human heart by ultrastructural recording of pre–ischemic degree of myo–cardial hypertrophy and degeneration. *Ann Thorac Surg* 1980;30:356–363.
5. Mirsky I, Pfeffer JM, Pfeffer MA, Braunwald E: The contractile state as the major determinant in the evolution of left ventricular dysfunction in the spontaneously hypertensive rat. *Circ Res* 1981;53:767–778.
6. Pfeffer MA, Pfeffer JM: Blood pressure and left ventricular dysfunction in the spontaneously hypertensive rat. *Fed Proc* 1983;42:2698–2702.
7. Malik A, Shapiro JM, Yanics J, Rojales A, Geha AS: A simplified method for producing rapid ventricular hypertrophy in rats. *Cardiovasc Res* 1976;8:801–805.
8. Snoeckx LHEH, Vusse GJ van der, Coumans WA, Willemsen PHM, Nagel T van der, Reneman RS: Myocardial function in normal and spontaneously hypertensive rats during reperfusion after a period of global ischemia. *Cardiovasc Res* 1986; 20:67–75.
9. Snoeckx LHEH, Schrijen JJW, Bilsen M van, Lammers WJEP, van der Nagel T, van der Vusse GJ, Reneman RS: A microcomputer system for haemodynamic measurements in isolated, working rat hearts. *Comp Biol Med* 1986;4:301–309.
10. Harmsen E, De Jong JW, Serruys PW: Hypoxanthine production by ischemic heart demonstrated by HPLC of blood purine nucleosides and oxypurines. *Clin Chim Acta* 1981;115:73–84.
11. Bergmeyer HU, Bernt E: UV–essay for lactate dehydrogenase with pyruvate and NADH. In: *Methods of Enzymatic Analysis*, Bergmeyer UH (ed.), Weinheim, Verlag Chemie GmbH 1974;2:727–733.
12. Wynants J, Belle H van: Single–run high–performance liquid chromatography of nucleotides, nucleosides and major purine bases and its application to different tissue extracts. *Anal Biochem* 1985;144:258–266.
13. Vusse GJ van der, Coumans WA, Veen FH van der, Drake A, Flameng W, Suy R: ATP, creatine phosphate and glycogen content in human myocardial biopsies: markers for the efficacy of cardioprotection during aorta–coronary bypass surgery. *Vasc Surg* 1984;18:127–134.
14. Snoeckx LHEH: Ischemia tolerance of the hypertrophied rat heart. *Ph.D. Thesis*, University of Limburg, Maastricht, 1987.
15. Snoeckx LHEH, Vusse GJ van der, Veen FH van der, Coumans WA, Reneman RS: Recovery of hypertrophied rat hearts after global ischemia and reper–fusion at different perfusion pressures. *Europ J Physiol (Pflügers Arch)* 1989; 413:303–312.

16. Phelan EL, Simpson FO: Vascular reactivity in genetic models of hypertension. *J Hypertension* 1986;4:S25–S27.
17. Abrahamsson T, Almgren O, Carlsson L: Ischemia induced noradrenaline release in the isolated rat heart: influence of perfusion substrate and duration of ischemia. *J Mol Cell Cardiol* 1983;15:821–830.
18. Carlsson L, Abrahamsson T, Almgren O: Release of noradrenaline in myocardial ischemia. Importance of local inactivation by neuronal and extraneuronal mechanisms. *J Cardiovasc Pharmacol* 1986;8:545–553.

CHAPTER 30:

CAPILLARY GROWTH IN NORMAL AND HYPERTROPHIC HEART

Olga Hudlicka and Margaret D. Brown

ABSTRACT

Capillary growth was elicited in normal hearts by long–term administration of coronary dilators, adenosine or propentofylline, in normal and hypertrophic hearts by long–term bradycardia induced by electrical pacing or i.p. administration of alinidine. The stimulus for angiogenesis (increased capillary wall tension due either to increased flow or to more vigorous contractions), as well as the relation of increased capillarization to heart performance, are explored.

INTRODUCTION

Capillary growth in normal adult heart occurs very rarely (see [1–3] for reviews). It has been described in young animals subjected to vigorous training, as a result of long–term administration of thyroxine [4] or coronary dilators such as dipyridamole [5], ethanol [6], adenosine or xanthine derivative propentofylline [7]. Pathological heart hypertrophy – due to either volume or pressure overload – results usually in less dense capillary network since capillaries do not grow while myocytes do, and the result are longer intercapillary distances, lower capillary density and lower capillary/fiber ratio (for review see [8]). Attempts to increase capillary supply in hypertrophic hearts were met with a relatively low degree of success: Tomanek et al. [9] reported a transient increase in 3H–thymidine labeling of capillary endothelium in 15 month old SHR rats, and Anversa et al. [10] showed some capillary growth in young SHR using morphometric techniques. Turek et al. [11] reversed the decreased capillary density in SHR by long–term treatment with nifedipine.

It is generally accepted that cardiac power output is dependent on oxygen consumption both in normal [12] as well as in hypertrophic [13] hearts. Higher oxygen consumption – a correlate of greater work output [14] is usually met by higher coronary blood flow entailing increased perfusion of the capillary vascular bed. Thus, a less dense capillary supply in hypertrophic hearts would severely limit heart power output. It is therefore of paramount importance to find means of stimulating capillary growth in both pressure and volume overload heart hypertrophy.

Since it is known that so called athletic animals (e.g. hare, wild rat) have a higher capillary density and a lower heart rate than more sedentary animals of similar species [15], we have tried to establish whether a drastic reduction of heart rate, achieved either by chronic electrical bradycardial pacing or by long–term administration of a bradycardic drug, alinidine, would induce growth of capillaries in normal as well as hypertrophic hearts, and, if so, by what mechanism.

METHODS

Experiments were performed in New Zealand red rabbits of either sex and in male Sprague–Dawley rats.

In rabbits, heart hypertrophy was induced (under fluothane anesthesia and aseptic conditions) by destruction of one cusp of the aortic valve using a specially constructed cannula introduced via the left carotid artery.

Bradycardia was achieved by electrical pacing (using bipolar electrodes introduced into the right atrium via the jugular vein). Pacing was carried out, usually for 4 weeks, as described previously [16, 17] and, in experiments with heart hypertrophy, started one month after the aortic valve damage.

In rats, pressure overload heart hypertrophy was produced by the one–clip two kidney Goldblatt method. Bradycardia was induced by alinidine (gift of Boehringer Ingelheim), given for 5 weeks (3 mg/kg ip twice daily); treatment in animals with renal hypertension started 2 weeks after the introduction of the clip.

Coronary blood flow was measured with radioactive labeled microspheres [7, 18], cardiac output by thermodilution and cardiac work was calculated from mean blood pressure and cardiac output. Maximal cardiac work was induced by infusion of noradrenaline give i.v. Heart hypertrophy was assessed as heart/body weight ratio. The extent of capillarization was estimated on the basis of staining for alkaline phosphatase in frozen sections taken from subendocardial and subepicardial regions of maximally contracted hearts, and, in several experiments in rabbits, in immersion fixed blocks prepared for electron microscopy. Values were expressed either as number of capillaries/mm^2 (capillary density, CD) or as capillary/fiber ratio (C/F). Student's t–test or analysis of variance were used for statistical evaluation. All values are given as means ±SEM.

RESULTS AND DISCUSSION

Effect of Bradycardial Pacing on Capillarization in Normal and Hypertrophic Rabbit Hearts

Chronic bradycardial pacing alone which decreased heart rate to about 50% of the original value (i.e. about 120 beats/min in conscious rabbits) did not result in heart hypertrophy: heart/body weight ratio was 0.221 ± 0.01% in paced and 0.216 ± 0.006% in control animals. Fiber diameters were 28.5 ± 1.8 μm and 28.5 ± 2.6 μm respectively.

Both C/F ratio (1.46 ± 0.07 in paced vs. 1.26 ± 0.07 in controls) and capillary density (2339 ± 148 paced, 1897 ± 95 controls in the subepicardium, 2360 ± 138 and 1844 ± 98 in the subendocardium) were significantly higher after bradycardial pacing.

Aortic valve lesion resulted in an increase in heart/body weight ratio by 33% and 25% in unpaced and paced hearts. Capillary density in hypertrophic hearts was lower, but pacing increased the values to such an extent that they were higher than in control hearts without hypertrophy (Tables 1 and 2). Capillary density in normal hearts is related to heart weight (Fig. 1). It is then possible to evaluate increase or decrease as a percentage with respect to a value predicted for a given heart weight. Figure 2 demonstrates a dramatic increase in capillarization in hypertrophic hearts as a result of bradycardial pacing. Such a high increase in capillary density can only be interpreted as due to capillary growth for two reasons. First, alkaline phosphatase is an enzyme that stains specifically capillary endothelium, and it thus demonstrates all capillaries present in the examined tissue. In a previous work [19], capillary density was compared in sections stained for alkaline phosphatase

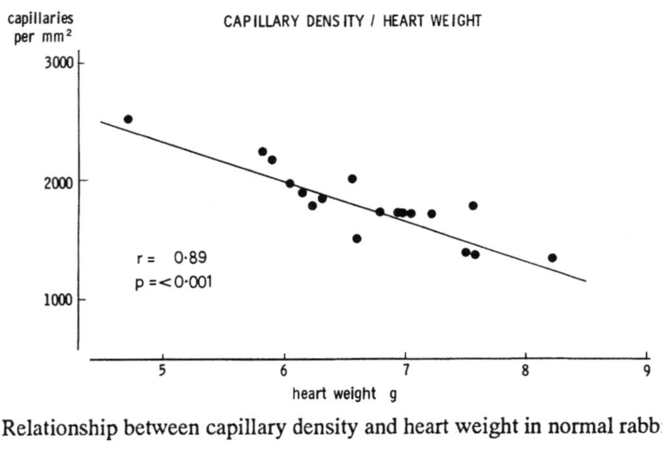

Figure 1: Relationship between capillary density and heart weight in normal rabbit hearts.

Table 1: Heart weight to body weight ratio, and capillary density in rabbit hearts.

	Heart Wt/Body Wt Ratio	Capillary Density/mm²	
		Subendocardial	Subepicardial
C (n=11)	0.228 ± 0.01	1,844 ± 0.98	1,897 ± 95
P (n= 7)	0.221 ± 0.09	2,360 ± 149	2,339 ± 160
VL (n=10)	0.275 ± 0.01	1,380 ± 99	1,386 ± 58
PVL (n= 6)	0.306 ± 0.02	2,469 ± 222	2,313 ± 149

C, control; P, paced; VL, valve–lesioned, PVL, valve–lesioned and paced; CD, capillary density. Numbers of animals in each group are in parenthesis.

Table 2: Analysis of variance between the four groups of animals.

Parameters	Comparison Groups	Probability Values
Body Weight	No significant differences	
Heart Weight	VL/PVL > C/P	< 0.001
Heart wt/body wt ratio	VL/PVL > C/P	< 0.001
CD Subendocardial	P/PVL > C/VL	< 0.001
CD Subepicardial	P/PVL > C/VL	< 0.001
CD Total	C/P > VL/PVL	< 0.05

See table 1 for legend.

with capillary density estimated in immersion fixed adjacent tissue stained with toluidine blue showed a very high degree of correlation ($r = 0.794$, $p<0.001$). Second, increased capillary density could result from decreased size of myocytes, but pacing did not affect fiber size in control hearts, neither did it decrease heart hypertrophy.

Effect of Alinidine on Capillarization in Normal and Hypertrophic Rat Hearts

Occlusion of one renal artery resulted in an increase in mean blood pressure (from 66 ± 13 mmHg to 143 ± 20 mmHg). Alinidine treatment significantly reduced heart rate by 14% in hypertensive and by 42% in normotensive rats. It did not affect the degree of heart hypertrophy in normotensive animals, but increased it in hypertensive rats (Table 3). Capillary supply was evaluated as C/F ratio to avoid complications of

Figure 2: Percentage change in capillary density in hypertrophic (valve lesioned) and hypertrophic bradycardially paced rabbit hearts. Zero represents predicted capillary density for a given heart weight as represented in Fig. 1.

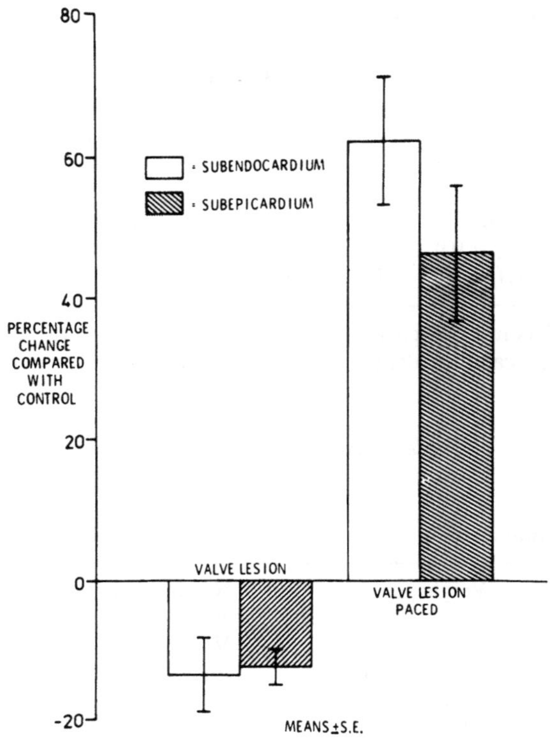

accounting for myocyte area changes. Both alinidine treated groups had a significantly higher C/F ratio (Table 3) [20]. The increase was similar to that found in bradycardially paced hearts. Thus, decrease in heart rate represents a stimulus for capillary growth.

Table 3: Heart/body weight ratio (H/B) and capillary fiber ratio (C/F) in rats with pressure overload hypertrophy. Means ± SEM. Number of animals in parenthesis.

	H/B	C/F
Control (4)	0.313 ± 0.19	1.31 ± 0.04
Hypertrophy (6)	0.372 ± 0.003+	1.30 ± 0.03
Alinidine control (9)	0.308 ± 0.012	1.54 ± 0.02*
Alinidine hypertrophy (4)	0.448 ± 0.013+	1.49 ± 0.03*

* denotes values significantly different between alinidine treated and control rats;
+ denotes values significantly different between hypertrophy controls ($p<0.05$).

Angiogenesis always starts with a disruption of basement membrane which enables migration of endothelial cell into the interstitial space and formation of capillary sprouts. It is traditionally ascribed to various growth factors, particularly in tumors or wound healing [21]. Low molecular growth factor was indeed demonstrated in hearts with myocardial infarction [22], and was found in all bradycardially paced hearts [23]. However, it was also present in 50% of normal hearts and there was no relationship between the degree of capillarization and presence or absence of this factor. We therefore hypothesized that capillary proliferation must be due to other causes, such as increased blood flow. This assumption was based on findings mentioned in the introduction, as well as on experiments where one drug – α 1 blocker prazosin – elicited capillary growth in organs where it increased blood flow (skeletal muscle) but not in the heart where flow was not changed [18]. Furthermore, it can be expected that coronary flow which is higher during diastole than during systole could be increased when the duration of diastole is prolonged. Long–term increase in flow leads to increased velocity of red blood cells and increased shear stress [24] as well as to wider diameters [25] and thus higher wall tension in capillaries in skeletal muscle. Although there are no data on either flow velocity or capillary diameters in hearts with long–lasting bradycardia, it is possible that wall tension would be increased since Tillmans et al. [26] found wider diameters in diastole than in systole. Increased capillary wall tension could result in disruption of the basement membrane as a direct effect of stretch and possibly release fibroblast growth factor which is present there (see [27]). Shear stress could damage the luminal side of capillaries and lead to a release of proteases that would disrupt the basement membrane with similar consequences.

Coronary Blood Flow in Hearts with Chronic Bradycardia

Maximal coronary blood flow during dilatation induced by propentofylline (which was used instead of adenosine because it did not lower blood pressure, [28]) was significantly lower in hypertrophic than in control rabbit hearts (Table 4), but pacing did not lead to any increase above the control levels. Although flow/heart beat was higher in paced than in control hearts [23], the difference in flow/beat between hypertrophic and paced hypertrophic hearts (2.86 ± 0.53 ml/100g/min vs. 2.91 ± 0.33) was not significant. Hearts with pressure overload hypertrophy also had lower blood flow than controls (Table 4), but alinidine treatment did not result in higher values in either control or hypertrophic hearts, and there was no difference in flow/beat. These results do not support the hypothesis that high shear stress would initiate capillary growth in hearts with long lasting bradycardia. However,it is possible that the wall tension is increased not only because of a longer duration of the total

Table 4: Maximal coronary blood flow (ml/g/min) in rabbit and rat hearts, measured by microspheres labeled with ^{46}Scandium. Means±SEM, number of animals in parenthesis.

	Rabbits	Rats
Control	11.16 ± 1.16 (9)	11.68 ± 1.93 (4)
Bradycardic	9.22 ± 1.72 (6)	6.73 ± 1.1 (4)
Hypertrophic	7.77 ± 0.65 (6)*	6.54 ± 1.62 (4)*
Hypertrophic bradycardic	9.32 ± 0.99 (6)	7.83 ± 1.61 (6)

* denotes significantly different values against controls ($p<0.05$).

time for which capillary diameters are wider during bradycardia, but possibly also because more forceful contractions of myocytes during each systole resulting from greater stretch during prolonged diastole. Stretch by surrounding tissues was indeed demonstrated as a potent stimulus for vessel growth during development many decades ago [29]. Bradycardial pacing increased significantly stroke work [16] not only in control but also in hypertrophic hearts while volume overload hypertrophy decreased it [17]. Moreover, long-term administration of inotropic drug, dobutamine (20 μg/kg/min iv for 2 weeks), resulted in an increase in stroke work and increased capillary density in the absence of heart hypertrophy and changes in coronary blood flow [30]. It can thus be concluded that long term drastic decrease in heart rate induces capillary proliferation very probably due to stretch of the capillary wall and possible release of some growth factors.

Acknowledgement

This work was supported by grants from the British Heart Foundation.

DISCUSSION

Dr. Strauss: Was the catecholamine damage restricted to the ventricles or did it also occur in the atria?

Dr. Hudlicka: We do not know yet. We only looked at the left ventricle and we are looking now at the atria as well.

Dr. Reneman: With heart hypertrophy in SHR rats, the endocardial capillary density is lower than the epicardium; if you increase the density with nithedipine, do you increase the density in endocardial layers more than in the epicardial layers? In other words, was there more homogeneous distribution of capillary density?

Dr. Hudlicka: The data on nithedipine are not from my experiments so I am not too sure. I believe the increase was similar in both regions. The pacing of the hypertrophied hearts increased capillary density more in the subendocardium than in the subepicardium.

Dr. Kammermeier: You mention that the capillary wall tension might induce capillary growth. If so, an increase in venous pressure should increase the intercapillary pressure and also induce capillary growth. Did you test that?

Dr. Hudlicka: Your deduction is absolutely right. I put this idea in a grant application and I did not get the grant because they thought it was a completely crazy idea, but I am going to do it anyway.

Dr. Noble: In the paired pacing way you get a slower heart rate but more forceful contraction which is presumably why the flow goes up. Would you be prepared to accuse Bassingthwaighte's baboons as being unhealthy nonexercised slobs?

Dr. Hudlicka: I suggest Dr. Bassingthwaighte answers this question.

Dr. Bassingthwaighte: They were not very heavily exercised. We tried to increase their coronary flows and cardiac outputs by exercise and also by heat stress. Their normal myocardial flows were around 1.5 ml/(g·min) but they increased to only a little above 2 ml/(g·min) with the exercise stress, peddling a bicycle, less than a factor of 2, so they were pretty slothful.

Dr. Marban: Did the paced hearts become hypertrophic?

Dr. Hudlicka: No, there is no hypertrophy in paced hearts and this is something which I do not understand because the stroke work is greater, the stroke volume is greater yet there is no hypertrophy.

Dr. Marban: Have you ever found a condition where hypertrophy occurs and capillary density is increased?

Dr. Hudlicka: No. There is a paper by Tomanek *[Tomanek et al., Circ Res 1982;51:295–304]* on SHR and he found some increase when the animal was about 15 months old. There was a decrease before and a decrease after and he cannot really explain it why at 15 months there would be some growth.

Dr. Kammermeier: Did you use stimulation to increase heart rate instead of reducing it? If so, what effect did you get?

Dr. Hudlicka: We have done it when we started to work on this problem. We only had about 3 or 4 experiments but capillary density went down.

REFERENCES

1. Hudlicka O, Tyler KR: *Angiogenesis*. Academic Press, London 1986.
2. Tomanek R: Response of the coronary vasculature to myocardial hypertrophy. *J Am Coll Cardiol* 1990;15:528–533.
3. Hudlicka O, Brown MD, Egginton S: Growth of capillaries in skeletal and cardiac muscle. *Physiol Rev* 1991;in press.
4. Chilian WH, Wangler, RD, Peters KG, Tomanek RJ, Marcus ML: Thyroxine-induced left ventricular hypertrophy in the rat: anatomical and physiological evidence for angiogenesis. *Circ Res* 1985;57:591–598.
5. Ljungqvist A, Tornling G, Unge G Jurdahl EB, Larsson B: Capillary growth in the heart and skeletal muscle during dipryridamole treatment and exercise. *Prog Appl Microcirc* 1984;4:9–15.
6. Mall G, Mattfeldt T, Volck B: Ultrastructural morphometric study on the rat heart after chronic ethanol feeding. *Virchow's Arch Abt A Path Anat* 1980;389:59–77.
7. Ziada AMAR, Hudlicka O, Tyler KR: The effect of long–term vasodilatation on capillary growth and performance in rabbit heart and skeletal muscle. *Cardiovasc Res* 1984;18:724–732.
8. Rakusan K: Microcirculation in the stressed heart, in Legato MJ (ed): *Growth of the Heart in Health and Disease*. Boston, Martinus Nijhoff Publishing, 1987, pp 107–123.
9. Tomanek RJ, Searl JC, Lachenbruch PA: Quantitative changes in the capillary bed during developing peak and stabilized hypertensive rat. *Circ Res* 1982; 51:295–304.
10. Anversa P, Melissari M, Beghi C, Olivetti G: Structural compensatory mechanisms in rat heart in early spontaneous hypertension. *Am J Physiol* 1984;246: H739–746.
11. Turek Z, Kubat K, Kazda S, Hoofd L, Rakusan K: Improved myocardial capillarisation in spontaneously hypertensive rats treated with nifedipine. *Cardiovasc Res* 1987;21:725–729.
12. Katz LN, Feinberg H: Relation of cardiac effort to myocardial oxygen consumption and coronary blood flow. *Circ Res* 1958;6:656–669.
13. Strauer BE: Ventricular function and coronary haemodynamics in hypertensive heart disease. *Am J Cardiol* 1979;44:999–1006.
14. Rooke GA, Feigl EO: Work as a correlate of canine left ventricular oxygen consumption and the problem of catecholamine oxygen wasting. *Circ Res* 1982; 50:273–286.
15. Poupa O, Rakusan K, O'tadal B: The effect of physical activity upon the heart of the vertebrates. *Medicine Sport* 1970;4:202–235.
16. Wright AJA, Hudlicka O: Capillary growth and changes in heart performance induced by chronic bradycardial pacing in the rabbit. *Circ Res* 1981;49:469–478.
17. Wright AJA, Hudlicka O, Brown MD: Beneficial effect of chronic bradycardial pacing on capillary growth and heart performance in volume overload heart hypertrophy. *Circ Res* 1989;64:1205–1212.
18. Ziada AMAR, Hudlicka O, Tyler KR: The effect of long–term administration of α_1–blocker prazosin on capillary density in cardiac and skeletal muscle. *Pflugers Arch* 1989;415:355–360.
19. Brown MD, Egginton S: Capillary density and ultrastructure in rabbit papillary muscle after a high dose of norepinephrine. *Microvasc Res* 1988;36:1–12.
20. Brown MD, Cleasby MJ, Hudlicka O: Capillary supply of hypertrophied rat hearts after chronic treatment with the bradycardic agent alinidine. *J Physiol (Lond)* 1990;427:40

21. Folkman J, Klagsbrun M: Angiogenic factors. *Science* 1987;235: 442–446.
22. Kumar S, West D, Shahabuddin S, Arnold F, Haboubi N, Reid H: Angiogenesis factor from human myocardial infarcts. *Lancet* 1983a;ii:364–368.
23. Hudlicka O, West D, Kumar S, El Khelly F, Wright AJA: Can growth of capillaries in the heart and skeletal muscle be explained by the presence of angiogenic factor? *Br J Exp Pathol* 1989;70:237–246.
24. Dawson J, Hudlicka O: The effects of long term administration of prazosin on the microcirculation in skeletal muscle. *Cardiovasc Res* 1989;23:913–920.
25. Myrhage R, Hudlicka O: Capillary growth in chronically stimulated adult skeletal muscle as studied by intravital microscopy and histological methods in rabbits and rats. *Microvasc Res* 1978;16:73–90.
26. Tillmans TH, Ikeda S, Hansen H, Sarma JS, Fauvel JH, Bing RJ: Microcirculation in the ventricle of the dog and turtle. *Circ Res* 1974;34:561–569.
27. D'Amore PA, Orlidge A: Growth factors and pericytes in microangiopathy. *Diabete, Metabolism* 1988;14:495–504.
28. Hudlicka O, Komarek J, Wright AJA: The effect of a xanthine derivative, 1-(5'oxohexyl)-3-methyl-7-propylxanthine (HWA 285), on heart performance and regional blood flow in dogs and rabbits. *Br J Pharmac* 1981;72:723–730.
29. Thoma R: Uber die Histomechanik des Gefassystems und die Pathogenese der Angiosklerose. *Virchow's Arch Path Anat* 1911;204:1–74.
30. Brown MD, Hudlicka O: Cardiac performance in vivo and anatomical capillary supply in the rabbit after prolonged dobutamine infusion. *J Physiol (Lond)* 1989;413:19P.

Myocardial Metabolism and Energetics

CHAPTER 31:

MYOCARDIAL BLOOD FLOW AND OXIDATIVE METABOLISM FINDINGS IN NORMAL AND DISEASED MYOCARDIUM

Heinrich R. Schelbert

ABSTRACT

Regional myocardial oxygen consumption in humans can now be estimated noninvasively in humans with positron emission tomography (PET) and Carbon–11 (C–11) labeled acetate. This agent traces the flux of substrates through the tricarboxylic acid (TCA) cycle as the final oxidative pathway common to all substrates. The activity of this particular pathway is synchronized with the rate of oxidative phosphorylation. Therefore, measurements of rates of substrate fluxes through the TCA cycle provide accurate estimates of myocardial oxygen consumption. The study describes (a) how regional myocardial oxygen consumption can be measured noninvasively with C–11 acetate; and (b) how this novel approach can be employed to the study of normal and altered function of the human heart.

INTRODUCTION

The availability of a noninvasive approach for the quantifying of regional rates of oxidative metabolism in human myocardium offers intriguing possibilities for studying cardiovascular physiology. Oxygen consumption can directly and quantitatively be related to myocardial blood flow and oxygen supply. Rates of substrate utilization, for example of glucose or free fatty acid can be compared to TCA cycle activity and the relative contributions of each substrate to overall oxidative metabolism can be estimated. If mechanical work is compared to oxidative metabolism, it should be possible to derive information on cardiac energy efficiency.

MEASUREMENTS OF MYOCARDIAL OXYGEN CONSUMPTION

Pure oxygen–15, a positron emitting isotope of oxygen, is available for imaging with PET. It has proved useful for measurements of the oxygen extraction and, when combined with tracers of blood flow, oxygen consumption of the brain. However, technical limitations have precluded its use for similar measurements in the heart. In contrast, C–11 acetate yields high myocardial to background activity ratios on serially acquired PET images. The resulting signal is suitable for accurate measurements of the tracer uptake in and clearance from myocardium and thus of the tissue tracer kinetics from which estimates of myocardial oxygen consumption can be obtained.

Following intravenous injection, C–11 labeled acetate rapidly clears from blood and accumulates in myocardium. High tracer first pass extraction fractions of about 0.63 at control flows of 1 ml/min/gr myocardium imply that C–11 acetate readily exchanges across the capillary and sarcolemmal membranes [1]. The fraction of tracer that exchanges from blood into myocardium during a single capillary transit depends also on blood flow. The inverse relationship between the first pass extraction fraction E and blood flow F is described by

$$E = 1 - 0.58 \exp -0.44/F \qquad (1)$$

where 0.44 reflects the permeability surface product. The high first pass extraction fraction of C–11 acetate indicates further that blood flow largely accounts for the initial uptake or the net extraction of tracer (as the product of blood flow and extraction fraction) in myocardium. However, this net extraction of C–11 acetate increases with high flows in a nonlinear fashion as a result of the nonlinear, inverse relationship between first pass extraction fractions and blood flow [1].

Once extracted, C–11 labeled metabolites begin to clear from myocardium with a several minute delay. The latter has been attributed to the time required for the tracer or its metabolites to equilibrate with intercellular metabolic pools, of for example, TCA cycle intermediates and bicarbonate. The subsequent bi–exponential clearance of tracer label from myocardium corresponds to the metabolic fate of the tracer and its labeled metabolites. The bi–exponential clearance curve morphology implies that the tracer label distributes between at least two metabolic pools of different sizes and different turnover rates. Regional myocardial time activity curves are typically obtained from rapidly acquired serial images on PET by assigning regions–of–interest to the left ventricular myocardium. Pertinent parameters derived from these curves include the relative sizes of both clearance curve components and their slopes (Figs. 1 and 2). The latter are described by the clearance rate constant k_1 for the rapid and the clearance rate constant k_2 for the slow

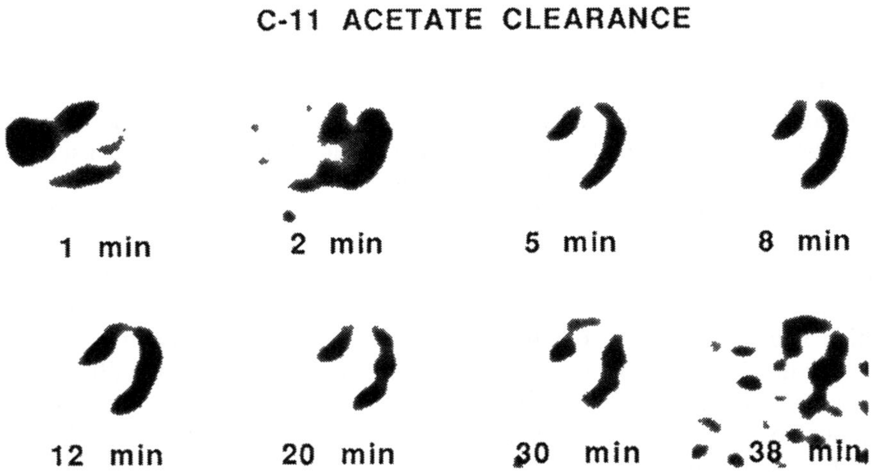

Figure 1: Serial cross–sectional PET images recorded following intravenous injection of C–11 acetate in a normal human volunteer. Note the C–11 activity in the right and left ventricular cavities early after injection (1 and 2 min). C–11 activity subsequently clears from arterial blood into the myocardium which is visualized best at 5 to 8 min. C–11 activity concentration subsequently declines in the myocardium. At 30 min after injection, tracer activity in the myocardium has decreased to mere background levels.

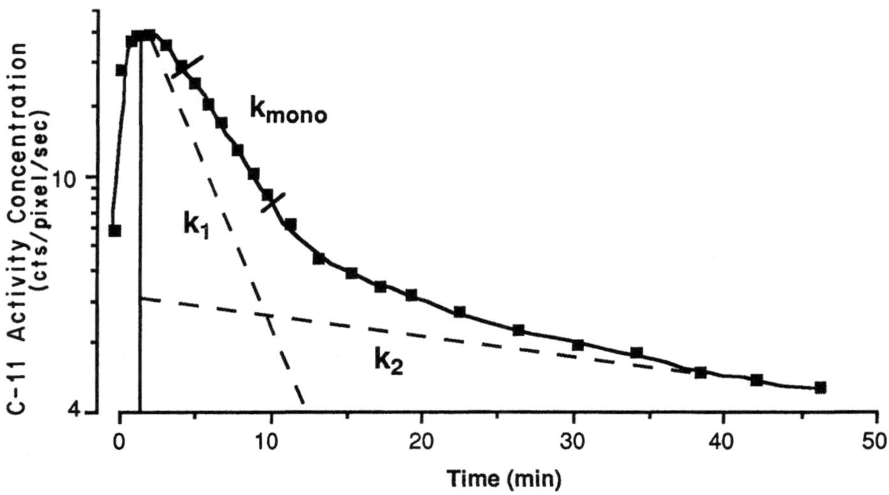

Figure 2: Myocardial time activity curve obtained from a region of interest assigned to the serially acquired PET images. Note the initial rapid uptake of C–11 acetate into myocardium and the subsequent bi–exponential clearance. The slope of thc two clearance curve components are determined by by–exponential least square fitting. k_1 defines the slope of the early rapid and k_2 the slope of slow clearance curve component. k_{mono} was derived by mono–exponentail fitting of the early portion of the time activity curve.

clearance curve component. The ratio of the relative sizes of the two clearance curve components describes the fractional distribution of tracer between these two pools. Rate constants k_1 and k_2 correspond to the turnover rates of these pools.

The nature of the slow clearance curve component and its significance for the assessment of oxidative metabolism remain unclear. Preliminary findings in animal experiments suggested that the slow clearance rates corresponds to the fraction of C–11 labeled that branches off the TCA cycle and becomes deposited in a pool of amino acids in the form of C–11 glutamate [2, 3]. Other possibilities include transfer of the label to long chain fatty acid. Overall, the relative size and the slope of this clearance curve component appear to remain relatively constant even in the presence of marked alterations in oxidative and substrate metabolism.

In contrast, the slope of the rapid clearance curve component closely changes with cardiac work and, importantly, with myocardial oxygen consumption [2–6]. Its rate constant, defined as k_1, changes in direct proportion to the heart rate blood pressure product and to myocardial oxygen consumption as demonstrated in animal and human experiments (Fig. 3). However, clearance rates of radiotracers from tissue depend on several factors. Foremost are (1) the rate of substrate flux through the metabolic pathway to be studied; (2) the volume of distribution or the size of the metabolic pool; and (3) backdiffusion of nonmetabolized tracer. The relationship between the rate of substrate flux and the distribution volume is described by $k_1 = F/V$ where F is the flux rate and V the volume of distribution. Thus, higher rates of substrate fluxes will increase k_1 whereas increases in V will lower k_1. Many of the features of the time activity curves relate directly to the metabolic fate of the label. After C–11 acetate has exchanged across the capillary and sarcolemmal membranes, presumably via passive diffusion, it is esterified to C–11 acetate CoA and proceeds into the mitochondria where the TCA cycle metabolizes the tracer to C–11 labeled CO_2 and H_2O.

There appear to be several metabolic pools with which C–11 acetate or its metabolites equilibrate before tracer label begins to clear from the myocardium and which accounts for the initial delay. These pools may include acyl–CoA or TCA cycle intermediates as well as of intercellular bicarbonate. For example, as demonstrated in rat hearts by Randle et al. [7] and subsequently Chance et al. [8], labeling of the small acyl–CoA pool requires only about 10 s while equilibration with the TCA cycle intermediates may amount to several minutes. Changes in the size of these pools could dramatically alter tissue clearance rates of tracer. Such a possibility has in fact been demonstrated in isolated arterially perfused rat hearts during unphysiologically high concentrations of acetate in the perfusate and, presumably, an increase in the intercellular acetate pool [2]. It appears, however, that under physiologic conditions, the size of these metabolic pools remains relatively constant.

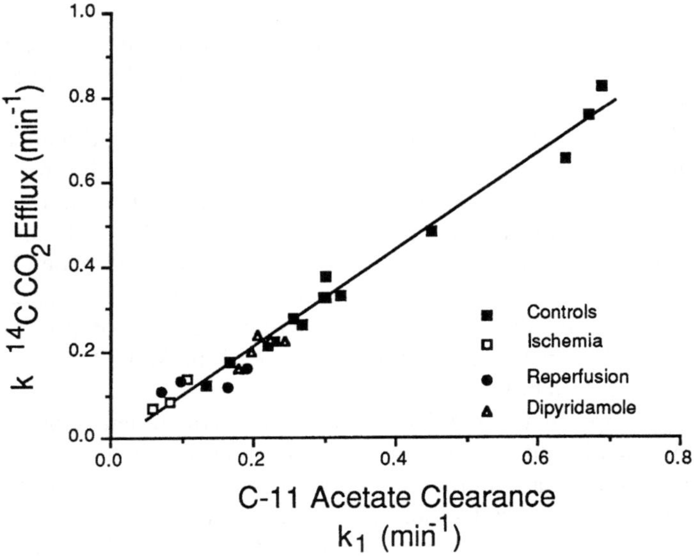

Figure 3: Relationship between myocardial oxygen consumption as determined by the coronary sinus catheter technique and the mono-exponential clearance rate constant k_{mono} for C-11 acetate in the dog's myocardium.

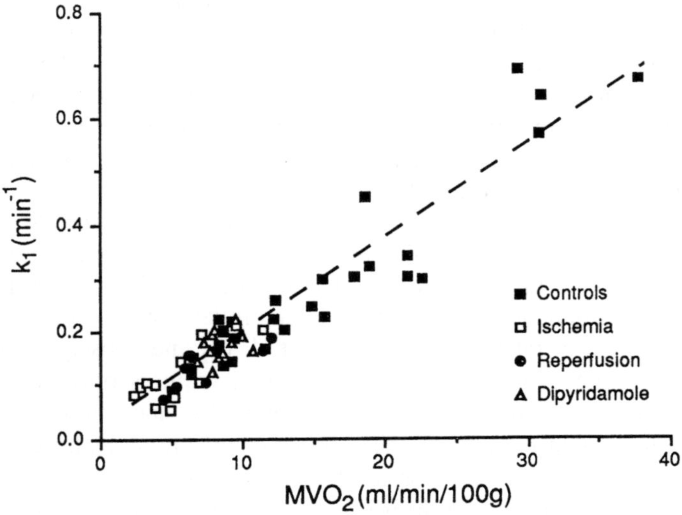

Figure 4: Relationship between the rate of efflux of C-14 labeled carbon dioxide from the myocardium and the rate of activity clearance from the myocardium after coinjection of C-11 and C-14 labeled acetate in a dog's myocardium.

Given the relatively constant size of the metabolic pools, the rate of substrate flux through the TCA cycle appears to be the prime determinant of the slope of the tissue clearance rate and its changes in response to altered oxygen consumption. Several animal experimental investigations have indeed described a close relationship between the efflux rate of C–11 labeled CO_2 from the myocardium and the externally measured rate of clearance of total C–11 activity from the myocardium (Fig. 4) [1]. However, the C–11 label in the coronary sinus effluent is not exclusively bound to CO_2. Most studies have found a 5 to 10% fraction of the total C–11 activity was unrelated to CO_2 and was bound to other chemical species [1–3]. Non–metabolized tracer returning from the myocardium into the venous blood probably accounts for most of the non–CO_2 related C–11 activity. Backdiffusion of the tracer can influence the tissue clearance rate and limit the utility of the clearance slopes as indices of metabolic flux rates. Variable degrees of tracer backdiffusion have, for example, limited the utility of C–11 palmitate as a tracer of fatty acid metabolism. A substantial fraction of C–11 activity in the venous effluent is related to backdiffusion of nonmetabolized C–11 palmitate. As the degree of backdiffusion markedly changes with metabolic alterations, for example with ischemia, the slope of the clearance phase becomes an unreliable indicator of fatty acid oxidation [9, 10]. This limitation however does not apply to C–11 acetate. The fraction of non CO_2 related C–11 activity and presumably of C–11 acetate in the myocardial effluent has been found to be rather constant. For example, changes in myocardial substrate selection or changes in myocardial blood flow or myocardial ischemia failed to significantly alter the ratio of CO_2 to non–CO_2 related C–11 activity. Therefore, the rate of clearance of C–11 activity from myocardium closely reflects the rate of clearance of C–11 labeled CO_2 as the end product of C–11 acetate oxidation by the myocardium.

Effects of substrate selection on the myocardial clearance of C–11 acetate have been demonstrated [3]. In canine myocardium, for example, tissue clearance rates were found to be about 15% faster for the same rate of myocardial oxygen consumption when myocardium preferentially oxidized carbohydrate relative to free fatty acid. The most plausible explanation for this observation are differences in the fraction of oxygen consumed by the TCA cycle for each substrate. For glucose and lactate, for example, 67% of overall myocardial oxygen consumption occurs through TCA cycle oxidation whereas for palmitate and oleate, the main circulatory plasma fatty acids, 70 to 72% of the oxygen consumption occurs via the TCA cycle oxidation [3]. On the other hand, as the myocardium rarely relies exclusively on one single substrate, these differences due to substrate selection are likely to be rather small. Although a statistically significant difference in clearance rates could be demonstrated noninvasively in dog experiments with serial PET imaging after intravenous C–11 acetate, this difference was relatively small and was not observed in normal human volunteers [3, 5]. The latter probably resulted from a less prominent shift in substrate oxidation and was further due to relatively high data noise in the recorded image data.

Oxygen consumption is typically measured in humans by intravenous injection of 15 to 20 mCi of C–11 acetate and serial imaging with PET at framing rates of one to two minutes for a total time of 30 to 40 minutes. Regional myocardial time activity curves are derived through regions of interest from the serial images. Different from experimental animals, the myocardial clearance curves of C–11 acetate in humans are usually difficult to fit bi–exponentially. This is because of disproportionately slower tracer clearance during the rapid clearance curve component, which especially under baseline conditions, obscures the slow clearance curve component. Because adequate bi–exponential fitting often cannot be achieved, an index of oxygen consumption is usually derived by fitting mono–exponentially only the early portion of the clearance curve. The derived clearance rate constant, frequently referred to as k_{mono}, has been validated as an accurate index of oxidative metabolism in both animal and human studies.

CLINICAL INVESTIGATIONS WITH C–11 ACETATE

Several studies have defined the tissue kinetics of C–11 acetate in normal human myocardium and thus provided a framework of reference. In normal human volunteers at rest, Armbrecht et al. [5] observed average clearance rates of 0.068 ± 0.006 (SEM) per min which were similar to the average values of 0.054 ± 0.014 per min reported by Henes et al. [6]. Converted into rates of oxygen consumption, these values correspond to 246 mmol O_2/min/gr myocardium. In both investigations, C–11 acetate was found to clear homogeneously from normal human myocardium. Regional clearance rates varied by approximately 9%. A more recent investigation has suggested some degree of regional variations [11]. Carbon–11 acetate was found to clear more rapidly from the anterior wall and the interventricular septum than from the remainder of the left ventricular myocardium. While these differences achieved statistically significance levels they were only small and amounted to about 5 to 10%.

Similar to findings in animal experiments, the tissue clearance rate of C–11 acetate in human myocardium correlated closely with cardiac work. For example, the clearance rate constants increased linearly with increases in cardiac work induced by either supine bicycle exercise or by intravenous dobutamine infusion [5, 6]. Mono–exponential clearance rates of C–11 acetate (k_{mono}) were found to correlate with the heart rate systolic blood pressure product (RPP) by $k_{mono} = 0.014 + 5.9 \cdot 10^{-5}$ bpm/mmHG. Converted into units of oxygen, these relationships predict an oxygen consumption of 4.3 ml per min per 100 gram for a heart rate x blood pressure product of 6,500 bpm x mmHg at rest which increases to 11.2 ml O_2 per 100 gr myocardium for a heart rate blood pressure product of 17,000 bpm x mmHg/ml during supine bicycle exercise. Both values are lower than those previously obtained noninvasively with the coronary sinus catheter technique and the Fick principle. The reasons for this underestimation remain uncertain at present and need further clarification.

The possibilities offered by C–11 acetate for examining cardiovascular function have thus far been explored in patients with coronary artery disease. These studies combined C–11 acetate with tracers of blood flow as well as of glucose utilization. The homogeneous clearance, characteristic for normal myocardium, was typically lost in patients with ischemic heart disease and regional perfusion and wall motion abnormalities. As demonstrated in patients with stable coronary artery disease by Brunken et al. [12] tissue clearance rates and, by inference, myocardial oxygen consumption were relatively normal in myocardium with apparently normal blood flow. Average rate constants of 0.07210.016 per min as observed in such segments were similar to those obtained in normal volunteers. Segments with evidence of ischemic injury, however, revealed markedly reduced clearance rates of C–11 acetate. The magnitude of these reductions appeared to be related to the severity of the ischemic injury. Simultaneous evaluation of myocardial blood flow with N–13 ammonia, and glucose utilization with F–18 deoxyglucose, in these patients permitted the categorization of segments with decreased blood flow but with preserved glucose utilization as "ischemic" and segments with proportionate decreases of both, blood flow and glucose utilization, as "infarcted". Relative to normal myocardium, blood flow was decreased by 29 ± 12% in "ischemic" segments and by 46 ± 18% in "infarcted" myocardium. These segmental flow reductions were associated with similar decreases in C–11 acetate tissue clearance rates. k_{mono} averaged 0.058 ± 0.010 per min in "ischemic" and 0.045 ± 0.019 per min in "infarcted" myocardium and thus were 20% and 38% lower than in apparently normal myocardium. Thus, in these patients with stable coronary artery disease, regional rates of oxidative metabolism were decreased in proportion to regional myocardial blood flow. As the ratios of oxidative metabolism and of blood flow in both, ischemically injured and normal myocardium were virtually identical, oxygen consumption must have remained closely linked to myocardial oxygen delivery and blood flow.

Because animal experiments had demonstrated a dissociation of blood flow and oxygen consumption from contractile function in the stunned myocardium, several studies explored whether a similar dissociation existed in patients early after an acute myocardial infarction and thrombolysis. Melin et al. [13] studied 8 patients several weeks after reperfusion while Czernin et al. [14] examined 7 patients within 5 days of an acute myocardial infarction. In both studies, patients had undergone thrombolysis within hours of onset of acute symptoms. Both groups of patients were studied with N–13 ammonia and F–18 deoxyglucose in addition to C–11 acetate. Both studies however reported that oxidative metabolism correlated closely with myocardial blood flow. In the patients studied several weeks after the acute event, blood flow in the "infarcted" myocardium was reduced by 35 to 40% below that in the normal myocardium. Regional clearance rates of C–11 acetate in these patients were reduced in proportion to myocardial blood flow. The findings of Czernin et al. [14] were similar. Blood flow and C–11 acetate clearance rates in the normal myocardium averaged 0.8 ± 0.3 ml/min/gr and 0.05 ± 0.06 per min, respectively. Infarcted segments revealed more sever reductions in both blood flow (0.03 ± 0.01 ml/min/gr) and acetate clearance rates (0.03 ± 0.01 per min) and a 50% decrease in glucose utilization. Again, comparable ratios of clearance rates to blood flow in normal, ischemic, and infarcted myocardium indicates that oxygen extraction remained constant and that oxygen consumption remained tightly coupled to blood flow in ischemic and in "infarcted" tissue.

Different from these observations, Henes et al. [15] did in fact observe a dissociation between flow and oxygen consumption in patients studied within 24 h after successful thrombolysis. While blood flow in reperfused myocardium was normal at 24 h after thrombolysis, and remained normal when reexamined at 48 h and again at the time of hospital discharge, oxygen consumption was significantly decreased. Compared to normal myocardium, it averaged at 24 h and at 48 h only 43 ± 24% and 60 ± 15%. This decrease persisted at least until the time of hospital discharge.

The reasons for the different findings in early post–infarction patients between the two studies remain uncertain. In both studies, patients had been submitted to coronary thrombolysis within hours of admission. Use of N–13 ammonia in the studies by Czernin et al. [14] as compared to the metabolically inert Oxygen–15 water as used by Henes et al. [15] is unlikely to account for this discrepancy. Similar proportionate reductions in both blood flow and oxidative metabolism were recently observed in acutely post–ischemic canine myocardium where blood flow was quantified with the microsphere technique [16]. It is possible that a dissociation between blood flow and oxygen consumption existed only very early after reperfusion and subsequently resolved.

More recent investigations have demonstrated the possibility of determining changes in oxidative metabolism in stunned myocardium during inotropic stimulation [16]. A transient, 20 min intercoronary balloon occlusion in acute dog experiments resulted in a segmental depression of regional wall motion and blood flow. Clearance rates of C–11 acetate in such segments were reduced in proportion to blood flow. Dobutamine infusion resulted in a marked improvement in regional wall motion and in increases in C–11 acetate clearance rate in both, stunned and control myocardium. In fact, the relative decrease in clearance rates present at the baseline study was largely abolished during dobutamine stimulation. Clearance rates remained only slightly though significantly lower in stunned than in control myocardium. Interestingly, relative increases in oxidative metabolism in stunned myocardium exceeded concomitant increases in blood flow.

SUMMARY

Several investigations have confirmed the possibility of noninvasive measurements of regional oxygen consumption in human myocardium. Consistent with animal experimental findings, oxygen consumption generally is homogeneous although some subtle regional differences may exist. Coronary artery disease can profoundly alter the pattern of homogeneity. Significant regional reductions in oxidative metabolism can occur, usually associated with proportionate decreases in blood flow. Superimposed on these changes may be regional alteration in glucose and in fatty acid metabolism. Combined evaluation of oxygen and of substrate metabolism can delineate the relationships between oxygen supply, oxygen consumption and substrate reference and substrate oxidation. Furthermore, correlative measurements of blood flow and of oxygen consumption unsuitable for studies of the relation of blood flow, oxygen supply and extraction and oxidative metabolism. It further would seem possible to elucidate with C–11 acetate adaptable mechanisms to substrate extraction and metabolism to limited oxygen supply in human ischemic heart disease.

Acknowledgements

Studies carried out for the U.S. Department of Energy, Washington, D.C. by the University of California under Contract #DE–AC03–76–SF00012. This work was supported in part by the Director of the Office of Energy Research, Office of Health and Environmental Research, Washington, D.C., by Grants #HL 29845 and #HL 33177 from the National Institutes of Health, Bethesda, MD, and by an Investigative Group Award from the Greater Los Angeles Affiliate of the American Heart Association, Los Angeles, CA.

The author wishes to thank Barbara Watson for her skillful secretarial assistance in preparing this manuscript and Wendy Wilson for the illustrations.

DISCUSSION

Dr. Gettes: Your meaning of ischemia in post–infarction patients is of interest. Does that imply that there are regions which have some form of altered metabolism with and without blood flow? How do you define that region?

Dr. Schelbert: Ischemia in the context of these studies was defined as a disparity between myocardial blood flow and glucose utilization as determined by F–18 2–deoxyglucose. Typically, such regions exhibit reduced blood flow, but with preserved or elevated rates of glucose utilization. This metabolic pattern, as assessed noninvasively with PET and N–13 ammonia as a tracer of blood flow and F–18 deoxyglucose as a tracer of exogenous myocardial glucose utilization, differs distinctly from the pattern of infarction which typically exhibits a concordant reduction in both blood flow and glucose utilization. The significance of both patterns in terms of ischemia and infarction has been confirmed by us and others. Myocardial regions in which both perfusion and glucose metabolism were concordantly reduced failed to improve contractile function following revascularization. In contrast, regions with reduced blood flow but preserved glucose metabolism usually improved contract function after the therapeutic restoration of blood flow.

These patterns have been defined previously in patients with chronic coronary artery disease. They were also observed in patients early after an acute myocardial infraction in whom, according to initial follow–up studies, they appeared to similarly identify ischemic and truly infarcted myocardium.

Dr. van der Vusse: I relate to the experiments performed with C–labeled acetate. Interestingly, the release of the C–labeled degradation products surpass the amount of the labeled oxygen released. Did you identify the nature of the other C–labeled break–down products of acetate?

Dr. Schelbert: Your question is an interesting and potentially an important one. However, there are no studies that have specifically examined the chemical species in which the label may be released from myocardium. Preliminary information obtained in our laboratory in dogs suggest that the overwhelming fraction of the carbon label not associated with CO_2 is released from myocardium in the form of acetate.

Dr. Bassingthwaighte: It seems to me that you have an ideal opportunity to test the hypothesis that I was raising earlier, namely, that there are local close relationships or nearly constant ratios between metabolism, flow, and transport capacity regionally. In your analysis you did not talk about analysis of the capacity for the glucose uptake or for the acetate metabolism region by region. Do you have observations that would give you measures of those transport rates or local metabolic rates relative to flow locally.

Dr. Schelbert: We have a number of observations in animal experiments which indicate a fairly close relationship between blood flow and the initial transmembraneous exchange of C–11 acetate, as defined by first pass extraction fractions as well as the myocardial acetate clearance rates and, by implication, myocardial oxygen consumption. Consistent with the Renkin–Crone model, the first pass extraction fractions were found to be inversely related to blood flow. The least square fitting of the data in these dogs indicated, as one would expect with most diffusible indicators, a change in the permeability surface product at higher blood flows. Furthermore, there was a fairly close linear relationship between blood flow and the clearance rate constant of C–11 acetate indicating a tight coupling of myocardial oxygen consumption to myocardial blood flow. This relationship held in normal and acutely ischemic myocardium but not in reperfused myocardium where blood flow exceeded oxygen consumption.

Dr. Corr: You report data with ammonia in which there was a good correlation between the microspheres at low and high flow. Others have found that ammonia is not able to predict flow at low flow levels. Are your findings a consistent phenomena?

Dr. Schelbert: I am not certain what particular studies you refer to. In our animal experiments, first pass extraction fractions and, consequently, myocardial net uptake of N–13 ammonia was tightly coupled to myocardial blood flow, although in a non-linear fashion. In these studies we observed that acute experimentally induced ischemia or lowering of plasma pH decreased first pass extraction fractions of N–13 ammonia. However, the magnitude of these changes was small and did not appear to invalidate the utility of N–13 ammonia as a tracer of blood flow. One study in isolated arterially perfused hearts reported, however, striking changes in the first pass extraction fraction when the perfusate was enriched with red blood cells.

Dr. Corr: Is that because of trapping of the ammonia?

Dr. Schelbert: No, I do not think so. Flow rates required to adequately oxygenate these isolated hearts were markedly lower with perfusate with than without red blood cells. As first past extraction fractions of N–13 ammonia are a function of blood flow, they were markedly higher when hearts were supported with perfusate containing red blood cells as compared to perfusate without red blood cells. Thus, the reported difference in first pass extraction fraction therefore appeared related to the blood flow rather than to the changes in metabolic trapping.

REFERENCES

1. Armbrecht JJ, Buxton DB, Schelbert HR: Validation of [1–11C] acetate as a tracer for noninvasive assessment of oxidative metabolism with positron emission tomography in normal, ischemic, post–ischemic and hyperemic canine myocardium. *Circulation* 1990;81:1594–1605.
2. Buxton DB, Schwaiger M, Nguyen A, Phelps ME, Schelbert HR: Radiolabeled acetate as a tracer of myocardial tricarboxylic acid cycle flux. *Circ Res* 1988;63:628–634.
3. Buxton DB, Nienaber CA, Luxen A, Ratib O, Hansen H, Phelps ME, Schelbert HR: Noninvasive quantitation of regional myocardial oxygen consumption in vivo with [1–11C] acetate and dynamic positron emission tomography. *Circulation* 1989;79:134–142.
4. Brown M, Marshall DR, Burton BS, Sobel BE, Bergmann SR: Delineation of myocardial oxygen utilization with carbon–11–labeled acetate. *Circulation* 1987;76:687–696.
5. Armbrecht JJ, Buxton DB, Brunken RC, Phelps ME, Schelbert HR: Regional myocardial oxygen consumption determined noninvasively in humans with [1–11C] acetate and dynamic positron tomography. *Circulation* 1988;80: 863–872.

6. Henes CG, Bergmann SR, Walsh MN, Sobel BE, Geltman EM: Assessment of myocardial oxidative metabolic reserve with positron emission tomography and carbon-11 acetate. *J Nucl Med* 1989;30(#9):1489–1499.
7. Randle PJ, England PJ, Denton RM: Control of the tricarboxylate cycle and its interaction with glycolysis during acetate utilization in rat heart. *Biochem J* 1970;117:677–695.
8. Chance EM, Seeholzer SH, Kobayashi K, Williamson JR: Mathematical analysis of isotope labeling in the citric acid cycle with applications to 13CNMR studies in perfused rat heart. *J Biol Chem* 1983;258:13785–13794.
9. Schön HR, Schelbert HR, Najafi A, Hansen H, Robinson, G, Huang SC, Barrio J, Phelps ME: C–11 labeled palmitic acid for the noninvasive evaluation of regional myocardial fatty acid metabolism with positron computed tomography. II Kinetics of C–11 palmitic acid in acutely ischemic myocardium. *Am Heart J* 1982;103:548–561.
10. Fox KA Abendschein DR, Ambos HD, Sobel BE, Bergmann SR: Efflux of metabolized and nonmetabolized fatty acid from canine myocardium. Implications for quantifying myocardial metabolism tomographically. *Circ Res* 1985;57:232–243.
11. Hicks R, Herman W, Wolfe E, Kotzerke J, Kuhl D, Schwaiger M: Regional variation in oxidative and glucose metabolism in the normal heart: comparison of PET–derived C–11 acetate and FDG kinetics. *J Nucl Med* 1990;31(5):774.
12. Brunken RC, Chan SY, Armbrecht JJ, Phelps ME, Schelbert HR: ^{11}C–acetate kinetic analysis reveals impaired oxidative flux in hypoperfused human myocardium with maintained glucose metabolism. *Circulation* 1989;80:II–377.
13. Melin JA, Vanoverschelde JL, Bol A, Wijns W: Regional oxidative metabolism in patients with reperfused infarction: Relation to regional blood flow and glucose uptake. *Circulation* 1989;80:II–378.
14. Czernin J, Porenta G, Brunken R, Chan S, Kuhle W, Phelps M, Schelbert HR: Oxidative and glycolytic metabolic tissue characterization in patients with acute infarction using dynamic PET. *J Nucl Med* 1990;31:774.
15. Henes CG, Bergmann SR, Walsh MN, Geltman EM: Recovery of myocardial perfusion and oxygen consumption after thrombolysis delineated with positron emission tomography (PET). *Circulation* 1989;80:II–312.
16. Hashimoto T, Nienaber CA, Buston DB, Armbrecht JJ, Krivokapich J, Phelps ME, Schelbert HR: Disparity between blood flow and oxygen consumption in reperfused canine myocardium. *Circulation* 1989;80:II–240.

CHAPTER 32:

OXIDATIVE SUBSTRATE METABOLISM IN REPERFUSED MYOCARDIUM

Rene Lerch, Gunter Goerge and Richard Benzi

ABSTRACT

The oxidative metabolic rate in retrogradely perfused isolated rat hearts is exceedingly high relative to contractile performance early after postischemic reperfusion. This hypermetabolism is not only observed in hearts that were reversibly "stunned" but also in hearts that were irreversibly injured after one hour of reperfusion. The dissociation between contractile function and oxygen consumption decreases during administration of ruthenium red, compatible with the involvement of mitochondrial calcium transport in the mechanisms of postischemic hypermetabolism.

INTRODUCTION

The advance of thrombolytic therapy in clinical cardiology during the past decade has provided the impetus for research on metabolism in reperfused myocardium. Recent evidence suggested that the relationship between overall oxidative metabolic rate and contractile work as well as the substrate pattern for oxidative metabolism may be altered after postischemic reperfusion.

Oxygen consumption in the normal myocardium is closely related to contractile work. Several studies indicate that this relationship may be altered in the postischemic myocardium. Employing isolated ferret hearts, Neubauer et al. [1] observed that 40 min after reperfusion, following 60 min of complete cessation of myocardial perfusion, the rate–pressure product recovered to only 9% of the preischemic value, whereas oxygen consumption still averaged 41%. Stahl et al. [2] subjected anesthetized dogs to ten 5 min coronary occlusions separated by 10–min reperfusion intervals. After the last reperfusion period, systolic shortening of the ischemic segment was markedly reduced to 3% of the control. However myocardial oxygen consumption was even higher (+ 13%) in the reperfused region. Similar observations have been made in chronically instrumented awake dogs. In the study of Laxton et al. [3], after three 10–minute coronary occlusions followed by reperfusion, regional myocardial oxygen consumption was increased by +8% despite a reduction of systolic shortening to 37%.

Several hypotheses have been put forward to explain the dissociation between external work and myocardial oxidative metabolic rate. Schott et al. [4] have argued that, despite the absence of wall motion, energy is required for "contractile efforts to maintain akinesis". Other authors suggest uncoupling between oxidative metabolism and energy production [5]. Since mitochondrial calcium transport is associated with electron transfer in the respiratory chain and stimulation of intermediary metabolism without synthesis of ATP, it is conceivable that activation of this process by postischemic calcium overload in the cytoplasm may be implicated in oxidative hypermetabolism after reperfusion [6].

Initial evidence for alterations in the substate pattern after postischemic reperfusion has been provided by studies employing radionuclide imaging with positron emission tomography (PET) and radiolabeled metabolic substrates. Schwaiger et al. [7, 8] observed in dogs with transient coronary occlusion that clearance of radioactivity from regionally extracted (1–^{11}C)–palmitate was slower after reperfusion in the postischemic region compared to a control region. Delayed clearance of the radionuclide normalized within 3 hrs after a short occlusion of 20 min duration [7]. However, in chronically instrumented dogs subjected to a 3–hour occlusion, the abnormal clearance pattern persisted for up to 7 days. Furthermore, 24 hrs after reperfusion, the postischemic region exhibited increased accumulation of ^{18}F–2–deoxy–2–fluoro–D–glucose (^{18}FDG) [8]. These observations are compatible with a shift from preferential fatty acid utilization to glucose utilization in postischemic myocardium. In support of this interpretation Myears et al. [9] observed in dogs with a 60 minute coronary occlusion that formation of $^{14}CO_2$ from ^{14}C–palmitate, injected intracoronarily 10 min after reperfusion, was reduced compared to control conditions. Conversely, the relative contribution of glucose to oxidative metabolism was enhanced. The concept of an altered substrate pattern after reperfusion has recently been challenged by Liedtke et al. [5, 10]. These authors reported that in the *in vivo* working swine heart model oxidation of both palmitate and glucose normalized rapidly after reperfusion following coronary flow reduction for 45 min. Similarly, Lopaschuk et al. [11] observed a rapid return of the oxidative rate of both substrates to control levels after reperfusion in isolated working rat hearts perfused with medium containing 11 mM glucose and 1.2 mM palmitate after subjection to 25 min of no flow ischemia. The apparent inconsistency in the observed substrate pattern for oxidative metabolism after reperfusion may be related, among other factors, to differences in animal species, substrate levels, and severity of ischemia.

An isolated perfused rat heart model is used in the present study to assess: (1) the effect of different degrees of ischemic injury on oxidative substrate metabolism after postischemic reperfusion; and (2) the possible involvement of mitochondrial calcium transport in postischemic hypermetabolism.

METHODS

Hearts were removed from anesthetized male ZUR:SIV rats (200–300 g) fasted for 24 hrs. They were perfused retrogradely in a non–recirculating system at constant flow with Krebs–Henseleit buffer containing albumin 0.4 mM, palmitate 0.4 mM and glucose 11 mM. Washed erythrocytes were added to the perfusate at a hematocrit of 30 % to assure sufficient oxygenation at a physiological flow rate of 2 $ml \cdot min^{-1} \cdot g^{-1}$. The perfusate was warmed to 37°C and equilibrated with 95% O_2 – 5% CO_2. Heart rate was maintained at 280 min^{-1} by atrial pacing. A latex balloon connected to a pressure transducer was advanced into the left ventricular cavity to continuously monitor isovolumic pressure. Caval and pulmonary veins were ligated and the pulmonary artery was cannulated for collection of the coronary effluent. Myocardial oxygen consumption was calculated by multiplying the perfusate coronary effluent difference of total oxygen content (hemoglobin bound and dissolved) with myocardial blood flow per gram tissue. For the measurement of the oxidative metabolic rate of individual substrates trace amounts of either (1–^{14}C)–palmitate or (U–^{14}C)–glucose were added to the perfusate and myocardial release of $^{14}CO_2$ was measured, as previously described [12].

RESULTS AND DISCUSSION

Effect of Different Degrees of Ischemic Injury on Myocardial Substrate Oxidation after Reperfusion

To produce different degrees of ischemic injury, hearts were subjected to either 60 min of low flow ischemia (5% of control; n = 15), 60 min of no–flow ischemia (n = 12) or 120 min of no flow ischemia (n = 6) [13]. Control hearts were perfused during 140 min without intervention (n =14).

Hearts reperfused after 60 min of low flow ischemia exhibited early after reperfusion a moderate reduction of developed left ventricular pressure to an average of 55% ($p<0.05$) of the mean value measured in control hearts (Table 1). Subsequently, contractile function recovered to 77% (NS) after 60 min of reperfusion. Myocardial creatine phosphate content was only slightly reduced at the end of the experiment to 78 ± 5% ($p<0.05$). Sustained recovery of both creatine phosphate and contractile function suggest that the bulk of the myo–

cardium was reversibly injured by ischemia. Myocardial oxygen consumption returned to normal immediately after the onset of reperfusion when contractile function was still significantly depressed. However, oxidation of (1–^{14}C)-palmitate was reduced early after reperfusion, whereas oxidation of (U–^{14}C)-glucose was increased. This change of substrate pattern tended to return to baseline after 60 min of reperfusion. Thus, in transiently "stunned" hearts, oxygen consumption was disproportionately high compared to contractile performance early after reperfusion. There was a transient change in the relative contribution of palmitate and glucose to oxidative metabolism.

Table 1: Contractile function and oxidative metabolism 5 and 60 min after the onset of reperfusion in 3 groups of hearts with different degrees of ischemic injury.

	60 min Low Flow		60 min No Flow		120 min No Flow	
	5 min	60 min	5 min	60 min	5 min	60 min
DLVP	55 ± 9**	77 ± 7	7 ± 2**	29 ± 9**	1 ± 1**	2 ± 1**
MVO	97 ± 8	93 ± 10	92 ± 3	68 ± 9	28 ± 4**	19 ± 5**
PaOx	45 ± 6**	91 ± 9	69 ± 6**	94 ± 9	22 ± 8**	11 ± 1**
GluOx	268 ± 39*	169 ± 25**	194 ± 33	153 ± 17**	117 ± 16	57 ± 7

Values are expressed in % of mean value in control hearts. DLVP: developed left ventricular pressure; MVO: myocardial oxygen consumption; PaOx: oxidation of palmitate; GluOx: oxidation of glucose. * $p<0.05$, ** $p<0.01$ vs control hearts.

After 60 min of noflow ischemia, recovery of contractile function was poor, reaching only 28% after 60 min of reperfusion. Myocardial content of creatine phosphate measured at the end of the experiment was severely depressed to 10 ± 4% ($p<0.01$). Thus, irreversible injury predominated in this group. Surprisingly, immediately after reperfusion myocardial oxygen consumption returned to levels similar to those measured in control hearts. Although oxygen consumption subsequently gradually decreased it still averaged 68% after 60 min of reperfusion. Similarly to hearts subjected to low flow ischemia, palmitate oxidation was transiently reduced and glucose oxidation increased early after reperfusion. Thus, recovery of oxidative metabolism early after reperfusion was not limited to hearts that ultimately restored contractile function, but also occurred in hearts that were irreversibly injured after the ischemia–reperfusion period. The experiments do not distinguish between irreversible injury occurring during ischemia or instead during the reperfusion interval.

Hearts subjected to 120 min of no flow ischemia exhibited no recovery of contractile function and oxidative metabolism remained severely depressed. Glucose was the predominant substrate of residual metabolism.

Role of Mitochondrial Calcium Transport in Oxidative Hypermetabolism After Reperfusion

To study the possible involvement of mitochondrial calcium transport in the mechanisms underlying oxidative hypermetabolism, we assessed the effect of ruthenium red on substrate metabolism of postischemic rat hearts [14]. Ruthenium red is an inhibitor of mitochondrial calcium transport [15]. Hearts were subjected to 60 min of no flow ischemia followed by 60 min of reperfusion. Ruthenium red was infused into the aortic root during the initial 40 min of reperfusion to give a final concentration of 6 μM. In untreated postischemic hearts myocardial oxygen consumption averaged 1.8 ± 0.2 $\mu mol \cdot min^{-1} \cdot g^{-1}$ at 15 min after reperfusion, which was similar to nonischemic control hearts (2.1 ± 0.1 $\mu mol \cdot min^{-1} \cdot g^{-1}$; NS). Ruthenium red elicited a reduction of myocardial oxygen consumption to 1.2 ± 0.2 $\mu mol \cdot min^{-1} \cdot g^{-1}$ ($p<0.05$ vs no flow experiments without ruthenium red). Oxidation of both palmitate and glucose were reduced to a similar extent, by 38 ± 7 % (NS) and 42 ± 16 % (NS) respectively. Despite reduced oxidative metabolism, developed left ventricular pressure was higher in the presence of ruthenium red (21 ± 7 vs 6 ± 2 mmHg; $p<0.05$). Thus, oxygen consumption per unit of contractile work was reduced by ruthenium red. The results are consistent with the hypothesis that enhanced mitochondrial calcium transport is at least in part responsible for the dissociation between contractile function and oxidative metabolism after reperfusion.

Acknowledgement

This work was supported by the Swiss National Science Foundation Grant 32–26373.89.

DISCUSSION

Dr. Spaan: With a diastolic left ventricular pressure of 50 mmHg, there can hardly be any coronary flow.

Dr. Lerch: Total coronary flow was kept constant by a pump. We did not measure regional distribution of myocardial blood flow. However, we did experiments in which we deflated the balloon to decrease the compression of the subendocardial tissue. In those experiments, metabolic data had not been changed. Thus, subendocardial compression was not a major factor that influenced overall metabolic recovery.

Dr. Strauss: Did any of these hearts exhibit cardiac arrhythmias during the reperfusion period as diastolic pressure started to go up?

Dr. Lerch: There were some arrhythmias. However, ventricular fibrillation was rare and they always reverted spontaneously. But, yes, these were arrhythmias.

Dr. Mowbray: Your results show that these postischemic hearts carry out a lot of lipid synthesis. This presumably requires oxygen uptake and energy. Could you make an estimate from the amount of extra-lipid synthesis whether that would account for a large part of the oxygen uptake?

Dr. Lerch: The incorporation of fatty acids into triglycerides utilizes energy. This pathway may therefore contribute to enhanced oxygen consumption.

Dr. Mowbray: Do the fatty acids produce more triglycerides in the postischemic period?

Dr. Lerch: Yes, but we have not studied this point in detail so far. We are presently looking at the extent of futile metabolic cycling during reperfusion to get an idea of the energy used for triglyceride resynthesis.

Dr. van der Vusse: Liedtke et al. *[Circ Res, 1988;62:535–542]* have recently shown that the postischemic oxidation of fatty acids is not impaired in pig hearts. Can you explain the discrepancy between your experimental findings and theirs?

Dr. Lerch: This discrepancy is only apparent from the discussion. If you look at the data of Liedtke and co-workers, it appears that even in the swine heart the production of $^{14}CO_2$ recovers within 15 minutes. So, during the first 15 minutes of reperfusion, values of $^{14}CO_2$–production are also reduced.

Dr. van der Vusse: Your experiments with ruthenium red suggest that blockade of mitochondrial Ca^{2+}–uptake is beneficial in the reperfused heart. I thought that mitochondrial accumulation of calcium ions was a safety measure to prevent high cytoplasmic levels of Ca^{2+} following reperfusion, a condition known to impair functional recovery. Do you have a mechanistic explanation why ruthenium red protects the reperfused heart?

Dr. Lerch: This was a great surprise to us as well. We have thought that when we block mitochondrial calcium transport, an overload of cytoplasmic calcium occurs, which is expected to damage the cell. However, similar observations have been made by Ferrari *[J Mol Cell Cardiol 1982;14:737–740]* on recovery of mitochondrial function. These authors did not look at contractile function, but mitochondrial function and ATP content were improved by ruthenium red. I am not sure about the mechanism by which we got improvement of contractile function and less cellular damage. Note that the concentration of ruthenium red that we used was 6 micromolar.

Dr. Mowbray: Ruthenium red is a very charged molecule and one might guess that because of its negative charges it would be a placebo experiment, as you might never expect it to penetrate normal cells, but evidently it does. Do you think it gets in by a specific transport mechanism?

Dr. Lerch: I cannot say how it gets in. The only group which has looked at this problem is the group of Dhalla *[Am J Physiol 1988;255:1413–1420]*. We worked first with 1.5 micromolar before their paper appeared and we had very little effect. In the Dhalla paper there was an effect at above 5 micromolar. After 5 minutes perfusion ruthenium red could be detected within the cell. It has been speculated by Ferrari *[J Mol Cell Cardiol 1982;14:737–740]*, who used a lower concentration, that damaged cells let it in easier, but this is purely speculative.

Dr. Marban: The question relates to specificity. It is worth considering SR calcium release. Yet, I believe you are interpreting your findings solely in terms of the mitochondria. Maybe you could speculate on the relative importance of these two factors.

Dr. Lerch: The fact that high energy phosphate content increases markedly is a strong argument in favor of a mitochondrial site of action.

Dr. Marban: Is SR calcium release important for energetic recovery?

Dr. Lerch: Yes, it is conceivable that you may have less contracture and therefore less activation of mitochondrial respiration. However, in the Ferrari study *[ibid]*, they measured mitochondrial and total calcium and although they had no change in total calcium, mitochondrial calcium decreased. I agree, the exact mechanism of the protective effect of ruthenium red is not clear.

REFERENCES

1. Neubauer S, Hamman BL, Perry SB, Bittl JA, Ingwall JS: Velocity of the creatine kinase reaction decreases in postischemic myocardium: A 31P–NMR magnetization transfer study of the isolated ferret heart. *Circ Res* 1988;63:1–15.
2. Stahl LD, Weiss HR, Becker LC: Myocardial oxygen consumption, oxygen supply/demand heterogeneity, and microvascular patency in regionally stunned myocardium. *Circulation* 1988;77:865–872.
3. Laxson DD, Homans DC, Dai XZ, Sublett E, Bache RJ: Oxygen consumption and coronary reactivity in postischemic myocardium. *Circ Res* 1989;64:9–20.
4. Schott RJ, Rohmann ER, Schaper W: Ischemic preconditioning reduces infarct size in swine myocardium. *Circ Res* 1990;66:1133–1142.
5. Liedtke AJ, DeMaison L, Eggleston AM, Cohen LM, Nellis SH: Changes in substrate metabolism and effects of excess fatty acids in reperfused myocardium. *Circ Res* 1988;62:535–542.
6. Vercesi A, Reynafarje B, Lehninger A: Stoichiometry of H+ ejection and Ca2+ uptake coupled to electron transport in rat heart mitochondria. *J Biol Chem* 1978;253:6379–6385.
7. Schwaiger M, Schelbert HR, Keen R, Vinten–Johansen J, Hansen H, Selin C, Barrio J, Huang SC, Phelps ME: Retention and clearance of C–11 palmitic acid in ischemic and reperfused canine myocardium. *J Am Coll Cardiol* 1985;6:311–320.
8. Schwaiger M, Schelbert HR, Ellison D, Hansen H, Yeatman L, Vinten–Johansen J, Selin C, Barrio J, Phelps ME: Sustained regional abnormalities in cardiac metabolism after transient ischemia in the chronic dog model. *J Am Coll Cardiol* 1985;6:336–347.

9. Myears DW, Sobel BE, Bergmann SR: Substrate use in ischemic and reperfused canine myocardium: quantitative considerations. *Am J Physiol* 1987;253: H107–H114.
10. Renstrom B, Nellis SH, Liedtke AJ: Metabolic oxidation of glucose during early myocardial reperfusion. *Circ Res* 1989;65:1094–1101.
11. Lopaschuk GD, Spafford MA, Davies NJ, Wall SR: Glucose and palmitate oxidation in isolated working rat hearts reperfused after a period of transient global ischemia. *Circ Res* 1990;66:546–553.
12. Luthy P, Chatelain P, Papageorgiou I, Schubiger A, Lerch RA: Assessment of myocardial metabolism with iodine–123 heptadecanoic acid: Effect of decreased fatty acid oxidation on deiodination. *J Nucl Med* 1988; 29:1088–1095.
13. Goerge G, Papageorgiou I, Lerch R: Fatty acid metabolism in postischemic myocardium: Effect of different degrees of ischemic injury (abstract). *J Mol Cell Cardiol* 1988;20 (supplement V):S.67.
14. Lerch R, Papageorgiou I, Marquez E, Benzi R: Role of mitochondrial calcium transport for myocardial hypermetabolism during reperfusion (abstract). *J Mol Cell Cardiol* 1990;22(Supp III):S–107.
15. Ferrari R, diLisa F, Raddino R, Visioli O: The effects of ruthenium red on mitochondrial function during postischaemic reperfusion. *J Mol Cell Cardiol* 1982; 14:737–740.

CHAPTER 33:

THE ROLE OF DRIVING FORCES IN MYOCARDIAL ENERGY METABOLISM AND TRANSPORT PROCESSES

Helmut Kammermeier

ABSTRACT

Energy fluxes from ATP synthesis to ion pumping can not only be considered under the aspect of amount synthesized and transported respectively, but also under the aspect of free energy gradients. Owing to the high efficiencies of energy transformation, some 60 to 90%, major changes in the free energy levels of driving and driven processes must affect the rate and even the direction of the processes. Moreover, the transport/transformation capacities of the different processes and their interplay with driving forces may be determinant for some pathophysiological processes. This is demonstrated with the consequences of an alteration of the mitochondrial ATP/ADP translocator and with the intracellular changes of sodium and calcium concentration occurring at the ischemia–reperfusion transition.

INTRODUCTION

Flux of energy and molecules is governed by driving forces and regulatory processes. However, in most instances regulatory processes are considered to dominate and the role of the driving forces is more or less neglected. This is also the case in energy metabolism where, for instance, ATP concentration and regulation of ATPases is considered to describe the substantial core of energy metabolism. On the other hand it is also well known that the major pathways of biological energy transformation exhibit high efficiency, and high efficiency requires delicate matching of driving forces and fluxes. In this context one has to keep in mind, first, that 100 % efficiency of metabolic processes is only possible with zero net flux and, second, that a gradient is required to generate energy fluxes.

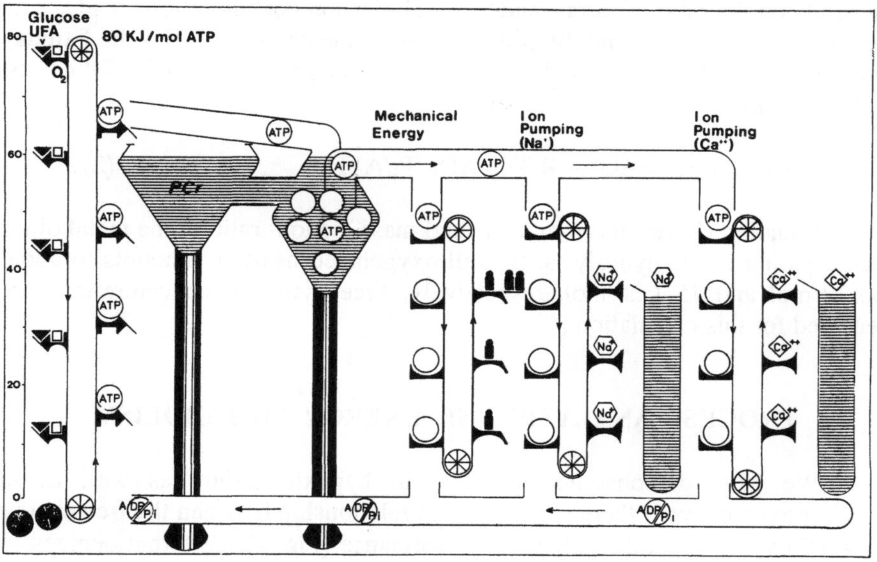

Figure 1: Schematic depiction of interplay of free energy gradients and chemical potentials of the main energy supplying and consuming processes of the myocardium, respectively.

THE BASIC ENERGY TRANSPORT MECHANISMS

Figure 1 demonstrates the importance of energy gradients. On the one side substrate and oxygen, dropping along an energy gradient, lift the ATP synthesized from ADP and Pi to an energy level of about 70 % of that available from glucose oxidation. The free energy level of ATP is buffered on this high level by the equilibrium reaction with phosphocreatine (symbolized by the width of the reservoirs). The major ATP consuming processes, those providing mechanical energy and ion pumping of sodium and calcium, utilize ATP dropping along an energy gradient to provide energy for liberation of mechanical energy and for lifting sodium and calcium ions against an electrochemical gradient. These electrochemical gradients with the given stoichiometries correspond to energy levels of more than 70 to 80 % of that available from ATP hydrolysis in well oxygenated hearts. It is obvious from the scheme that the energy level to which ions are lifted cannot exceed that of ATP, and that to maintain flux or pumping a certain energy gradient must exist.

The energy which is delivered or required in ion transport processes according to the following equation,

$$\Delta G = (\Delta\mu) = R \cdot T \cdot \ln(C_i/C_e) + z \cdot F \cdot \Psi \quad (1)$$

depends on the ratio of intra– and extracellular ion concentrations, C_i/C_e, and on the membrane potential, Ψ (volt); z denotes the charge and F is the Fareday constant; R, T, have the usual meaning. The free energy ΔG of ATP can be calculated from

$$\Delta G = (dG/d\xi) = \Delta G^\circ + R \cdot T \cdot \ln ADP(P_i/ATP) \qquad (2)$$

Equation (2) results from the actual mass–action–ratio of the metabolites participating in the hydrolysis. In well oxygenated hearts it amounts to about 60 kJ/mol and 15 kcal/mol, respectively. Free cytosolic concentrations are required for this calculation [1].

PROCESS ANALYSIS AND ENERGY METABOLISM

We have demonstrated [2] that in hypoxic failure as well as in posthypoxic recovery there is an apparent relationship between the free energy of ATP hydrolysis and mechanical performance (Fig. 2). Different degrees of hypoxia were induced by different oxygen content of the perfusion medium of isolated hearts. A comparable relationship does not exist between ATP

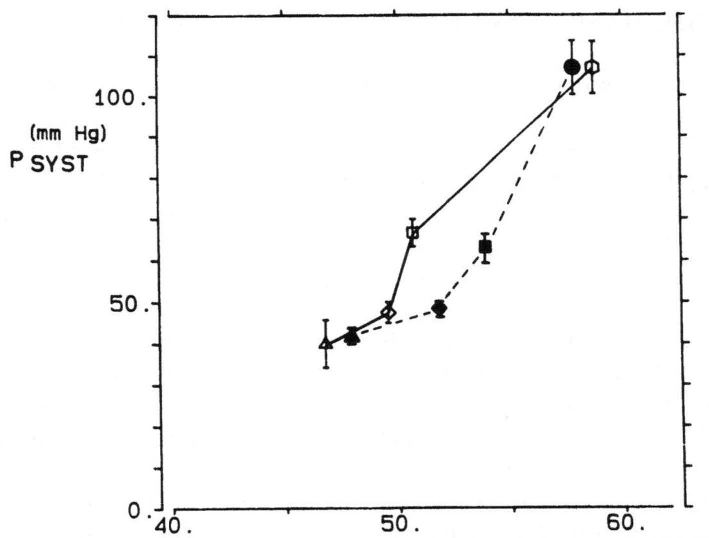

Figure 2: Peak isovolumic systolic pressure of isolated perfused rat hearts vs. free energy of ATP hydrolysis in graded hypoxia (dashed line) and graded reoxygenation (solid line) (PO_2: 60 to 700 mmHg).

concentration and contractile performance, according to numerous investigations (c.f. [1]). However, apparent correlation does not necessarily imply causal relationship. Moreover, there might be different mechanisms involving ATP consuming processes. Nevertheless, the high efficiencies mentioned, 80 % and more, indicate that free energy of ATP can not be reduced beyond a certain level without feedback to the ATP dependent processes.

The energy transformation of ATP and its yield to cytosolic ATP consuming processes includes one interesting step, namely the lifting of free energy of ATP at the transport step across the matrix membrane of mitochondria into the cytosol. Inside the matrix space the free energy of synthesized ATP is about 5 kJ/mol lower than in the cytosol. The energy required to lift ATP on the higher energy level is provided by the proton electrochemical gradient in the coupled transport of the ATP translocase reaction. This translocase process nicely reveals the importance of energy gradients. Dr. Schultheiss's group [3] has demonstrated that inhibition of this translocator by specific antibodies does not allow to utilize the proton gradient to lift free energy of ATP when passing the matrix membrane. So the energy gradient is almost abolished. When these hearts are exposed to workload, ATP can only go down along the energy gradient i.e. free energy of cytosolic ATP becomes still lower [4], which includes lower efficiency. Moreover, the hearts are failing with lower workload [4]. Earlier change to an anaerobic energy supply by glycolysis is a second index of impaired oxidative phosphorylation in those hearts.

Energy gradients are of similar importance in ion exchange. The quantitatively most important ion transport processes of the myocardium are depicted in Fig. 3. In these schemes of coupled transport, length of arrows indicates energy required/provided for transport. In some cases, either component can mutually act as the dominating direction determining step. For example, the lactate/proton symporter and the sodium/calcium antiporter can

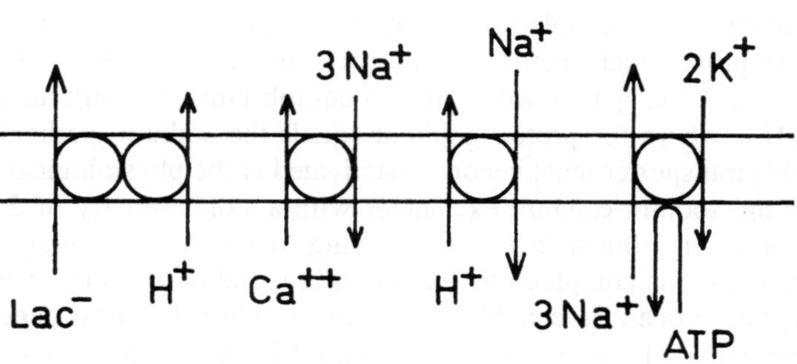

Figure 3: Ion transporters involved in myocardial ion homeostasis.

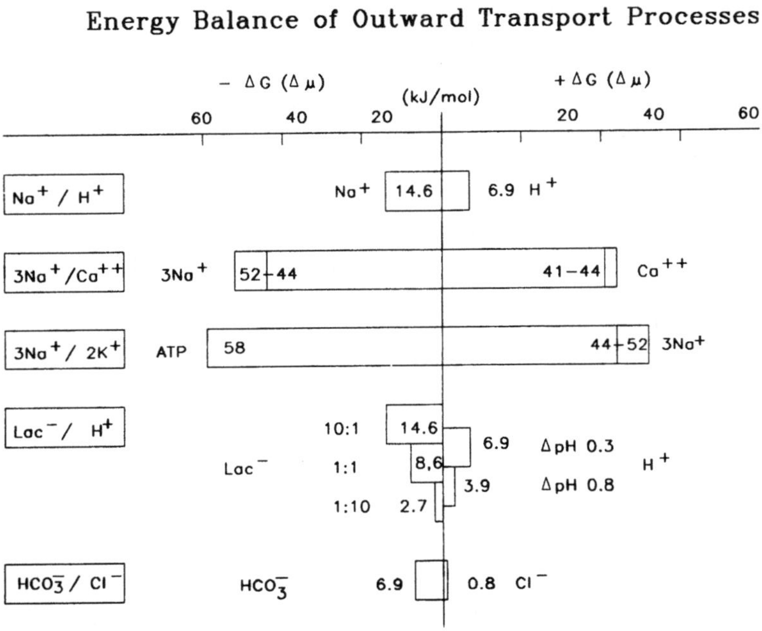

Figure 4: Balance of driving forces of ion transporters involved in ion transfer of the myocardial sarcolemma. The ratios indicate $[Lac^-]i : [Lac^-]e$.

also run in the opposite direction; determinants for the direction of transport are again the energy gradients.

The balance of driving forces resulting from concentration gradients, stoichiometries and membrane potentials are compared in Fig. 4. At the left hand side with $-\Delta G$ the bars represent the energy yielding, that is thermo-dynamic passive transport components. At the right hand side the lengths of the bars represent the coupled thermodynamic uphill processes. In the case of the sodium proton exchanger there is always an energy excess of the sodium gradient. If this transporter would run to equilibrium, intracellular pH would exceed pH 8 where the proton gradient equals the sodium gradient. Conse-quently this transporter must become inactivated at the physiological pH.

For the sodium calcium exchanger with a stoichiometry of 3 to 1, the driving forces are almost in balance. Owing to the varying energy obtained from 3 sodium ions compared to 1 calcium ion, and depending on membrane potential, this process is considered to run in either direction depending on membrane potential. The sodium /potassium ATPase utilizes a free energy of ATP of about 58 kJ/mol, to lift 3 sodium ions to a sum of energy of 44 to 52 kJ/mol depending on the intracellular sodium concentration. The energy demand for potassium pumping can almost be neglected at high resting membrane potential. It is obvious that the energy level of ATP hydrolysis, and that required to maintain physiological ion gradients, are rather close together. This includes high efficiency as well as a certain dependence of the ion distribution on the energy level.

Different states have to be considered for the lactate proton symporter. The first state is that of a well oxygenated heart, where intracellular lactate concentration is about 1/10 of extracellular and the ΔpH across the sarcolemma is 0.3. Here the driving forces obviously favor lactate uptake, as it is observed in the well oxygenated state. The second state is that of hypoxic perfusion, where intracellular and extracellular lactate concentration are similar and ΔpH is also about 0.3. Here the driving force of lactate release is almost in balance with that required to expel protons. Thus, the net driving force for lactate and proton release is rather small. A different state is reached during postischemic reperfusion: the high driving force resulting from more than 10 fold intracellular lactate accumulation exceeds that required for proton extrusion, resulting from ΔpH of about 0.8 substantially, and, at least according to the driving forces, major lactate and proton release should occur. But as is shown below, the capacities of the different transporters have a major impact on the actual exchange. The bicarbonate chloride exchanger (Fig. 4) can presumably be neglected in the myocardium.

Table 1 summarizes the amounts of lactate and protons formed and accumulated in hypoxia and ischemia and the capacity of the transporters involved. According to different sources, and depending on species, lactate formation amounts to 1 – 5 μmol / g·min. There is no major difference between hypoxia and ischemia except in the later states of ischemia, where glycolysis is suppressed. In contrast, major differences occur in lactate and

Table 1: Lactate and proton rate of formation and amount accumulated during hypoxia and ischemia. Capacity of lactate H^+ symporter and Na^+/H^+ antiporter.

	Hypoxia		Ischemia	
Lactate(Proton) formation (μmol/g·min)	< 1.5 – 2	[5]	< 1.5 – 2	[5, 12, 13]
	3 – 5	[9–11]		
Lactate (Proton) amount (μmol/g*)	< 1	[5]	20 – 60	[5, 14]
Lactate/H^+ transporter (μmol/g·min)	1.2 – 2.6 (10 mM)	[6, 7]	4 – 8 (40 mM)	[6, 7]
	0.7 – 1.6 (5 mM)	[6, 7]	2.7 – 5.3 (20 mM)	[6, 7]
Na^+/H^+ transporter	170 (pH_i 7.0)	[8]	170 (pH_i 6.5)	[8]

V_{max} = 8.0 μmol/g·min; K_m = 20 mM [6]; V_{max} = 4.0 μmol/g·min, K_m = 19 mM [9]; V_{max} = 170 μmol/g·min; K_m $H^+ = 10^{-7.35}$ M [8], K_m $Na^+ \approx$ 20 mM [8].

proton accumulation: whereas minor accumulation occurs in hypoxia because of the continuous removal, the content of lactate in ischemic tissue can accumulate to 20 to 40 μmol/g and more. Upon reperfusion and reoxygenation, two mechanisms compete for lactate and proton removal, i.e. the lactate proton symporter and the sodium proton antiporter (Figs. 3 and 4). Driving forces of both favor proton and lactate removal, respectively (Fig. 4). However, the transport capacity, as expressed in V_{max}/g tissue, differs by a factor of 20 to 40. The high capacity of the sodium proton antiporter permits removal of accumulated protons within seconds. In contrast, because of the low capacity of the lactate proton symporter, 5 to 10 minutes are required to restore control values. However, this fast removal of protons achieved by the sodium proton exchanger takes place at the cost of equimolar sodium gain.

The interplay of these transport mechanisms can be summarized as follows: during hypoxia the supply and removal from the minor lactate and proton pool is in balance. Because of the minor pH changes, the sodium proton antiporter contributes little to proton removal and consequently sodium gain will be negligible. Conversely, upon reperfusion following ischemia, the large proton/lactate pool will be drained to a major part by the sodium proton exchanger due to its high transport capacity and this phenomenon must result in a substantial sodium load.

This concept has recently been well demonstrated by Tani and Neely [14, 15]. According to Fig. 5, drawn from their data, intracellular sodium concentration 2 minutes following postischemic reperfusion depends on lactate and intracellular proton accumulation. This is obvious for the 2 data points with higher lactate and proton concentration. The data points with low Na^+ concentration belong to experiments with hypoxic perfusion with pyruvate containing medium. Here, excess lactate and proton are not correlated. Excess proton amount in hypoxic perfusion (cf.1) is very small and sodium gain by the Na/H antiporter is negligible. The increase in cytosolic lactate concentration resulting from hypoxia and pyruvate supply is not correlated with proton accumulation and consequently is also not accompanied by sodium gain.

Tani and Neely [14, 15] also demonstrated convincingly that sodium gain results in calcium accumulation with a certain delay because of the sodium calcium exchanger. According to Fig. 6,also drawn from their data, sodium accumulation in the cytosol can be manipulated by a couple of interventions: increase in sodium uptake can be accomplished by administration of the sodium ionophor monensin and sodium gain can be reduced by inhibition of the sodium proton exchanger by amiloride and by preventing lactate and/or proton accumulation by anoxic perfusion and glycogen depletion, respectively.

A good relation exists in Fig. 6 between sodium gain and the subsequent calcium accumulation (right inverted ordinate). Furthermore, mechanical

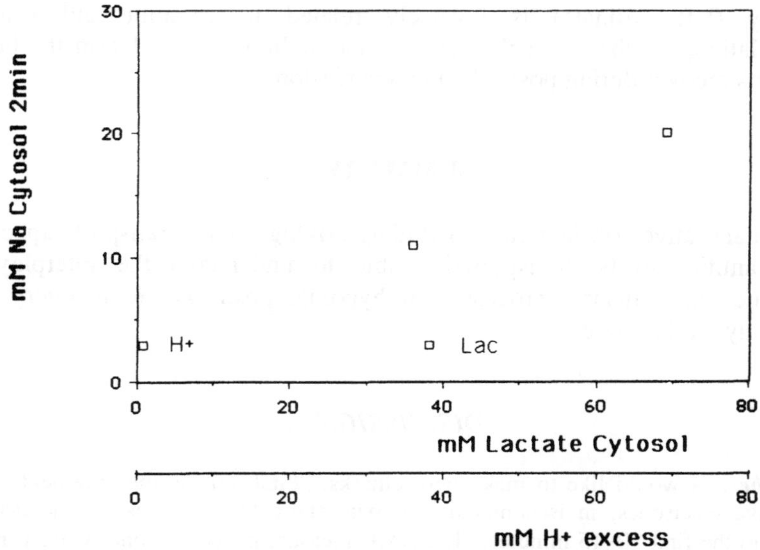

Figure 5: Myocardial cytosolic Na^+ concentration 2 minutes following postischemic reperfusion vs. ischemic lactate and proton accumulation. The upper two data points coincide for H^+ and lactate. In the lower range, due to anoxic perfusion with pyruvate containing medium, lactate accumulation is not accompanied by proton accumulation (c.f. [1]). Data of Tani and Neely, [14, 15], with permission.

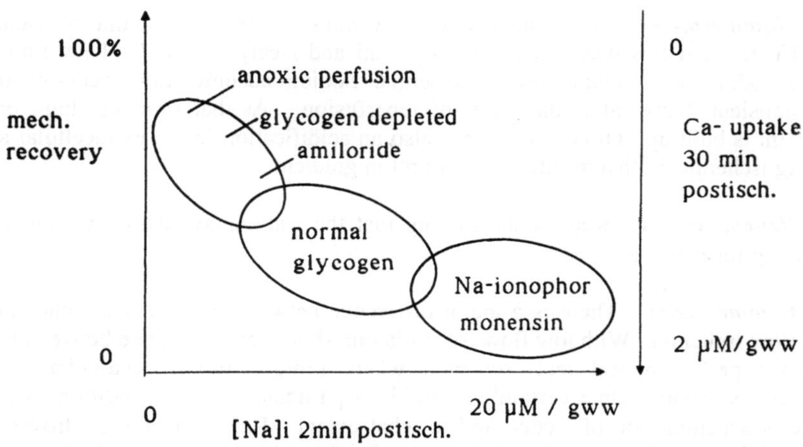

Figure 6: Schematic depiction of Ca^{++} uptake and mechanical recovery following 30 minutes postischemic perfusion vs. Na^+ accumulation after 2 minutes postischemic perfusion. Ellipses indicate the approximate range of data points. Data of Tani and Neely [14, 15], with permission.

recovery (left ordinate) is inversely related to calcium and sodium accumulation, i.e. the lower the calcium and sodium accumulation the better the hearts recover during postischemic reperfusion.

SUMMARY

Quantitative considerations including driving forces, transport capacities and quantities to be transported enable to understand the interplay of metabolic and transport processes in hypoxia, postischemic recovery and certain myopathic states.

DISCUSSION

Dr. Kléber: I would like to make two remarks: First, all the measurements with insensitive electrodes, in ischemia or hypoxia, show that there is no increase of sodium in the first 10–15 minutes. The NMR measurements are controversial in rat. There is a fast, immediate increase, but in rabbits or guinea pigs there is also an initial stable phase when sodium remains low. Second, there is evidence that the extracellular pH change is much more expressed than the decrease of intracellular pH. The transmembrane gradient of pH may even reverse. This will actually block the proton exchange through the sodium–proton antiporter. Therefore, regulation of intracellular pH in ischemia might be more complex than described in your presentation.

Dr. Kammermeier: I consider mainly a transient from ischemia to ischemic reperfusion and this was investigated by Tani and Neely *[14, 15]*. They find only minor sodium accumulation during ischemia. Sodium accumulation occurs mainly at the transient 2 min after the onset of reperfusion. At that time the high proton gradient is built up. Of course there is also an acidification in the extracellular space during ischemia, with a resulting minor proton gradient.

Dr. Reneman: You were demonstrating that the sodium hydrogen exchanger was effecting reperfusion.

Dr. Kammermeier: There is a major difference between total or real ischemia and low flow ischemia. With low flow ischemia one should expect a state between that of hypoxic perfusion and real ischemia, where protons might accumulate in the extracellular space. Once certainly should look particularly at the conditions, whether there is accumulation of lactate and to what extent. Regarding the controversy on Neely's data on glycogen depleted hearts, one has simply to look whether sodium accumulation took place or not. A real discrepancy exists only if there is no sodium accumulation. Each particular case has to be investigated as to whether or not there is sodium accumulation.

REFERENCES

1. Kammermeier H, Schmidt P, Juengling E: Free Energy change of ATP–hydrolysis a causal factor of early hypoxic failure of the myocardium. *J Mol Cell Cardiol* 1982;15:267–277.
2. Griese M, Perlitz V, Juengling E, Kammermeier H: Myocardial performance and free energy of ATP–hydrolysis in isolated rat hearts during graded hypoxia reoxygenation and high K^+ perfusion. *J Mol Cell Cardiol* 1988;20:1189–1201.
3. Schulze K, Becker BF, Schultheiss HP: Antibodies to the ADP/ATP carrier an autoantigen in myocarditis and dilated cardiomyopathy penetrate into myocardial cells and disturb energy metabolism in vivo. *Circ Res* 1989;64:179–192.
4. Schultheiss HP: *Personal Communication.* 1990.
5. Kuebler W, Spieckermann PG: Regulation of Glycolysis in the ischemic and the anoxic myocardium. *J Mol Cell Cardiol* 1970; 1:351–377.
6. Mann GE, Zlokovic BV, Yudilevich DL: Evidence for a lactate transport system in the sarcolemmal membrane of the perfused rabbit heart: kinetics of unidirectional influx carrier specificity and effects of glucagon. *Biochim Biophys Acta* 1985;819:241–248.
7. Kammermeier H, Wein B, Gerards P, Lang U, Wendtland B, Schmitz D, Rose H: Barriers in cardiac substrate supply. *Basic Res Cardiol* 1985;80:57–60.
8. Lazdunski M, Frelin C, Vigne P: The sodium/hydrogen exchange system in cardiac cells: Its biochemical and pharmacological properties and its role in regulating internal concentrations of sodium and internal pH. *J Mol Cell Cardiol* 1985; 17:1029–1042.
9. Rose H, Schnitzler N, Kammermeier H: Electrically stimulated cardiomyocytes: Evaluation of shortening oxygen consumption and tracer uptake. *Biomed Biochim Acta* 1987;46:622–627.
10. Williamson JR: Glycolytic control mechanisms: Kinetics of intermediate changes during aerobic–anoxic transition in perfused rat heart. *J Biol Chem* 1966;214:5026–5036.
11. Rovetto MJ, Whitmer JT, Neely JR: Comparison of the effects of anoxia and whole heart ischemia on carbohydrate utilization in isolated working rat heart. *Circ Res* 1973;32:669–711.
12. Rovetto MJ, Lamberton WF, Neely JR: Mechanisms of glycolytic inhibition in ischemic rat hearts. *Circ Res* 1975; 37:742–751.
13. Neely JR, Morgan HE: Relationship between carbohydrate and lipid metabolism and the energy balance of the heart. *Ann Rev Physiol* 1974;36:413–459.
14. Tani M, Neely JR: Role of intracellular Na^+ in Ca^{2+} overload and depressed recovery of ventricular function of reperfused ischemic rat heart. *Circ Res* 1989;65:1045–1056.
15. Tani M, Neely JR: Na^+ Accumulation increases Ca^{2+} overload and impairs function in anoxic rat heart. *J Mol Cell Cardiol* 1990;22:57–72.

CHAPTER 34:

THE EFFECT OF EXOGENOUS LACTATE AND PYRUVATE ON TRIACYLGLYCEROL TURNOVER IN THE NORMOXIC HEART

Ger J. van der Vusse and Monique J.M. de Groot

ABSTRACT

Both lactate and pyruvate (5 mM) reduce the decline in functional stability of isolated rat hearts, normoxically perfused with a glucose (11 mM) containing Krebs–Henseleit buffer. Unlike pyruvate, lactate considerably stimulates the release of glycerol from the heart. As cardiac triacylglycerol (TG) levels remain unaltered, the conclusion must be drawn that lactate stimulates the triacylglycerol–fatty acid (TG–FA) cycle in the normoxic heart. It is unlikely that this effect is caused by mitigation of the supposed feedback inhibition of FA on TG lipase since tissue FA are found to be elevated in the lactate perfused hearts. The present findings are in favor of an important role of glycerol–3–phosphate (G–3–P) in control of the TG–FA cycle.

INTRODUCTION

The normoxic heart is able to utilize a variety of oxidizable substrates to produce energy for electrical and mechanical processes. Fatty acids, glucose, lactate, pyruvate, ketone bodies and amino acids can be consumed by the heart [1, 2]. The actual amounts and type of substrate used strongly depend on the extracellular concentration of these substances, workload of the heart, the presence of competing substrates, and circulating and locally released hormones. In isolated rat hearts, glucose (in the absence of insulin) is a poor substrate during normoxic perfusion [3–5]. In order to fulfill the energy requirements, the glucose–perfused organ starts to deplete its endogenous substrate stores. Both glycogen (GLYC) and triacylglycerols (TG) are used to maintain an adequate supply of oxidizable substrates.

The present study was conducted to investigate whether addition of either lactate (5 mM) or pyruvate (5 mM) to a glucose–containing perfusion buffer was sufficient to meet the energy needs of the heart during prolonged, normoxic perfusion. To this end, mechanical performance of the isolated, ejecting rat heart was monitored, and the effect of both cosubstrates on endogenous levels of GLYC and TG assessed. Furthermore, this study focussed on the stimulation of the TG–FA cycle, if any, by lactate or pyruvate. To elucidate a possible mechanism of action, the influence of the cosubstrates on parameters most likely involved in TG turnover, such as intracellular levels of FA, TG, G–3–P and glycerol, was measured.

MATERIALS AND METHODS

Animals: Hearts were obtained from male Lewis rats (10 weeks old, average body weight 280 g). The animals were fed *ad libitum* (Standard diet 2, Hope Farms).

Preparation: After light ethyl ether anesthesia and thoracotomy, the hearts were rapidly excised and immediately immersed in an ice–chilled perfusion medium. Lung and fat tissue was removed and the hearts were ligated to the aortic cannula of the perfusion system and retrograde perfusion was started at a perfusion pressure of 8 kPa. A second cannula was connected to the left atrium to allow antegrade perfusion of the hearts. The perfusion system has been described previously [6]. The perfusion medium, a modified Krebs–Henseleit buffer, contained (in mM): NaCl (130.0), KCl (5.6), $CaCl_2$ (2.2), $MgCl_2$ (1.0), NaH_2PO_4 (1.2), $NaHCO_3$ (28.6). D(+)–Glucose (11 mM) or D(+)–glucose (11 mM), in combination with either L(+)–lactate (5.0 mM) or pyruvate (5.0 mM), was added as substrate. The buffer was equilibrated with a mixture of 95% O_2 and 5% CO_2 (pO_2>75 kPa) and continuously filtered (Millipore filter; 1.2 μm pore size) during the perfusion period. The pH range was adjusted to 7.35–7.45. Temperature was kept at 37°C.

Experimental Protocol: The hearts were preperfused at a pressure of 8 kPa according to the Langendorff technique. After a stabilization period of 10 min, hearts were freeze–clamped (t=0 min) or subjected to antegrade perfusion for 120 min. Left atrial filling pressure was 1 kPa. Diastolic aortic pressure was kept at 8 kPa. The hearts were freeze–clamped after 120 min of antegrade perfusion (t = 130 min).

Assessment of Functional Parameters: Variables of interest were registered on a six–channel recorder (Schwarzer). Aortic pressure was measured at the entrance of the aortic cannula via a catheter connected to a CTC pressure transducer. Left ventricular (LV) pressure was measured via a catheter inserted into the LV through the apex. Left ventricular developed pressure (LVDP) was calculated as the difference between LV systolic and end diastolic pressure. Aortic flow (AF) was measured by an electromagnetic flow probe (Skalar) in the aortic outflow tract connected to a sine wave electromagnetic flowmeter (Transflow 601). The coronary flow (CF) was determined by timed collection of the coronary effluent. Cardiac output (CO) was assessed by adding AF and CF. Heart rate (HR) was calculated from the pressure registrations.

Collection of Tissue and Effluent Samples: At the end of preperfusion (t = 0 min) and at 120 min of antegrade perfusion (t = 120 min) the ventricular tissue was rapidly dissected from atrial tissue with a surgical blade and immediately freeze–clamped between aluminum tongues, cooled in liquid nitrogen. Coronary effluents were sampled throughout the experiment. Samples were stored at –80°C until analysis.

Tissue – Biochemical Analysis:

Glycerol 3–phosphate, glycerol and glucose: Small parts of the frozen ventricles were freeze–dried at –30°C. Subsequently, G–3–P, glycerol and glucose were extracted by use of a mixture of perchloric acid (3 M) and dithiothreitol (5 mM) as described previously [6]. Glycerol–3–phosphate and glycerol were determined fluorometrically according to Laurell and Tibbling [7]. Glucose was measured fluorometrically as described by Van der Vusse et al. [8].

Glycogen: Aliquots of ventricular tissue were freeze–dried. After adding 1 M HCl to the freeze–dried material, GLYC was hydrolyzed at 100°C for 3 h. The samples were neutralized with Tris (0.12 M)/KOH (2.1 M), saturated with KCl, and subsequently glucose residues were measured fluorometrically [8]. The values obtained were corrected for the content of free glucose in the tissue.

Fatty acids and triacylglycerols: Part of the frozen tissue was powdered at the temperature of liquid nitrogen and subsequently extracted with chloroform/methanol (2:1, by volume) [9, 10]. Neutral and polar lipids were separated by silica–gel column chromatography [9]. The neutral lipid fraction was further separated using thin–layer chromatography. The lipid spots, corresponding to FA and TG, were scraped from the plates. After elution from the silica–gel powder and methylation, the FA methyl esters were quantitated by gas chromatography [9, 10].

Coronary effluent: Glycerol was determined fluorometrically as described by Laurell and Tibbling [7].

Normalization of Data: Measurements in tissue samples are expressed in moles per gram dry weight. The dry weight of cardiac tissue was determined by freeze–drying overnight. Data concerning the release of substances in the coronary effluent are expressed as moles per gram wet weight. Therefore it is accepted that release per heart is equivalent to release per gram wet weight, as earlier findings demonstrated that hearts of rats of this age weigh approx. 1 g (0.97 ± 0.07; mean ± S.D.; unpublished observations). For conversion of wet weight to dry weight a factor 6.25 is used as calculated by the ratio of dry and wet weight data.

Statistical Analysis: Differences within groups were evaluated for significance using Wilcoxon's matched–pairs signed ranks test. Differences between groups were tested for significance using th Mann–Whitney U test and the multiple comparison test. P values less than 0.05 were considered to be statistically significant.

RESULTS

Using glucose as sole substrate, cardiac performance had slightly but significantly decreased during 120 min of antegrade perfusion (Table 1). Left ventricular developed pressure declined by 10.2%; cardiac output (CO) fell by about 11.1%. Addition of either lactate (5.0 mM) or pyruvate (5.0 mM) to the perfusion medium reduced the decline in function. In hearts perfused with the glucose–containing perfusion medium, endogenous substrate stores were found to become depleted (Table 2). During 120 min of normoxic perfusion the content of TG and GLYC decreased by 78 and 75%, respectively. When

Table 1: Functional performance of isolated, ejecting hearts during prolonged normoxic perfusion. The effect of lactate or pyruvate added to a glucose–containing perfusion medium.

	Glucose (11 mM)		Glucose (11 mM) + Lactate (5.0 mM)		Glucose (11 mM) + Pyruvate (5.0 mM)	
Time (min)	15	120	15	120	15	120
n	15	15	13	13	6	6
HR	278	280	278	286	272	270
	(266 –323)	(250 –306)	(248 –294)	(252 –300)	(256 –294)	(248 –316)
LVDP	12.8	11.5*	12.3	12.5	12.4	12.1*
	(12.1–13.3)	(11.1–12.1)	(11.6–12.9)	(11.3–12.8)	(11.2–13.5)	(10.4–12.8)
CO	64.5	57.4*	63.7	62.1	60.4	56.8*
	(60.6–67.2)	(54.9–63.1)	(58.8–67.4)	(55.5–67.4)	(53.6–68.0)	(50.0–65.0)

Heart rate (HR), LV developed pressure (LDVP) and cardiac output (CO) are expressed in beats/min, kPa and ml/min, respectively. Data are presented as median values and 95% confidence limits; n refers to number of experiments. *Indicates values significantly different from those monitored at t = 15 min.

Table 2: Myocardial tissue content of GLYC, TG, FA and G–3–P, of hearts freeze–clamped immediately after stabilization (t=0min) and hearts that were normoxically perfused for 120 min.

	Glucose (11 mM)		Glucose (11 mM) + Lactate (5.0 mM)		Glucose (11 mM) + Pyruvate (5.0 mM)	
Time (min)	0	120	0	120	0	120
GLYC	139	35*	145	184*	n.m.	232
	(114 –169)	(25 –71)	(108 –149)	(144 –197)		(183 –307)
TG	15.2	3.3*	21.6	19.0	n.m.	23.2
	(13.7–25.7)	(2.6 –6.6)	(15.1–31.9)	(11.5–34.4)		(21.1–25.3)
FA	72	70	115	217*	n.m.	254
	(40 –241)	(61 –114)	(64 –174)	(168 –267)		(241 –266)
G–3–P	0.12	0.35*	0.61	0.94*	n.m.	0.24
	(0.06–0.30)	(0.20–0.79)	(0.40–0.90)	(0.67–1.41)		(0.00–1.23)

GLYC is expressed as μmol glucose equivalents per g dry weight; TG as μmol FA equivalents per g dry weight; FA as nmol per g dry weight and G–3–P as μmol per g dry weight. *Indicates significant difference from values measured in hearts supplied with the same substrate(s) at t=0 min ($p<0.05$);n.m., not measured.

either lactate or pyruvate was added as cosubstrate to the perfusion medium both TG and GLYC stores were maintained. The level of FA remained unchanged in glucose perfused hearts, but significantly increased in hearts perfused with lactate as cosubstrate. The FA levels in pyruvate perfused hearts were comparable to those in lactate perfused hearts after 120 min of normoxic perfusion (Table 2). The tissue content of G–3–P increased in glucose and glucose plus lactate perfused hearts during 120 min of normoxic perfusion. The increase in the latter group surpassed significantly the value in glucose perfused hearts (Table 2). Measurements in coronary effluents, collected after the initial stabilization period, showed that the cumulative release of glycerol in glucose plus lactate perfused hearts was about 7 and 2.6 times higher than in hearts perfused with glucose plus pyruvate or with glucose as sole substrate, respectively (Table 3).

DISCUSSION

The present study shows that glucose as sole substrate, in the absence of insulin, is unable to meet the energy requirements of the isolated, ejecting rat heart during prolonged perfusion under normoxic circumstances. This observation supports earlier findings reported by other investigators [3, 4]. Endogenous substrate stores are depleted to maintain the supply of oxidizable

Table 3: Cumulative release of glycerol during 120 min of normoxic perfusion. Hearts were perfused with glucose (11 mM), glucose (11 mM) plus lactate (5.0 mM) or glucose (11 mM) plus pyruvate (5.0 mM).

Group	Glycerol Release (μmol/g wet weight, 120 min)		
	Median	95% Confidence Limits	n
Glucose	0.8	0.2–1.4	13
Glucose + Lactate	2.1*	1.3–2.6	13
Glucose + Pyruvate	0.3	0.0–1.0	6

*Indicates significantly different from values measured in glucose–perfused hearts ($p<0.05$);n refers to number of experiments.

compounds at the desired level. Both GLYC and TG, providing glucose and FA moieties, respectively, serve this purpose. When either lactate or pyruvate, both at the concentration of 5 mM, were added to the perfusion medium decline in function was less during prolonged normoxic perfusion, indicating that both substances perfectly serve as oxidizable substrates. This notion is substantiated by the observation that both substrates completely block the consumption of substrates, released from the endogenous stores, i.e. GLYC and TG. The exact site and mechanism of inhibition are incompletely elucidated. Pyruvate is known to stimulate its own intracellular conversion by keeping pyruvate dehydrogenase in an active state [11]. Since CoA is an important cofactor for the conversion of pyruvate into acetyl CoA, competition with FA activation, a reaction step that also requires CoA, might take place at the level of CoA utilization. The same may hold for lactate, once converted into pyruvate by action of lactate dehydrogenase. Moreover, conversion of lactate into pyruvate will enhance the cytoplasmic $NADH/NAD^+$ ratio. This, in turn, will enhance the mitochondrial redox state creating an ideal situation to inhibit the oxidation of FA. Bielefeld and coworkers [12] have suggested that lactate inhibits the enzyme carnitine acyl transferase I (CAT–I), located at the inner side of the mitochondrial outer membrane [13]. Their conclusion is based on the findings that the cytoplasmic concentration of acylcarnitine drops while the incorporation of fatty acyl moieties into the endogenous TG pool is enhanced in lactate perfused hearts. The nature of the inhibition of CAT–I remained unresolved. The present findings might indicate that lactate perfusion does not necessarily block CAT–I activity in a direct manner. It is quite feasible, that increased G–3–P levels in the cardiac cell will compete for acylCoA to produce TG. Reduced supply of acylCoA to CAT–I will subsequently result in a reduced supply of acyl moieties to the mitochondrial matrix.

The present study clearly indicates that the isolated, ejecting rat heart continously produces glycerol. Earlier studies of Denton and Randle [14] have shown that the release of glycerol from isolated rat hearts represents a reliable parameter to estimate the degradation rate of the endogenous TG pool. In this respect it is interesting to note that the total amount of glycerol released from hearts, perfused with glucose as sole substrate for 120 min, is about 0.8 μmol/g wet weight. This value corresponds well with the decrease of the TG pool (about 0.6 μmol/g wet weight, when expressed on glycerol content). Since the total amount of TG did not undergo a measurable change in pyru-vate or lactate perfused hearts, release of glycerol in these hearts indicates that degradation of the endogenous TG pool is fully compensated by concomitant resynthesis of TG molecules. This situation will create an operative TG–FA cycle, in which the rate of TG synthesis is in equilibrium with TG hydrolysis. Compared to pyruvate perfused hearts the turnover rate of the TG–FA cycle is about 7 times higher in the lactate group.

In order to elucidate the mechanism underlying the lactate–induced turnover of the endogenous TG pool both the hydrolytic and synthetic pathway have to be considered. Hydrolysis of the intracellular TG pool is mainly achieved by action of the lysosomal lipases in addition to intracellular neutral lipases [15]. The actual activity of these lipolytic enzymes is regulated by a variety of hormonal and metabolic factors. Recently, the suggestion has been made that FA are important regulators of cardiac TG hydrolysis [16, 17]. In vitro experiments have shown that FA inhibit lipase activity, suggesting that they regulate their own production rate by feedback inhibition. However, the present paper questions the physiological role of FA as inhibitory agents in the cardiac cells. Firstly, in lactate–perfused hearts endogenous FA levels are enhanced rather than decreased. This finding is difficult to reconcile with the notion that FA exert feedback inhibition of TG lipase as the production rate of glycerol is also enhanced in lactate perfused hearts. Secondly, Stam and Hülsmann [18] observed inhibition of lipase activity at FA levels of 250 to 2000 μM. Severson and Hurly [17] reported inhibition at a concentration of 100 μM FA. It is uncertain whether these concentrations will be reached in the normoxic cardiac cell. In the present study the total FA concentration (about 220 nmol/g dry weight, which corresponds to about 60 μmol/l intracellular fluid) in the lactate–perfused hearts does not reach those *in vitro* levels.

The increased levels of G–3–P in lactate–perfused hearts are most likely caused by an enhanced NADH/NAD^+ ratio in the cytoplasmic compartment. This will shift the equilibrium between dihydroxy–aceton phosphate and G–3–P towards the latter. It can be hypothesized that high cellular G–3–P levels stimulate the formation of TG from acylCoA and G–3–P by mass action. Enhanced TG production explains only part of the increased turnover rate of the TG–FA cycle. The increased degradation rate of TG still remains to be clarified. We would like to hypothesize that enhanced esterification of acyl moieties due to an increased intracellular content of G–3–P may convert part of the endogenous TG pool into a metabolically more active state. This newly formed pool of neutral lipids might, in turn, be more susceptible to lipolytic enzymes and as a consequence degradation of TG may occur at an increased rate [5].

Since TG–FA cycling is an energy consuming process the question can be raised whether activation of this cycle poses a burden on the energetic state of the lactate perfused heart. The release of glycerol in lactate perfused hearts amounted to 17.2 nmol/min per g wet weight (Table 3). This value corresponds to an intracellular release of FA of about 0.3 μmol/min per gram dry weight. Since the total FA content of the endogenous TG pool is in the order of 20 μmol/gram dry weight, it can be calculated that each minute about 1.5% of the total TG pool is cycled. For the conversion of 3 fatty acyl moieties into 3 acylCoA 6 mol ATP is used. As the production of 1 mol G–3–P requires the hydrolysis of 1 mol ATP, it can be inferred that 7 mol ATP is hydrolyzed for each mol of glycerol released from the heart. Hence, the production of 17.2 nmol glycerol/min per g wet weight of tissue corresponds with the consumption of 120 nmol ATP/min, g wet weight. Using the formula of Hütter and coworkers [19], the oxygen consumption by the isolated, working rat heart in the present study is about 7 μmol O_2/ min, g wet weight. Assuming an ATP over O_2 ratio of 6, the total amount of ATP produced by rephosphorylation of ADP is in the order of 42 μmol/min, g wet weight. This value indicates that about 0.3% of ATP generated is used for the lactate–stimulated "futile" TG–FA cycle. This calculation implies that increased production of glycerol in lactate perfused hearts has most likely no consequences from an energetic point of view. Pilot studies performed in our laboratory have revealed that lactate greatly stimulated glycerol release in the post–ischemic heart. It cannot be excluded that TG–FA cycling under these pathophysiological conditions adds to the impaired functional recovery of the heart after a transient ischemic insult.

ACKNOWLEDGEMENTS

The authors are indebted to Karlijn Dickison, Bianca Lahoye and Emmy van Roosmalen for their help in preparing the manuscript and to Peter Willemsen and Will Coumans for their skillful help in assessing the biochemical substances.

DISCUSSION

Dr. Marban: This is an interesing work. I wonder if you could speculate on why lactate vs. pyruvate might be important in fibrillation. Secondly, given that all the hearts perfused with glucose and lactate are fibrillating, it may be difficult to compare these data to control since fibrillation itself has such dramatic metabolic effects under certain conditions.

Dr. van der Vusse: I can not offer a satisfactory explanation for the lactate–induced fibrillation of post–ischemic hearts. Maybe, lactate creates a high redox–state in the heart. High levels of NADH (and concomitantly low levels of NAD^+) may favor the production of oxygen free radicals. In this respect, it is interesting to note that Vandeplassche et al. *[J Mol Cell Cardiol 1989;21: 383–392]* found increased activity of NADH–oxidase in the post–ischemic heart. Increased activity of this enzyme complex in combination with elevated NADH levels will likely result in enhanced production of oxygen free radicals.

Dr. Marban: Given the fibrillation phenomenon, how can you control for arrhythmia vs. substrate dependence?

Dr. van der Vusse: This is very difficult to do in our experimental set–up. We found in the present study that pyruvate perfused hearts did not fibrillate during reperfusion after 45 minutes of ischemia, whereas lactate perfused hearts did. In addition, the glycerol release in lactate perfused hearts was increased many fold in the post–ischemic hearts. If enhanced glycerol production is a sign of an increased turnover of the endogenous TG pool, then a futile energy–consuming cycle is present under these conditions.

Dr. Mowbray: Do you know whether, if you use pyruvate instead of lactate, you could also shut off FA oxidation if you use high enough levels? You showed that Dr. Neely and others showed that CPT_1 was inhibited by lactate. Would you like to speculate how the lactate works?

Dr. van der Vusse: Theoretically, the most important difference between lactate and pyruvate as exogenous substrates is in the first metabolic conversion step inside the cardiac cell. Lactate has to be converted into pyruvate prior to mitochondrial consumption in the citric acid cycle via acetyl CoA. The conversion of lactate into pyruvate most likely increases the cytoplasmic NADH level, which, in turn, may stimulate the conversion of dihydroxy aceton phosphate into G–3–P. Increased levels of the latter compound may stimulate the TG–FA cycle. Pyruvate is directly converted into acetyl CoA by the pyruvate dehydrogenase complex.

Lactate inhibits FA oxidation at the level of carnitine palmitoyl transferase as has been shown by Bielefeld and coworkers *[J Moll Cell Cardiol 1985;17:619–625]*. The precise mechanism of this inhibitory action is unknown. Pyruvate may act as an inhibitor of FA oxidation in a similar fashion.

Dr. Mowbray: Let us take it another stage further. If you lower the free FA release then you will allow the free FA synthesis and the first thing you get is Malonyl CoA production which is an extremely powerful inhibitor of Palmitoyl Carnitine transfer. Now you have got no FA oxidation in the mitochondria leading to the increase you see.

Dr. van der Vusse: Yes, that is correct.

Dr. Bassingthwaighte: I am a little uncertain about your use of the glycerol production as a measure of futile cycling. If I regard glycerol production to be a combination of lactate metabolism and triglyceride breakdown, then 2 things are going on. One is the use of lactate to produce glycerol in this situation and the other is a net breakdown of triglyceride. Would it be possible to do an experiment in triglyceride depleted hearts in order to see the effect of lactate as distinguished from the effect of net triglyceride breakdown.

Dr. van der Vusse: The tissue level of TG did not change throughout the experimental period in lactate perfused hearts. Therefore, release of glycerol can be considered as a sign of enhanced TG degradation and resynthesis, if the assumption is correct that TG is the sole source of glycerol in the heart.

Dr. Bassingthwaighte: In the absence of glucose?

Dr. van der Vusse: In the absence of glucose, GLYC might be the source of G–3–P, via glycolytically derived dihydroxy aceton phosphate.

Dr. Corr: Have you ever tried using norepinephrine treatment of hearts to stimulate the triglyceride lipase and see if you could increase glycerol release?

Dr. van der Vusse: This is a good suggestion. Pyruvate perfused hearts should be perfused in the presence or absence of norepinephrine or another lipolysis–stimulating hormone.

REFERENCES

1. Taegtmeyer H: Carbohydrate interconversions and energy production. *Circulation* 1985;72 (Suppl IV):1–8.
2. Bieber LL: Fatty acid and ketone metabolism. *Circulation* 1985;72 (suppl IV):9–12.
3. Neely JR, Morgan HE: Relationship between carbohydrate and lipid metabolism and the energy balance of heart muscle. *Annu Rev Physiol* 1974;36:413–459.
4. Van Bilsen M, Van der Vusse GJ, Snoeckx LHEH, Arts T, Coumans WA, Willemsen PHM, Reneman RS: Effects of pyruvate on post–ischemic myocardial recovery at various workloads. *Europ J Physiol* 1988;413:167–173.
5. De Groot MJM, Willemsen PHM, Coumans WA, Van Bilsen M, Van der Vusse GJ: Lactate–induced stimulation of myocardial triacylglycerol turnover. *Biochim Biophys Acta* 1989;1006:111–115.
6. Snoeckx LHEH, Van der Vusse GJ, Coumans WA, Willemsen PHM, Van der Nagel T, Reneman RS: Myocardial function in normal and spontaneously hypertensive rats during reperfusion following a period of global ischemia. *Cardiovasc Res* 1986;20 67–75.
7. Laurell S, Tibbling G: An enzymatic fluorometric micromethod for the determination of glycerol. *Clin Chim Acta* 1966;13:317–322.
8. Van der Vusse GJ, Coumans WA, Van der Veen FH, Drake A, Flameng W, Suy R: ATP, creatine phosphate and glycogen content in human myocardial

biopsies: markers for the efficacy of cardio–protection during aortacoronary bypass surgery. *Vasc Surg* 1984;18:127–134.
9. Roemen THM, Van der Vusse GJ: Application of silicagel column chromatography in the assessment of non–esterified fatty acids and phosphoglycerides in myocardial tissue. *J Chromatogr* 1985;344:304–308.
10. Van der Vusse GJ, Roemen THM, Prinzen FW, Coumans WA, Reneman RS: Uptake and tissue content of fatty acids in dog myocardium under normoxic and ischemic conditions. *Circ Res* 1982;50:538–546.
11. Bünger R, Mallet RT, Hartman DA: Pyruvate–enhanced phosphorylation potential and inotropism in normoxic and post–ischemic working heart. Europ *J Biochem* 1989;180:221–233.
12. Bielefeld DR, Vary TC, Neely JR: Inhibition of carnitine palmitoyl–CoA transferase activity and fatty acid oxidation by lactate and oxfenicine in cardiac muscle. *J Mol Cell Cardiol* 1985;17:619–625.
13. Murthy MSR, Pande SV: MalonylCoA binding site and the overt carnitine palmitoyltransferase activity reside on the opposite sides of the outer mitochondrial membrane. *Proc Natl Acad Sci USA* 1987;84:378–382.
14. Denton RM, Randle PJ: Measurements of flow of carbonatoms from glucose and glycogen glucose to glyceride glycerol and glycerol in rat heart and epididymal adipose tissue. *Biochem J* 1967;104:423–434.
15. Hülsmann WC, Stam H, Breeman WAP: Acid and neutral lipases involved in endogenous lipolysis in small intestine and heart. *Biochem Biophys Res Commun* 1981;162:440–448.
16. Schoonderwoerd K Broekhoven–Schokker S, Hülsmann WC, Stam H: Enhanced lipolysis of myocardial triglycerides during low flow ischemia and anoxia in the isolated rat heart. *Basic Res Cardiol* 1989;84:165–173.
17. Severson DL, Hurley B: Regulation of rat heart triacylglycerol ester hydrolases by free fatty acids, fatty acylCoA and fatty acylcarnitine. *J Mol Cell Cardiol* 1982;14:467–474.
18. Stam H, Hülsmann WC: Regulation of lipases involved in the supply of substrate fatty acids for the heart. *Europ Heart J* 1985;6:158–167.
19. Hütter JF, Piper HM, Spieckermann PG: An index for estimation of oxygen consumption in rat heart by hemodynamic parameters. *Am J Physiol* 1985;249:H729–H734.

General Discussion

CHAPTER 35:

GENERAL DISCUSSION

Samuel Sideman and Rafael Beyar, Moderators

PROPAGATION AND ARRHYTHMIAS

Dr. Sideman: Whether experimental or theoretically, we aim to determine the activation processes which affect the normal and abnormal behavior of the heart. The approaches vary in complexity and accuracy as well as in the level or scale of involvement; varying from ionic channel transport to the global 3D activation patterns and the body surface potential. Could we identify some of the particularly promising avenues of research?

Dr. Wit: We are all trying to get to an answer to what causes certain kinds of arrhythmias, with the final goal to be able to intervene and prevent the various kinds of arrhythmias. The particular kinds of arrhythmias that I have been interested in, which I think are the most important kinds of arrhythmias, are arrhythmias caused by reentrant excitation. If we could learn what causes these arrhythmias, we might be able to learn how to treat these arrhythmias effectively. In order to approach that sort of goal we have to go back to the very basics of impulse propagation because the characteristics of propagation, both normal and abnormal, are what eventually leads to reentrant excitation. We can approach the study of propagation on many different levels and the beauty of the first part of this symposium is that we saw the different levels that we can look at propagation and understand what the characteristics are. Hopefully we may be able to figure out how to modify the propagation in a controlled way, at will. Examples of that, beginning with membrane channels, Dr. Fozzard's wonderful expose on how the basic membrane physiologists are dissecting the properties of the membrane channel, particularly the sodium channel, to learn more about its characteristics. One of the primary goals of studying the characteristics of the sodium channel is to understand it enough so that, maybe, one could design specific chemical agents that might interact with the sodium channel in such a way as to influence propagation and influence arrhythmias. We can also study propagation from the point of view of intracellular connections and we have learned how very important it is to look at intracellular connections, and how they affect propagation. One of the many excellent presentations was the presentation of Dr. Weingart, on the properties of the channels of the gap junctions. Again, if we understand more about how these properties work, we may some way intervene and alter these properties and eventually influence reentrant excitation and various kinds of arrhythmias. We have also learned that a very useful approach to study propagation is by computer modeling and there are many things that many of us experimentalists would like to do in the experimental laboratory but are not capable of doing. We can learn more about propagation and the causes of arrhythmias through models. We have heard some excellent presentations with different points of view, different approaches, all aimed towards understanding propagation more, from Bob Plonsey, Ron Joyner and Yoram Rudy and Franz van Cappelle. It is impossible for me to summarize what we have learned from their methods, but we have learned that to get the kinds of conduction delays that are necessary for reentrant arrhythmias, it is likely that something has to happen to cause abnormalities in intracellular connections. Finally, on the next level we have seen what is going on with the studies on propagation in the intact heart and how can using electrophysiological mapping show us what is going on during different kinds of cardiac arrhythmias, and during acute ischemic arrhythmias, during more chronic kinds of arrhythmias. We have learned that tissue anisotropy may, perhaps, be an important factor in causing reentry. I have learned that antiarrhythmic drugs do not affect conduction in real reentrant circuits in the way that I would like to see them affect conduction in order to stop reentrant arrhythmias. I am hoping to learn from the people who are studying channels and those who are doing modeling, what has to be done in order to stop these very important rhythm disturbances.

To summarize, a multi–faceted approach towards understanding propagation, that is from channel studies, computer modeling, and whole animal studies, is crucial in terms of what the final goal of many of the people who are studying propagation, and that is to be able to effectively treat arrhythmias in the future.

Dr. Sideman: The understanding of the "cell to system" relationships is one of the major goals of these Henry Goldberg meetings. We wish to identify and hopefully control the interaction and signal transfer from the elementary cell to the integrated performance of the cardiac system.

Dr. Noble: There has been a lot of literature recently about calcium overload causing a transient inward current following repolarization, which causes after–depolarizations and arrhythmias from that mechanism, quite different from the reentrant mechanism. That has not been mentioned at all in this symposium and I wonder whether this is unimportant in real life or whether it was just a subject that could not be fitted in.

Dr. Wit: On a cellular basis, the study of after–depolarizations has provided many fascinating hours for many people, including myself. In as much as we can deduce mechanisms of arrhythmias from clinical studies, and certainly there is a limit on that, the evidence seems to suggest that at least the delayed after–depolarizations have minimal importance in terms of causing life threatening arrhythmias. They have been identified as a certain cause of cardiac arrhythmias resulting from digitalis intoxication and they have been identified as a likely cause of a type of ventricular tachycardia that is not associated with any specific kind of cardiac disease and not associated with sudden death. But I do not believe that the clinical cardiologists who are interested in arrhythmia mechanisms think at this time that triggered arrhythmias are an important cause of life threatening rhythm disturbances.

Dr. Rosen: From the point of view of proving that clinical arrhythmias, or a subset of them, are triggered, I would agree with Dr. Wit. There is very little proof that a major subset of arrhythmias is induced by triggered activity. Certainly that is true in the ischemic setting and, as Dr. Wit said, with a few exceptions, like digitalis toxicity or a subset of digitalis toxicity, and a subset of exercise induced arrhythmias, you could say that the rules for reentry are followed. But then you have the question of what triggers arrhythmia, what induces arrhythmia, even if it is reentry. There is still room for argument, discussion and research. People who have been working with various types of recording techniques to look at monophasic action potentials have been able to see after–potentials both in animal preparations, whole animals and in patients. One of the problems with many of the measureing techniques is that it is still very hard to differentiate artifact from delayed after–depolarization. Thus, I agree with Dr. Wit that triggered activity has not been shown to be a major cause of lethal arrhythmias, but we have to look further.

Dr. Bassingthwaighte: The analyses presented by Dr. Goldberger and Dr. Jalife are very important. These are nonlinear processes, and the simplest expression of these was given by Lorenz in 1963 by 3 non–linear differential equations which give rise to an apparent chaotic and quite unpredictable process, but limited. Cellular reactions of a biochemical or physicochemical nature are low order reactions. It is very important to appreciate the power of their approach; you can use it to discover new things and look at phenomena in a way that is not obvious in our traditional ways. In the Belousov–Zhabotinsky reaction this kind of thinking is applicable because the system is so simple and you can understand its kinetics thoroughly. That is not true in a more complex system. For example, calcium transport in contraction involves SR release, reuptake and so on. The process has all the prerequisite features, namely nonlinear excitability, regenerative depolarization, energy–dependent relaxation, etc. Such processes should be considered from the point of view of non–linear dynamics, a new view that helps develop new insight.

Dr. Strauss: Once you have this kind of analysis, how you approach and go about trying to figure out what the underlying mechanism is. For instance, heart rate variability. Is there anything in that analysis which tells you what the mechanism underlies the reduction in heart rate variability with advanced age? People have done frequency analysis of heart rate variability and have correlated the change in frequency analysis with change in vagal tone. That has been done through characterization in a variety of different ways and they have deduced that when heart rate variability is reduced, then the vagal tone is reduced. How would you treat this similarly in a system or how would you go about trying to deduce underlying mechanism if you were just given an analysis of the records, without having access to prior data.

Dr. Goldberger: Vagal tone, by itself, has been correlated previously with higher frequency fluctuations (e.g. about 0.2 Hz) associated with respiratory sinus arrhythmias. However, the normal heart rate spectrum is broadband, including very low frequency (<0.0001 Hz), as well as higher frequency fluctuations. This spectrum has a 1/f–like distribution since there is an inverse relationship between frequency and power when the data are plotted on double–log axes. The respiratory peak is only one small component of overall heart rate variability.

The traditional approach has been to say that much of physiological heart rate variability, aside from respiratory sinus arrhythmia or oscillations associated with baroreceptor control, is "noise" due to external influences. According to this standard *homeostatic* model, fluctuations or deviations from the mean are thought to be induced primarily by *external* influences and the heart rate would settle down to a steady state if left unperturbed. This is the classic view of Walter B. Cannon that is still prevalent. The key difference in the approach of those of us who subscribe to *chaos theory* as a potentially important physiologic mechanism is to say that you can get complex variability from an *intrinsic* type of nonlinear feedback process. Therefore, it should be possible to write out the equations of motion that could generate complex variability such as one might see with a healthy heart rate. In this model, the variability that generates the broadband 1/f–like spectrum would not be due to extrinsic factors but would be *deterministic*.

So far, no one has succeeded in writing down those equations of motion for the heart beat. That is one of our interests, to see if one can in fact model the types of variability, that have previously been attributed to noise, as well as complex periodic patterns that have thus far defied mechanistic interpretation. The next step would be to alter the parameters of that model to see if we can replicate the loss of variability that one sees under certain pathologic conditions, including aging and patients at high risk of sudden death. Further, chaos theory offers new ways of analyzing time series that go beyond t–tests or spectral analysis. These new methods of assessing complexity, including measurements of Lyapunov exponents, entropy and dimension, should provide new ways of determining whether data sets are, in fact, generated by chaotic mechanisms and guide the mathematical modeling.

Dr. Jalife: My laboratory has demonstrated that cardiac Purkinje fibers can undergo chaotic dynamics when stimulated repetitively with electrical stimuli of rhythmic frequency. We have developed a unidimensional model that accurately describes the dynamics and properties of the system. It only looks at behavior. And it looks at behavior in terms of recovery of excitability, action potential duration and latency to excitation. If you just take those three parameters, analyze their frequency dependence and include them in a difference equation, you can reproduce the dynamics very accurately and you can demonstrate that at certain stimulation frequencies, the behavior corresponds to low dimensional chaos in the mathematical sense of the word. So cardiac tissues do indeed have intrinsic behavior which may be under certain conditions chaotic with high sensitivity to initial conditions.

MUSCLE FUNCTION AND CORONARY FLOW

Dr. Reneman: The nice thing in this meeting is that we started out with mechanical properties of fiber shortening and Dr. Pollack gave us a very nice survey on the pros and cons of the sliding filament theory. We have to read the basic concept in his book. I am not surprised there is shortening of the thick filaments because I know that data, but I am surprised with the statement that there is no stiffening in the lateral direction of the thick filaments. That might have some interesting consequences. This brings me to the 4 papers by Westerhof, Spaan, Sipkema and the modeling of Dr. Beyar. Although I decided 20 years ago not to spend more work on intramyocardial pressure anymore, because of all the frustrations we went through in the 60's trying to measure it, it is fashionable again and we have to think about it. What we have learned here is that we don't disagree anymore about intramyocardial pressure being determined by left ventricular pressure, but we now know, what we knew 20 years ago when the implications were there, that under certain circumstances there might be an uncoupling between the IMP and the LV cavity pressure. What we still do not agree upon is what is the mechanism that generates intramyocardial pressure and left ventricular pressure. Dr. Westerhof feels that compliance is the most important parameter. Alternatively you can think about the generation of pressure in a fluid compartment in the wall, or a fluid compartment being the left ventricular cavity. You need steps between the generator of the pressure in the fluid compartment and the intramyocardial pressure having the effect on the microvessels. I wish Drs. Westerhof and Spaan success in selecting the experiments needed to resolve that particular question.

We had several papers on flow and Dr. Bassingthwaighte presented the intriguing inhomogeneity in the permeability of the surface area. If you do the experiments as they are being done in Dr. Bassingthwaighte's department to increase flow and oxygen demand, and the first experiments indicate that inhomogeneity of flow is significantly diminishing when you increase the oxygen amount to the myocardium, it is very interesting to know whether the FS inhomogeneity is diminishing too. I agree with Dr. Bassingthwaighte that those are the experiments to look at.

Next we move to the flow in the hypertrophied heart. There we feel that the flow is the limiting factor following the ischemic insult and you have to find a way to maintain flow at low afterloads. Dr. Hudlicka's work relates to this. Last but not least, we had a flow study that was very intriguing to me. Dr. Welkowitz showed that you can generate enough energy at very low flows in the coronary arteries, sufficient to generate resonance of the arterial wall. You need a very significant constriction, down to 60% diameter reduction corresponding to about 85–90% surface area reduction, before you get enough energy in fluid turbulence to generate resonance in the wall. It might have something to do with the fact that the coronary arteries are basically muscular arteries, probably stiffer than the carotids, and are easier to generate resonance. But it is unlikely that you get turbulence at this degree of narrowing. I wonder where the sounds come from in the blood. Obviously, there is much to learn here.

CA^{2++} AND THE MEMBRANE

Dr. Marban: We have learned here about calcium in the cell and how it is handled. We have heard about the importance of voltage–dependent calcium cycling and the relative roles of calcium entry and release from Dr. Noble, whose work, I think, is built upon the work of Gibbons and Fozzard in the 70's in this area. Dr. Strauss has emphasized the importance of figuring out, in a rigorous way, how cation balance across the SR is handled during the process of calcium release; what counterbalances the enormous electrochemical gradient that must be generated by the release of calcium across the sarcoplasmic reticulum? His elegant permeation studies help us to understand how the cell manages this incredible feat. Dr. Gettes and I emphasized the importance of the regulation of calcium during hypoxia and ischemia and, particularly, how important it is to understand the interrelationship of calcium and energy metabolism if we are to understand what actually causes damage to the cell, and how contractile dysfunction is generated in post–ischemic myocardium. Dr. Rosen emphasized the α–adrenergic mechanisms and the importance of G–proteins in the pathways of arrhythmogenesis. Dr. Corr did not talk about that particular aspect of his work in this symposium but he has related α–adrenergic mechanisms to calcium handling in a very nice way elsewhere. Dr. Lerch emphasized the importance of mitochondrial calcium transport and post–ischemic recovery. Finally, Drs. Kammermier and Mowbray emphasize the importance of considering the energetics of ion transport, in any kind of scheme that we may come up with, in figuring out how the cell maintains its ionic gradients against all sorts of challenges. Dr. Kléber, although he did not emphasize this aspect in this particular presentation, has made major contributions in this area.

Dr. Strauss: It has been suggested that triggered fibrillation affects changes in calcium in and outside the cells. Have you looked at the VF cycle length over time following such changes?

Dr. Marban: Measurements of intracellular calcium and high energy phosphates during electrically induced ventricular fibrillation in isolated hearts show the calcium rises dramatically. Mean [Ca^{2+}] goes up by a factor of 10–12 from control levels; it has a rather quick rise in about 3 minutes and then it stays stable. If you unload the heart, i.e. if the heart is not pumping against the isovolumic load, ATP plus phosphocreatine, P_i and pH stay constant. But the increase in calcium would be expected to transfer into changes in membrane properties at different levels, including calcium mediated effects on potassium channels or on electrogenic transporters. Another point would be that you might expect changes in longitudinal resistance or the cell–cell coupling if you have dramatic changes in [Ca^{2+}]. The changes in [Ca^{2+}] during ventricular fibrillation are bigger than the changes we measure in total ischemia at 15 minutes.

Dr. Janse: It is very interesting to hear that the changes in intracellular calcium occur within 3 minutes and then remain stable. This is exactly what the VF interval does in our experiments. I do not know whether there is a change in intercellular coupling and whether this might affect conduction. What we would like to have, and what we do not have yet, is some sort of index of conduction velocity during ventricular fibrillation. This is very difficult to achieve in a three–dimensional event such as ventricular fibrillation. If we could use Dr. Allessie's method of freezing the inner 2/3rd or 4/5th of the ventricular wall and still have enough muscle mass to maintain fibrillation, you might be able to say something of the conduction velocity of reentrant wavelets during fibrillation. It would be very interesting to see if something changes in the conduction velocity in the first 3 minutes, which might be related to partial uncoupling. At this point, however, I do not know.

I agree that particularly in ischemic myocardium, the VF interval is not only determined by the excitability of the tissue but also by the stimulating efficacy of the wavefront, which is certainly reduced in ischemic tissue. It would be interesting to see whether there is an excitable gap in ischemic tissue during fibrillation, and to see whether one could capture the ischemic myocardium by very strong electrical stimuli during fibrillation. We have not done this yet, but it should be done.

METABOLISM AND TRANSPORT

Dr. Bassingthwaighte: Metabolism and transport touches on everything we talked about throughout this meeting, starting with ionic transport and ending up with metabolic transport. However, we missed a few topics and I will try to pin–point a few of them. One subject that we ought to have touched on is the relationship between the mechanical and the electrical aspects of the heart. It was brought up in passing but nobody talked much about it. On the other hand, the relationship between the ionic events and the metabolic events did come out very clearly throughout the meeting. These are obviously fundamental, not only in relating the energetics to the ion transport and the maintenance of the cells internal ionic milieu but also to the events in ischemia. We use the damaging of the membrane in ischemia to see what happens to the ionic transfer and also the substrate transfer. In ischemia, even large proteins escape from the cell and the environment for the ion channels is changed, as outlined by Dr. Corr.

The energetics of the cell membrane regulation of the ionic transfer was discussed in the papers by Drs. Kammermeier, Lerch, and van der Vusse. Dr. Kammermeier asked us to please put it all together, and that is what we here are trying to do. In trying to focus on the energetic balance and to examine each of the mechanisms for ionic transfer, he showed the role of the lactate/proton transfer and a variety of other transport mechanisms which are energy coupled. Those I regard as key events. Dr. van der Vusse made a most provocative observation on the relationship between the ions and the substrates. Why is it that lactate–supplied hearts do not recover after ischemia, but fibrillate? Why do they fibrillate whereas the pyruvate–perfused ones seem to recover normal rhythm?

Amongst the biochemical events, the one most interesting and most novel was that of Dr. Mowbray on the oligomeric ATP. Perhaps this solves the mystery of the NMR invisible phosphorous. His observations stimulate a broadening of the discussion of where and how these high energy phosphates operate in the cell. One of the things we did not get to here was intracellular micro–compartmentalization. We failed to talk about the sub–sarcolemmal ionic concentrations and the distinction between sub–sarcolemmal mitochondria and ATP generation for ionic transport in contrast to the ATP generated in deeper mitochondria and utilized by the myofilaments. We do not have clear data on these phenomena. We will have to do different experiments from what most of us try to do, in order to look at subcellular compartmentalization of substrate utilization and ion transfer. Those are going to be some of the puzzles that we have to solve in the future.

Dr. Schelbert discussed the relationships between the different substrates in oxidative metabolism; Drs. Lerch and Kammermeier carried it further. That trade–off is not very well characterized in normal or ischemic hearts. How this is regulated is going to be very important to our future discussions, an exciting thing to look at in the future. Coming out of the metabolism and transport aspects of this meeting were as many new problems as there were solutions.

We have also not talked here about the various kinds of receptors and what they mean in terms of local effects. A major factor in this conference has been arrhythmia and its generation. Dr. Noble brought up the question of local events. We know that arrhythmias are provoked by heterogeneities in the properties of the tissue, particularly in the post–ischemic situation, and, that might be a topic for further discussion in other meetings. There are these ischemia–induced heterogeneities, but heterogeneities exist normally. Are these related to heterogeneities in receptor densities or degrees of activation? There is much left to do!

Dr. Mowbray: Ischemic myocytes throw away adenine nucleotides, some of which ends up in endothelial cells. Do you suggest that these have a special role in organizing recycling nucleotides?

Dr. Bassingthwaighte: It is a big mystery why ATP is in such high abundance in endothelial cells in which the metabolic requirements would be zero for contractile purposes and relatively modest for other purposes. The fact is that the concentrations are very high there and I do not know why.

Dr. Strauss: It is even more so as far as the degradation of ATP following an hypoxic or ischemic insult.

Dr. Bassingthwaighte: It is intriguing to me how rapid these kinetics are. These events can be extremely rapid and the "why" is very provoking. It is true that endothelial cells are involved in huge numbers of reactions that govern both myocyte and vascular responses, but that is only pushing the question harder. I do not have answers as yet.

CLOSING REMARKS

Dr. Marban: I am sure that I speak for all of us here that one of the most exciting things in this meeting, under the auspices of the Henry Goldberg Workshop, is the opportunity to listen to subjects that one would ordinarily not have a chance to listen to; it expands our horizons to have done so. I would like to thank the organizers who made it all possible and express our appreciation to them for making this workshop a most enjoyable affair and a most exciting scientific encounter.

Dr. Sideman: I would like to thank the members of Scientific Committee and the Organizing Committee who contributed their time and wisdom to assure the success of this meeting. Particularly, I would like to thank my Co-Chairmen, Dr. Rafael Beyar for his continued pursuit of the goals of the Henry Goldberg Workshops and to Dr. André Kléber for his unique contribution to the success of this meeting and the making of the fine local arrangements. In memory of this beautiful occasion we would like to present him with the a medal, which reads:

To André Kléber:
A Medal of Friendship
Commemorating the Joint Technion Israel Institute of Technology,
and the University of Berne
7th Henry Goldberg Workshop."

Finally, thanks are due to each and every participant, whose stimulating presentations made this 7th Henry Goldberg Workshop a most memorable and exciting scientific event. This stimulating encounter of scientists from the different disciplines of cardiac research brings to mind the words of the Nobel Laureate Francis Jacob in his book "The Statue Within":

> ...What matters more than the answers are the questions and how they are formulated; for in the best of cases, the answers lead to new questions...

Last, but not least, heartiest thanks to Ms. Deborah Shapiro, Secretary of the Heart System Research Center in the Technion, for her diligence and painstaking efforts to bring all of these words together and make this book a reality.

Index

INDEX

A

A–band, 215, 219, 221, 225
Abdominal aorta, 300
Acetate (C–11 labelled), 319, 322, 325, 329
Acidosis, 110, 196, 200, 196, 197, 200
  intracellular, 110
Action potential, 14, 30, 39, 44, 47, 59, 68, 83, 84, 87, 88, 92, 94, 116, 138, 160, 199
  amplitude, 117, 120
  duration (APD), 48, 49, 52, 63, 122, 163, 164, 170, 197
  monophasic, 367
  reentrant, 68, 70
  upstroke, 109, 110, 111
Activation, 17, 56, 65, 107, 174, 188,
  electrical, 13
  front, 22, 74, 77
  map, 39, 42
    colorcode, 103
  pattern, 47, 102, 103
  premature, 20, 107
  ratio, 116, 118
  reentrant, 76
Acylcarnitine (LCK), 195, 200, 356
acylCoA (acyl–CoA), see Carnitine
Adenine, 205, 206, 208, 212, 213, 297, 302, 373
  nucleotides, 205, 373
Adenosine, 158, 205, 206, 208, 209, 210, 211, 213, 291, 297, 307, 312, 341, 358
  diphosphate (ADP), 342
    ribosylation, 170, 174
  monophosphate, 158
  triphosphate (ATP), 158, 190, 199, 205, 213, 290, 298, 334, 342, 344, 358, 371, 373
    hydrolysis, 190, 350
    myocardial, 301, 302, 303
    synthesis, 341
Adrenergic,
  α, 169, 171, 173, 174, 175, 371
  β, 169, 171, 173, 175, 180, 192, 197
Aequorin, 156, 158, 191, 192
After–depolarization, 58, 198, 366, 367
  delayed, 198
Aging, 126, 131, 135, 369
Alinidine, 307, 310, 312
Alkaline phosphatase, 308
Alkylating agent, 174
Amino acids, 140, 322, 351
Amiodarone, 179, 180
Ammonia, 292, 326, 327, 329, 330
  nitrogen–13, 327, 330
Anatomic block, 37, 41, 42, 44
Anatomical circuits, 44
Angiogenesis, 307, 312
Angiograms, 262, 271
Angioplasty, 261, 265, 266, 270, 272
Anisotropy, 7, 13, 23, 29, 33, 45, 101, 105, 106, 366
Antiarrhythmic drugs, 44, 137, 180, 366
Antibodies (Ab), 141, 143, 169, 173, 175, 344
Antidepressants, 181, 291
  tricyclic, 181
Aorta, see Aortic
Aortic,
  flow, 353
  pressure, 239, 263, 296, 352
  valve, 308, 309
APD, see Action potential duration
Arrhythmia, 15, 63, 115, 123, 169, 202, 289, 337, 365, 373, see also specific types
  functional, 101
  ischemic, 366
  lethal, 176, 196, 200, 367
  malignant, 195
  reentrant, 29, 73, 77, 366
  responses, 58
  sinus, 368
  ventricular, 198
Arrhythmogenesis, 195, 202, 371
Arterial,
  blood, 321
  collaterals, 246
  flow, 229, 274, 276, 277
  resistance, 256
  tree, 262
Arteries, 231, 247, 248, 251, 296, 370, see also Arterial, Arteriolar
  constricted, 261

pulmonary, 282
stenosed, 265, 266
Arteriolar, 231
resistance, 252
Arterioles, 231, 233, 256, 263, 286
Aspartate, 89, 90
Atrial,
cannula, 296
fibrillation, 32
filling pressure, 352
tissue, 47
Atrio–venous extraction, 284
Automaticity, 20, 25, 58, 169, 171, 174, 175, 176, 177
Autoradiography, 195, 203
Autoregressive, 265, 270, 272
Autoregulation, 233

B

Back pressure, 246, 273
Balloon, 186, 275, 327, 335, 337
angioplasty, 265
Basal activation loop, 40
Beeler–Reuter (BR), 14, 16, 25, 48, 49, 51, 64, 72
Bicarbonate, 320, 322
chloride exchanger, 346
Bifurcation, 117, 118, 121, 122
Blood flow, see Circulation; Coronary blood flow; Myocardial blood flow
Body surface potential, 9, 25, 365
Body weight ratio, 297, 300, 308, 310
BR, see Beeler–Reuter
Bradycardia, 307, 308, 312, 313

C

Cable theory, 96
Ca^{++}, see Calcium
Caesium, 153
Caffeine, 146, 224
Calcium, 111, 138, 155, 156, 159,
activation, 224
channel, 180
blocker, 180
concentration, 341
dependent, 193
entry, 370
exchange, 162, 186
extracellular, 96
intracellular, 89, 109, 110, 112, 160, 185, 198, 371
ions, 156, 174, 338
mitochondrial, 333, 334, 335, 338, 339, 371
overload, 193
release, 145, 146, 148, 152, 158, 160, 161, 165
transient, 185, 187, 188, 190
transport, 367
uptake, 348
Capacitance, 26, 54, 58, 83, 246, 248, 256, see also specific types
Capillaries, see Capillary
Capillarization, 310, 311, 312
Capillary, 107, 232, 247, 281, 291, 304, 309, 312, 314,
density (CD), 288, 308, 310, 311, 313
growth, 307
network, 286, 307
tissue, 285
Carbon–11 (C–11), 319,
acetate, 321, 325, 328
Carbon–14 (C–14),
labeled carbon dioxide, 323
Carbon dioxide, 323
C–14 (^{14}C), 335
Cardiac,
activation patterns, 47
arrhythmias, 101, 115
automaticity, 169
contraction, 160, 229, 237, 273, 278
electrophysiology, 125
excitation, 63, 115, 120, 121, 137, 160
fiber, 63
lymph, 198
muscle, 215, 234
output, 308, 354
physiology, 132
sarcolemmal membrane, 180
sudden death, 125, 195, 199
tissue, 4, 8, 115
work, 308, 322
Cardiomyopathy, 350
Carnitine acyltransferase (CAT I), (acylCoA), 197, 199, 322, 356, 358
Carotid artery, 261, 308, 370
Catecholemine, 158, 175, 196, 197, 303, 313, 315

Cationic flux, 145
Cations, 89, 135, 145, 159, 170, 174, 370
monovalent, 284
selective, 153
Cavity pressure, 235, 274, 277
Cell,
isolated, 58, 83
myocardial, 158, 236
smooth muscle, 198
surface, 288
two coupled, 57
Cell–to–cell, 13, 54, 63, 71, 83, 94, 97, 371
coupling, 63
Cellular,
activation, 3, 5, 9
membrane, 147
transport, 216
uncoupling, 63, 67, 68
uptake, 281
Cesium, 89
Channel selectivity, 141
Chaos, 115, 118, 120, 126, 130, 135, 368, 369
Chloride, 89, 90, 346
Chloroethylclonidine (CEC), 174, 178
Chordae tendenae, 126
Circulation, 72, 102, 135, 245, 246, 274, 277, 278, 286, see also Blood flow; specific types
Circumflex coronary artery, 287
Circus movement, 30, 63, 70, 106
Class IC Drugs, 44
Clearance curve, 322
Collagen, 253, 258
fibers, 245, 235, 255
Collaterals, 246, 247, 252, 257
Compliance, 230, 274, 277, 369, see also specific types
Conductance, 65, 67, 83, 84, 145, 148, 174
gap junctional, 48
single channel, 87, 139, 149, 150
Conduction, 4, 14, 26, 50, 53, 59, 64, 70, 72, 87, 91, 127, 133, 137, 366, 371
anisotropic, 29, 102
block, 30, 37, 40, 63, 101, 102, 105, 198
decremental, 63
longitudinal, 101
network, 128
transverse, 101, 102
velocity, 15, 17, 20, 23, 38, 39, 43, 66, 73, 78, 91, 95, 96, 97, 103, 104, 109, 110, 372
ventricular, 174
Conductivities, 5, 9, 19, 78
Congestive failure, 125
Contractile, 158, 162
activation, 189
elements, 145, 164
failure, 189
force, 188, 189
function, 335, 336
mechanism, 215
performance
work, 333
Contractility, 229, 240, 245, 273, 275
Contraction, 156, 158, 218
cardiac, 145, 273, 278
isobaric, 237
Core conductor, 92, 93
Coronary, 353, 369,
arterial tree, 126, 261, 262, 270
arteries, 109, 370
artery disease (CAD), 270, 326, 328
blood flow (CBF), 261, 273, 308, 312
circulation, 245
constricted arteries, 261
dilators, 307
flow, 96, 110, 186, 229, 233, 237, 261, 264, 276, 278, 295, 302, 304, 337, 369
impediment, 241, 275
reserve, 274
ligation, 109
occlusion, 334
pulsatile flow, 273
resistance, 230
sinus catheter, 323
stenosed segment, 263
thrombus, 195
vascular compartment, 230
vascular resistance,
venous tree, 126
vessels, 237
Creatine phosphate, 299, 300, 335
Cross–bridge, 215, 222
theory, 221
Current, 84, 93, see also specific types
clamp, 48, 53, 83, 84, 85

injection, 86
ionic, 64, 69
second inward, 155, 161, 163
single-channel, 139
threshold, 56
transient inward, 366
transmembrane, 147
voltage-peak, 138
Curvature, 8, 78
Cycle length, 45
Cyclic AMP, 196
Cytoplasm, 158, 334
Cytoplasmic, 16, 141, 143, 149, 158, 188, 199, 338, 356, 357, 359
resistance, 83, 94

D

Deformation, 258
Dehydrogenase, 290, 297, 356, 359
Depolarization, 39, 58, 75, 86, 127, 129, 133, 139, 140, 155, 157, 164, 181, 304
regenerative, 367
Deterministic behavior, 119, 122
Dextrose, 48, 170
Diastolic interval, 121, 122
Dichotomous branching, 127
Digitalis toxicity, 367
Dihydropyridine, 180, 183
Dipyridamole, 307, 315
Dobutamine, 313, 325, 328
Double-layer, 5, 7

E

Ectopic activity, 10, 21, 23, 58, 61
Elastance, 230, 240, 274, 276, 277
time varying, 233, 275, 278
varying, 237, 238, 240
Electrical analog, 246
Electrocardiogram, 10, 129, 170
Electrochemical, 344
gradient, 137, 342, 370
Electron microscope, 195
Electron spin resonance, 196
Electron transport, 199
Electrophysiologic, 35, 76, 195, 199, 200
abnormalities, 289
derangement, 196
Electrotonic, 48, 51, 87
coupling, 50, 89
response, 86
Ellipsoid, 20, 103, 104
shape, 17
ENDO, see Endocardial; Endocardium; Subendocardium
Endocardial, 4, 10, 19, 22, 45, 75, 91, 103, 231, 232, 233, 239, 249, 275, 300, 313
epicardial flow ratio, 299
myocytes, 98
vessels, 229
Endocardium, 98, 233, 249, 256
End-systolic, 240, 278
Energy, 112, 148, 210, 271, 290, 298, 328, 338, 351, 370
force, 190
free, 343, 344
metabolism, 289, 341, 343
transport, 342
utilization, 190
EPI, see Epicardial; Epicardium; Subepicardium
Epicardial, 10, 247, 252, 275, 279, 300, 313
border zone, 37, 39, 42, 43, 103
capacitance, 246
excitation, 108
Epicardium, 10, 98, 101, 233, 249, 313,
Ethanol, 208, 307, 315
Ether, 296, 352
Exchange pump,
sodium/calcium, 163
sodium/potassium, 170, 174
Excitability, 48, 66, 116, 118, 120, 122, 137, 197, 140, 367, 369, 372
Excitation, 40, 50, 70, 93, 107, 115, 120, 133, 162, 175, 262, 264, 365, 369
contraction coupling, 163, 185, 189
Extracellular, 7, 32, 58, 72, 83, 91, 105, 109, 138, 141, 161, 165, 182, 196, 199, 200, 343, 346, 351
bipolar electrogram, 95
conductivity, 5
resistance, 91, 95, 96, 110
space, 156
Extra-systolic potentiation, 241

F

Fareday constant, 343
Farey tree, 116, 118
Fatty acids (FA), 112, 196, 198, 287, 322, 328, 351, 353
  utilization, 334
Fatty acyl chain, 181
5–F–BAPTA, 185, 186, 190
^{19}F–NMR, 185
^{18}F–Deoxyglucose, 326, 327
Ferret hearts, 155, 186
Fetal distress, 125
Fiber,
  direction, 101, 107
  muscle, 219, 253
  shortening, 369
Fibrillation, 15, 32, 35, 44, 101, 199, 202, 301, 338, 358, 359, 371, 372
Finite–elements, 15, 25
Flecainide, 44
Flow measurement, see Blood flow
Flow reserve, see Blood flow
Fluid–fiber model, 246, 252, 255
Force generation, 185
Fractal, 126, 130, 135, 293
  mechanism, 125
Fractal–like, 127
Freeze–clamped, 297, 352, 355
Functional,
  block, 37, 38, 45
  circuits, 37, 44

G

Gap, 31, 34, 53, 80, 82, 85, 155, 156, 372
  channels, 88
  conductance, 48, 86
  junctions, 83, 86, 88, 91, 366
  resistance, 67, 72, 83, 84
Gating, 16, 69, 137, 141, 147
Gaussian distribution, 17, 74, 75
Glucose, 334, 353, 355
Glutamine, 142
Glycerol 3–phosphate (G–3–P), 353, 356
Glycogen, 347, 351, 353

H

Heart rate, 125, 128, 130, 132
  variability, 368
α–Helical, 141
Heparin, 48
Heptanol, 48, 87, 102
Heterogeneity, 105, 202, 282, 287, 290, 373
  intramitochondrial, 288
Hidden flow, 252
His–Purkinje, see Purkinje
Hodgkin–Huxley, 140, 143, 215, 220, see also Models
Homeostasis, 129, 135, 344
Homogeneous, 69, 71, 74, 219, 300, 313, 326, 328
  conducting layer, 74
Hydrogen, 142,
  ions, 142
Hypermetabolism, 333,
Hypertension, 295,
  renal, 308
Hypertrophy, 295, 297, 302, 307, 308, 370
  rabbit hearts, 309
  volume overload, 313
Hypoxanthine, 205, 297
Hypoxia, 199, 200, 201, 343, 346, 370, 373

I

Imaging, 320, 324, 334
Impedance, 288, 289
Impulse, 15, 31, 35, 37, 83, 85, 95, 96, 109, 170, 365
  conduction, 91, 101
  initiation, 169, 176
  propagation, 83, 85, 109, 111
  transfer, 87
Indicator dilution, 282
Infarction, 37, 38, 42, 44, 63, 101, 195, 261, 312, 326, 329, 339
  healing, 37
Infarcts, see Infarction
Inhomogeneities, see Inhomogeneity
Inhomogeneity, 9, 14, 15, 29, 31, 66, 67, 69, 72, 176, 190, 370
  functional, 70, 71
  regional, 104
Inhomogeneous, see Inhomogeneity

Injury current, 58
Inter–cellular, see Intercellular
Intercellular,
  resistive coupling, 47
  resistive values, 20
Interstitial cavity, 238
Interstitial volume, 237, 240
Intracellular, 138, 141, 199
  conductivity, 13
  connection, 365
  lactate, 346
  longitudinal resistance, 83
  norepinephrine, 303
  pH, 112
  propagation, 13
  resistance, 83, 91, 92, 109, 112
  space, 156
Intramyocardial, 231, 370, see also Myocardial
  arteries, 252, 255
  bed, 246, 247
  blood volume, 275
  pressure (IMP), 235, 237, 239, 241, 242, 245, 246, 249, 252, 254, 273, 276, 279, 369
  pump, 274, 276, 277
  vessels, 233, 234, 246, 255, 277
  volume, 277, 278
Ion, 146, 174, 196, 201, 341, 371, 373
  channels, 141
  transporters, 344, 372
Ischemia, 31, 185, 195, 252, 295, 326, 346
  injury, 335
  insult, 370
  metabolism, 96
  myocardium, 77, 185, 196, 197, 198, 200
  region, 15
  simulated, 109
  tissue, 48
Ischemia–reperfusion, 336, 341
Ischemic, see Ischemia
Isochronal lines, 31
Isochrones, 4, 6, 10, 39, 42, 103, 104, 106
  interpolated, 40
Isoproterenol, 172, 175, 203
Isovolumic, 186, 229, 230, 238, 240, 252, 279, 335, 343, 371
  contraction, 240
  relaxation, 252

K

K^+, see Potassium
Krebs–Henseleit buffer, 335, 351

L

Lactate, 351, 357, 359,
  /proton symporter, 344
  /proton transport, 372
  uptake, 346
LAD, see Left anterior descending artery
Langendorf, 352
  perfused rabbit hearts, 101, 102
  perfused rat hearts, 205
Late potentials, 14, 126, 133
Left anterior descending (LAD) artery, 38, 266, 276
  occlusion of, 266
Left circumflex artery, 266
Left coronary artery, 262, 266, 276
Left ventricle, see also Left ventricular (LV)
Left ventricular (LV), 15, 240, 245, 276, 283, 352, 369
  activation, 14
  cavity pressure, 249, 274
  cell, 58
  developed pressure, 354
  elastance, 252
  empty beating, 242
  epicardium, 39
  fibrillation, 32
  hypertropy, 295
  myocardium, 91, 283, 284, 325, see also Myocardial
  pressure (LVP), 229, 234, 235, 237, 273, 274, 277, 279, 337, 369
  pump, 274
  tachycardia, 14, 20, 39, 44, 101
    sustained, 25
  wall, see Wall
Length–tension relation, 217, 221
Lidocaine, 44, 198,
Limulus, 219, 224
Lipid, 146, 179, 182, 183, 353, 358
  bilayers, 147
  synthesis, 338
Longitudinal resistance, 110
LV, see Left ventricular, specific types

Lypunov exponent, 129
Lysophosphatidylcholine (LPC), 195, 197, 204

M

Magnetic resonance imaging (MRI), 9
Mapping, 101, 120, 366, see also specific types
Markov chain, 140
Mechanical restitution, 15, 158, 162, 163
Membrane, 4, 83, 181
  activation, 13
  bilayer, 179, 182
  capacitance, 83
  channels, 365
  lipids, 180
  potential, 16, 137, 343, 345
    resting, 110, 197
  properties, 33, 109, 111
  receptors, 179
  resistance, 197
Metabolic, see Metabolism
Metabolism, 96, 176, 198, 210, 326, 370, 373,
  fatty acid, 328
  oxygen, 210, 319, 328
  rate, 211, 333, 334
  requirements, 281
  transport, 372
Michaelis–Menten kinetics, 23, 150
Microcirculation, 278, 316
Microspheres, 216, 230, 252, 274, 283, 286, 290, 299, 308, 313, 330
Microvascular, 339
Microvessels, 274, 370
Mitochondria, 198, 201, 208, 212, 339, 344, 359, 373
  ATP/ADP translocater, 341
  calcium transport, 333, 337, 371
  outer membrane, 356
Models, 19, 23, 112, 140, 216, 235, 245, 261, 278, 281, 289, 296, 366, see also specific types
  analytical, 120
  cellular activation, 3, 9
  computer, 29, 63, 127, 132, 366
  coronary artery, 262
  coronary perfusion, 246
  discontinuous ring–shaped, 63
  3D finite–elements, 15
  fluid–fiber, 246, 252, 255
  3D ventricular, 13, 23
  Hodgkin–Huxley, 137, 138
  incremental network, 261
  isotropic myocardial, 75
  linear–core–conductor, 7, 9
  Markov chain, 137
  mathematical, 137, 278
  papillary muscle, 110
  pressure–dependent, 251
  ring–shaped, 71
  sound generation, 270
  source, 4
  stochastic, 139
  2D tissue, 15
  waterfall, 245, 250
Monoclonal antibody, 143
Monolayer tissue, 174
Monte Carlo Algorithm, 75
MRI, see Magnetic resonance imaging
Murmurs, 261, 272
Muscle bundles, 107
Muscle fibers, 219
Myocardial,
  activation, 15
  blood flow, 281, 282, 319, 326
  cells, 155, 158
  compression, 245, 252
  fibers, 43
  function, 245
  human, 328
  infarction, 44, 101, 326
  ischemia, 63, 109, 195, 197 specific types
  isolated sarcolemma, 196
  oxygen,
    consumption (MVO_2), 319, 320, 323, 326
    delivery, 326
  perfusion, 245, 255
  regional flow, 282
  reperfusion, 333
  structure, 103
  tissue packed, 97
Myocardium, see Myocardial
Myocytes, 26, 48, 77, 83, 89, 174, 196, 307, 313, 373
  isolated, 111, 199, 200
Myofilament, 185, 190, 373
Myosin, 191, 216, 220, 222

N

Na, see Sodium
NaCl, see Sodium Chloride
Natural frequency, 262
Nerve growth factor (NGF), 169, 170
Neutron diffraction, 181
Nifedipine, 307
Nimodipine, 179, 180
Noncontractile, 190
Nonlinear architecture, 126
Nonuniformity, see Anisotropy
Noradrenaline, 308
Norepinephrine, 175, 303,
Nuclear magnetic resonance, 183, 185, 206, 212, 372

O

Oligo-phosphoglyceroyl-ATP (oligo-PG-ATP), 207, 209
Oscillatory flow, 229, 234
β-Oxidation, 199
Oxidative, 212, 289, 322, 356, 359,
  metabolism, 210, 319, 328, 334, 336, 337, 373
  phosphorylation, 319
  substrate metabolism, 333
Oxygen, 112, 190, 211, 281, 282, 302, 307, 319, 326, 335, 359
  consumption, 190, 307, 319, 325, 328, 329, 333, 336
  content, 343

P

Pacemakers, 76, 169, 174
  modeling, 20
Pacing, 25, 38, 41, 43, 47, 52, 53, 56, 102, 105, 118, 122, 163, 187, 275, 296, 309, 313, 335
Papillary muscle, 48, 72, 95, 98, 110, 155, 156, 191, 315
Partition coefficient, 180
Patch clamp, 139, 144
Perfusion,
  pressure, 229, 239, 298, 299
  transmyocardial, 245
Permeability, 145, 151, 281, 282, 285, 320, 329, 370
  ratio, 151
PET, see Positron emission tomography
Pharmacological, 48, 87, 179,
Phenothiazine, 181, 183
Phosphate, 112, 162, 181, 206, 212, 297, 299, 300, 302, 335, 336, 351, 353, 357, 359, 372
  high energy, 371
Phosphocreatine, 342, 371
Phosphoglyceroyl-ATP, 205, 206, 213
Phospholipase, 176, 197, 201, 304
Phospholipids, 162, 165
  bilayer, 179
Poiseuille's law, 230
Positron emission tomography (PET), 319, 320, 321, 334
Posthypoxic recovery, 343
Postischemic, 186, 202, 333, 334, 335, 337, 338, 346–348
  myocardium, 371
  oscillations, 212
Potassium, 89, 101, 106, 107, 109, 141, 142, 147, 149, 151, 152, 282, 304, 345, 371
  channel blocker, 146
  channels, 288
  extracellular, 102, 106, 109, 111
  intracellular, 282
  permeability, 145, 146, 150
P–R interval, 171, 172
Premature activation, 20
Premature stimulation, 69
Premature stimulus, 65
Premature ventricular contractions, 19
Pressure–cross–sectional area, 231
Pressure–dependent, 246
Pressure–diameter, 235
Pressure–flow, 230, 273, 280
Pressure–volume, 236, 238, 240, 277, 279
Propagation, 4, 6, 7, 9, 10, 13, 19, 22, 23, 26, 29, 53, 53, 65, 77, 85, 91, 102, 111, 129, 365, 366
  circular, 73
  plane–wave, 9
  velocity, 7, 16, 17, 31, 92
  zig–zag, 9
Propranolol, 171, 179, 180, 183, 203
Proteins, 137, 143, 151, 158, 169, 173, 179, 198, 201, 206, 216, 372

contractile, 158, 162
G, 174, 175, 371
G_i, 172
G_s, 172
GTP regulatory, 169, 170, 171, 174
kinase, 190
Proton, 165, 344, 345, 372
accumulation, 347
gradient, 345
symporter, 347
Pulmonary, 335
artery, 282
vein, 282
Purine, 297, 305
nucleotides, 205, 210, 213
Purkinje, 17, 45, 59, 118, 119, 122, 133, 138, 204
fibers, 17, 116, 369
system, 46, 126, 128, 132
Pyruvate, 210, 296, 347, 348, 351, 352, 354, 355, 359, 372

Q

QRS, 127, 128, 133, 172
Q–T interval, 169, 171, 175
Quinidine, 60

R

Radioactivity, 206, 208, 299, 308, 334
Radionuclide, 334
Radiotracers, 322
Rats, 83, 140, 308, 322, 333, 351
aorta–banded, 303
hypertensive, 310
isolated hearts, 351
perfused hearts, 335
sham–operated, 302
spontaneously hypertensive (SHR), 296, 297
Wistar–Kyoto (WKY), 296, 297
Recirculation fraction, 159
Recovery, 48, 66, 87, 116, 122, 140, 155, 212, 295, 297, 300, 301, 335, 343, 348, 358, 369, 371
Reentrant, 24, 35, 37, 64, 68, 70, 73, 76, 101, 102, 198, 365, 366, 371
activation, 23, 24
circuit, 37, 38, 39, 41, 43, 44
excitation, 365, 366
tachycardia, 29, 40, 101
Reentry, 14, 19, 22, 37, 41, 43, 45, 63, 67, 77, 101, 366, 367
arrhythmias, 73, 109
loop, 20, 23, 25
single cycle, 77
Refractoriness, 29, 31, 32, 67, 69, 71, 163
Refractory period, 17, 32, 69
Relaxation, 158, 163, 252, 368
Reperfusion, 195, 301, 302, 333, 337, 347,
Repolarization, 68, 120, 163, 174
Resistance, 30, 47, see also specific types,
distribution, 34
input, 59
intercellular, 16
intracellular, 16
pressure–dependent, 246
reentrant, 24
vascular, 273
Resonant spectrum, 262
Reoxygenation, 347
Respiration, 129, 211, 368
Respiratory chain, 334
Resting potential, 44
Return maps, 120
Right ventricular, 155, 156, 283
Ring preparation, 106
Rubidium, 285, 288, 290, 292
Runge Kutta Verner solution, 249
Ruthenium red, 333, 337, 338,
RV, see Right ventricular
Ryanodine, 109, 111, 113

S

Sarcolemma, 145, 160, 180, 195, 196, 346
Sarcomere, 216, 217
Sarcoplasmic reticulum, 145, 147, 149, 158, 160, 339
Scorpion toxin, 142
Selectivity, 141, 148, 209
Servo–null method, 239
Shear stress, 312
Shunt pathway, 146
Shunt resistance, 95
Signal processing, 261

Simulation study, 63
Single–channel, 148, 150
  current, 139
  opening, 139
Skeletal muscle, 150
Sliding filament, 215
Slow inward current, 69
Sodium, 44, 67, 72, 102, 106, 110, 137, 141, 159, 160, 162, 163, 165, 199, 342, 344, 365
  activation gate, 67
  calcium, 112, 167
    antiporter, 344, 347
    exchange, 163, 345, 347
  channel, 137, 140, 365
  concentration, 341
  conductance (gNa), 67, 68
  current, 64, 65, 137
  cytosolic, 348
  exchange, 162
  extracellular, 111
  gradient, 345
  intracellular, 111, 112, 174
  inward current, 111, 112
  kinetics, 67, 138
  permeability, 150
  potassium pump, 170
Sodium Chloride (NaCl), 48, 102, 156, 170, 296, 352
Sound, 261, 370
  generation model, 270
  resonance, 263
  spectrum, 264
Space sickness, 125
Spatial inhomogeneity, 66
Spectral analysis, 127
SR, see Sarcoplasmic reticulum
Stellate ganglion, 32
Stenosis, 263, 264, 270, see also specific types
Step–wise, 221,
  phenomena, 216, 222
  shortening, 216, 222
Stiffness,
  measurements, 222
  transverse fibrous, 253
  varying, 238
Stimulation,
  inotropic, 327
  repetitive, 116
Stochastic properties, 139
Stress, see specific types
Striated muscle, 216, 220
Stroke work, 313
Stunned myocardium, 185, 187, 189, 326, 327, 328, 336
Subendocardial, 46, 98, 245, 249, 303, 308, 337
  flow, 230
  resistance, 230
Subendocardium, see Supendocardial
Subepicardium, 15, 245, 255, 309
Sucrose gap, 155
Sympathetic, 35, 210
  activity, 31
  innervation, 169, 175
  modulation, 170
Syncytium 4, 59, 91, 108

T

Tachycardia, 14, 15, 20, 25, 37, 39, 40, 46, 57, 58, 78, 101, 106, 199, 301, 367
Tetrodotoxin, 141, 142, 144
Thallium, 284, 288, 291
Thermodilution, 308
Thick filament, 215, 219, 222, 369
Thickening, 235, 254, 255
Thoracotomy, 352
Three–dimensional (3D), 13, 171
  activation, 365
  channel–structure, 140
  finite–elements model, 15
  ventricular model, 13
Thrombolysis, 326, 327
Thrombolytic therapy, 333
Thrombus, 195
Thymidine, 307
Time varying elastance, 230, 233, 273, 275, 278
Tissue,
  cardiac, 4, 18
  preparations, 23
Trabeculae, 93, 166
  isolated, 93
Tracer washout techniques, 252
Tracheo–broncheal tree, 125
Transmembrane, 7, 16, 19, 22, 44, 45, 92, 98, 118, 120, 137, 141, 142, 143, 147, 287
  potentials, 43
  resting, 39
  transport, 284

Transmural,
  pressure, 237, 246, 248
  ventricular mechancis, 246
Transport mechanisms, 347
Transverse,
  block, 105
  coupling resistance, 31
  resistance, 22
  resistivities, 8
Triacylglycerol (TG), 351
  fatty acids (TG–FA), 351
Tricarboxylic acid (TCA), 319
Troponin, 191
TTX, see Tetrodotoxin
Turbulent,
  flow, 115, 262
  pressure, 263
Twist, see also specific types
Twitch force, 155
Two–dimensional matrix, 29, 73
Two–dimensional models, 15
Tyrode, see Tyrode's solution
Tyrode's solution, 48, 94, 95, 98, 102, 116, 170, 171, 201, 296
  perfused hearts, 295
Tyrosine hydroxylase, 175

U

Unidirectional block, 15, 24, 29, 63, 66, 70
Uniform propagation, 4
Uric acid, 205, 297

V

Vagal tone, 368
Vascular,
  collapse, 274
  compliance, 230
  resistance, 273, 274
  volume, 230
Vasodilation, 248, 252
  maximal, 229, 234
Venous,
  effluents, 198
  flow, 273
Ventricular, see Left ventricular
Verapamil, 96, 111, 113
Vessel,
  endocardial, 229, 232
  epicardial, 229
  transmural, 232
VF–intervals, 32, 33
Viscoelastic effects, 263
Voltage–clamp, 83, 138, 141, 155, 157
Voltage dependent, 140, 148
  calcium cycling, 370
Voltage gated, 152
  ion channels, 141
Volume,
  conductor, 91, 97
  element, 6
  extracellular, 95
  fraction, 95
  intracellular, 95
Vortex element, 75
Vulnerable window, 64

W

Wall,
  ventricular, 234
Waterfall, 274
  flow, 251
  model, 245, 250, 251
Wavefront,
  conducting, 103
  excitation, 91
Window of vulnerability, 69, 71

X

Xanthine, 205, 297, 307, 316
x–ray diffraction, 151, 183, 223

Z

Zero–current, 148
Z–line, 219, 221
Zig–zag propagation, 9

MIX
Papier aus verantwortungsvollen Quellen
Paper from responsible sources
FSC® C105338

If you have any concerns about our products,
you can contact us on
ProductSafety@springernature.com

In case Publisher is established outside the EU,
the EU authorized representative is:
Springer Nature Customer Service Center GmbH
Europaplatz 3, 69115 Heidelberg, Germany

Printed by Libri Plureos GmbH
in Hamburg, Germany